Bacterial Cellulose

Bacterial Cellulose

Synthesis, Production, and Applications

Edited by
Sher Bahadar Khan and Tahseen Kamal

CRC Press
Taylor & Francis Group
Boca Raton London New York

CRC Press is an imprint of the
Taylor & Francis Group, an **informa** business

First edition published 2022
by CRC Press
6000 Broken Sound Parkway NW, Suite 300, Boca Raton, FL 33487–2742

and by CRC Press
2 Park Square, Milton Park, Abingdon, Oxon, OX14 4RN

© 2022 Taylor & Francis Group, LLC

CRC Press is an imprint of Taylor & Francis Group, LLC

ISBN: 978-0-367-63312-7 (hbk)
ISBN: 978-0-367-63313-4 (pbk)
ISBN: 978-1-003-11875-6 (ebk)

DOI: 10.1201/9781003118756

Typeset in Times
by Apex CoVantage, LLC

Contents

Preface

Bacterial cellulose (BC) refers to the cellulose produced by bacteria of specific genera through a series of enzymatic reactions using glucose, fructose, and other simple sugars as precursors. Besides bacteria, several other microbial species, including algae, have also shown success in BC production. In recent years, BC production through a cell-free enzymatic technique has also gained fame as the exclusion of live cells during the production process has relaxed the processing conditions, specifically temperature and pH. Since this discovery, BC production has been achieved in various forms, including sheets, films, granules, and so on through multiple synthetic strategies such as static, shaking, and agitation. Simple sugars, being the precursors, have opened the ways of producing BC from a variety of sources. To overcome the cost issues associated with commercial synthetic media, BC production has been achieved from a number of waste resources, including fruit waste, industrial waste, agricultural waste, and municipal waste.

Over the past few decades, BC has gained substantial attention due to its abundance, purity, reproducibility, outstanding physical properties, mechanical strength, crystallinity, special surface chemistry, and biological features. BC nanofiber networks develop a net-shaped 3D structure with a porous geometry that endows BC with impressive properties and subsequently numerous applications. Additionally, the presence of free −OH groups on the BC surface gives it hydrophilic behavior. Both porous geometry and hydroxyl sites enable BC to synthesize composites using multiple materials, including nanoparticles, metals, metal oxide, natural chemical extracts, polymers, and other materials. The porous fibrous network together with hydrogen bonding sites empowers it to immobilize and impregnate various materials on its surface and in its internal matrix for enhanced applications.

BC, due to its moldable nature, nontoxicity, and high mechanical properties has been used in medical fields as wound dressing material, as a drug carrier, for skin care, for tissue regeneration, and in the development of artificial organs. BC has also been used in various industries and commercial product development, including the textile industry, paper and pulp industry, filter membranes, food supplements, support surface for catalysis and pollutant degradation, development of facial masks, and other products. More recently, focus has been placed on altering the BC structure through chemical methodologies for incorporating additional features such as conductivity, magnetic, bactericidal, and biocompatibility for targeted applications.

Various aspects of BC, including its structure, production processes, synthetic and cost-cutting approaches, alternative culture media, pre- and post-synthetic moldability, chemical modification, and synthesis of its composites for a number of applications in diverse fields, have been reported. It has been felt that there is definite need to compile these important aspects in a single book that provides comprehensive information on BC. Therefore, in *Bacterial Cellulose: Synthesis, Production, and Applications* we provide an inclusive and timely overview of different aspects of BC, starting from its basic structure and moving to its chemical modification and targeted applications. This book consists of nine chapters and covers four major areas of BC research: (1) BC history, structure, and the pathways developed for its production, (2) various synthetic approaches developed for BC production, including cellular and cell-free systems, (3) surface modification of BC and its composite development with different materials, and (4) the wide range applications of BC in the biomedical field, food industry, environment, catalysis support, and other industries.

We hope that the topics covered in this book will be of great interest to BC researchers and general readers and will inform them of the multiple directions BC research has taken and its many achievements to date.

Sher Bahadar Khan and Tahseen Kamal

Editors

Prof. Dr. Sher Bahadar Khan is a Full Professor in the Chemistry Department, King Abdulaziz University, and is conducting research in nanochemistry and nanotechnology, which comprises catalysis, hydrogen production, catalytic reduction, drug delivery, solar cell, photocatalysis, and fabrication of perceptive bio-sensors and chemi-sensors using metal-based nanomaterials and their green environmental nanohybrids. He is the author of 350 research articles, 12 books, and six patents, has a 53 h-index, an impact factor of more than 1,200, and around 10,000 citations. He was awarded the Best Researcher award by King Abdulaziz University (2019) and Best Scientist award by KP, Pakistan (2018).

Dr. Tahseen Kamal is an Assistant Professor at King Abdulaziz University, Jeddah, Saudi Arabia (2015–2021). His research interests are modification of polysaccharides and synthetic polymers with metals for utilization in dealing with aqueous-based pollutants. Previously, he worked on the structures of polymeric materials during his doctoral studies, primarily focusing on structure–property relationships in the solid and solution states using X-ray and neutron scatterings at Kyungpook National University (Daegu, South Korea), HANARO (Daejeon, South Korea), and NSRRC (Hsinchu Science Park, Taiwan) (2009–2014).

Contributors

Laeeq Ahmad
Department of Earth Science
IIT Bombay, India

Md Wasi Ahmad
Department of Chemical Engineering,
    College of Engineering
Dhofar University
Salalah, Sultanate of Oman

Zubair Ahmad
Department of Chemistry
University of Swabi
Anbar, Pakistan

Kalsoom Akhtar
Division of Nano Sciences and
    Department of Chemistry
Ewha Womans University
Seoul, Korea

Youssef O. Al-Ghamdi
Department of Chemistry, College of
    Science Al-zulfi
Majmaah University
Al-Majmaah, Saudi Arabia

Jawad Ali
School of Environmental Science and
    Engineering
Huazhong University of Science and
    Technology
Wuhan, PR China

Atiya Fatima
Department of Chemical Engineering,
    College of Engineering
Dhofar University
Salalah, Sultanate of Oman

Adnan Haider
Department of Biological Sciences
National University of Medical Sciences
    (NUMS)
Rawalpindi, Pakistan

Muhammad Ismail
Department of Chemistry
Kohat University of Science &
    Technology
Kohat, Pakistan

Tahseen Kamal
Center of Excellence for Advanced
    Materials Research
King Abdulaziz University
Jeddah, Saudi Arabia

Ashi Khalil
Institute of Chemical Sciences
University of Peshawar
Peshawar, Pakistan

Mohd Shariq Khan
Department of Chemical Engineering,
    College of Engineering
Dhofar University
Salalah, Sultanate of Oman

M.I. Khan
Department of Chemistry
Kohat University of Science & Technology
Kohat, Pakistan

Murad Ali Khan
Department of Chemistry
Kohat University of Science &
    Technology
Kohat, Pakistan

Shahid Ali Khan
Department of Chemistry
University of Swabi
Anbar, Pakistan

Shaukat Khan
School of Chemical Engineering
Yeungnam University
Daegu, South Korea

Sher Bahadar Khan
Center of Excellence for Advanced
    Materials Research
King Abdulaziz University
Jeddah, Saudi Arabia

Waleed Ahmad Khattak
Hyperdrive Lubricants Limited
Redditch, UK

Sehrish Manan
Department of Biomedical Engineering
Huazhong University of Science and
    Technology
Wuhan, PR China

Noor Qahoor
Department of Chemical Engineering,
    College of Engineering
Dhofar University
Salalah, Sultanate of Oman

Fazal Qayyum
Department of Chemistry
University of Swabi
Anbar, Pakistan

Naveed Ur Rahman
Materials Science Institute, PCFM
    Lab and GDHPRC Lab, School of
    Chemistry
Sun Yat-sen University
Guangzhou, PR China

Sher Ali Shah
Department of Chemistry
University of Swabi
Anbar, Pakistan

Ajmal Shahzad
Department of Biomedical
    Engineering
Huazhong University of Science and
    Technology
Wuhan, PR China

Mazhar Ul-Islam
Department of Chemical Engineering,
    College of Engineering
Dhofar University
Salalah, Oman

Salman Ul Islam
School of Life Sciences, College of
    Natural Sciences
Kyungpook National University
Daegu, South Korea

Muhammad Wajid Ullah
Department of Biomedical
    Engineering
Huazhong University of Science and
    Technology
Wuhan, PR China

Guang Yang
Department of Biomedical
    Engineering
Huazhong University of Science and
    Technology
Wuhan, PR China

Sumayia Yasir
Department of Chemical Engineering,
    College of Engineering
Dhofar University
Salalah, Sultanate of Oman

1 Bacterial Cellulose
History, Synthesis, and Structural Modifications for Advanced Applications

Atiya Fatima, Sumayia Yasir, Noor Qahoor,
Tahseen Kamal, Mohd Shariq Khan,
Shaukat Khan, Muhammad Wajid Ullah,
Mazhar Ul-Islam, and Md Wasi Ahmad

CONTENTS

DOI: 10.1201/9781003118756-1

1.1 INTRODUCTION

Biomaterials, including biopolymers, are receiving immense interest thanks to their abundance, natural synthesis, effective physicochemical features, and widespread applications. A number of polymers, including polysaccharides, polyamides, and polycarbonates, are being explored through multiple synthetic routes for exclusive applications (1). Bacterial cellulose (BC), also termed as biocellulose and microbial cellulose, is among the most prevalent emerging biopolymers because it has shown impressive structural, physicochemical, and mechanical features. Its high crystallinity, absorbing capabilities, porous geometry, biocompatibility, and post-synthetic modification features have led to advanced application in environmental, biomedical, pharmaceutical, food, packaging, catalysis, and electromagnetic fields (2, 105). Exploration of novel applications is continued together with new synthetic media and cost-effective BC production processes (3, 106).

The medical and environmental sectors are currently facing significant challenges. The modern era of industrialization has eased human life by offering advanced and diversified technologies, but it has greatly affected environmental quality by incorporating toxic and poisonous emissions into the atmosphere (4). The highly toxic or carcinogenic organic contaminants are potentially dangerous to human health, causing cancers, physical birth defects, and mental disorders. Various chemical or biological materials have been utilized for the effective removal or degradation of a wide range of toxic contaminants. In this context, carbon nanotubes (CNTs), graphene and biochar, and metal-based adsorbents and catalysts have been extensively explored as potent pollutant removal materials (5). Various biological materials such as cellulose nanofibrils (CNFs), cellulose nanocrystals (CNCs), microfibrillated cellulose (MFC), and BC have emerged as environmentally friendly, cost-effective, and efficient materials for environmental applications.

As a matter of fact, pure materials lack certain important features, which restricts their applications in diverse fields. BC, despite its impressive biological, physical, and mechanical features, lacks antibacterial, magnetic, conducting, and antioxidant features. This lack consequently greatly reduces the applicability of BC in medical, environmental fields, in electromagnetic device synthesis and in the pharmaceutical industry. To cope with these limitations and add additional features to BC, its structure has been modified through various chemicals entities, including nanomaterials, polymers, acids, and alkalis (3, 6). The modified BC has shown tremendous enhancement in its physicomechanical and biological features and consequent applications in a variety of fields. The main factors contributing toward the structural modification of BC are its 3D porous geometry and availability of OH moieties. A number of materials have been combined with BC in the form of composites through inside, *ex situ*, and polymer blending techniques (2, 3). BC has been combined with chitosan, alginate, polyethylene glycol, polyaniline, silver oxides, zinc oxides, cobalt oxides, clay materials, natural products, and numerous other materials (2, 7). Indeed, every modification or composite development is achieved through a specific approach.

This chapter basically illustrates the BC structure, its synthetic pathways, production techniques, and various modifications made in its structure for enhanced applications. In upcoming sections, the main topics are all discussed in detail and are accompanied by a comprehensive literature study.

1.2 BACTERIAL CELLULOSE

BC is produced through a number of synthetic routes by various microbes, including bacteria, fungi, and algae (3). Glucose is the main precursor, whereas other sugars, including fructose and galactose, have also been converted to cellulose through multiple metabolic pathways. In recent years, a novel approach to developing BC through a cell-free enzymatic system has grabbed attention because it can eliminate the shortcomings associated with the cellular system (8, 9). Conventional BC is produced from synthetic media consisting of sugar and protein sources. However, considering the production cost, attention has shifted toward waste and naturally available cheap resources (10, 107). BC has been produced in various shapes; however, it is most commonly produced as BC sheets or pellicles on a media surface under a static cultivation strategy. BC appears as a gel-shaped semi-transparent structure consisting of microfibrils connected in a web-shaped structure (11). During microbial synthesis, specifically using a static cultivation strategy, BC is produced at the water-air interface, growing in the form of a gel and resulting in a thick sheet.

BC has exceptional physicomechanical features that include high crystallinity, water-holding capabilities, mechanical strength, nontoxicity, high porosity, moderate biocompatibility, biodegradability, and pre- and post-synthetic moldability (12, 13). The physical, mechanical, and biological properties of BC are better than those of plant cellulose are. The main precursor of BC production is glucose, but its production can be carried through a verity of monomer sugars, including fructose and galactose. Monosaccharides enter the bacterial body where, through series of enzymatic reactions, they are converted to cellulose chains. These chains come out of the bacterial cell wall through pores and unite through a hydrogen bonding interaction, forming microfiber ribbons and eventually a net-shaped sheet structure (14, 15).

BC has diverse applications, most prominently in wound healing, facial masks development, antimicrobial membranes, skin tissue repair, drug delivery, electronics, display devices, diaphragms, foods, and paper (16, 17). The biomedical field is the prime area for BC applications because of BC's high biocompatibility, porous geometry, nontoxic nature, and chemical nature (18, 19). As mentioned earlier, applications using pure BC were restricted by its lack of certain important features.

Pure BC lacks antimicrobial, antioxidant, biocompatible, conducting, and magnetic properties that partially diminish its competencies in biomedical and electronic sectors (16). An approach developed to cope with such limitations is its structural modification with a variety of materials to develop composites. Numerous composites of BC with polymers and nanomaterials have been produced to overcome such deficiencies and enhance its biological activities, mechanical strength, conduction, magnetic properties, biocompatibility, transparency, and biomedical applications (16).

1.3 HISTORY AND SYNTHETIC APPROACHES

Bacterial cellulose is a natural polymer produced by various bacteria (such as *Escherichia, Azotobacter, Agrobacterium, Rhizobium, Achromobacter, Aerobacter, Sarcina, Salmonella,* and many cell-free systems) using numerous synthetic and non-synthetic media (20–22). The first BC production was reported by A.J. Brown in 1886 using glucose as the main constituent; however, in time production was also reported from varied sugar sources (including galactose, fructose, and sucrose) (23). It is well established through numerous synthetic pathways that glucose and other sugars such as fructose are interconverted by enzymatic reactions. BC synthesis involves several routes differentiated by the different controlling enzyme, as illustrated in Figure 1.1. Bacterial species consume sugar media to formulate glucose chains, which protrude out of their bodies (in the form of an exoskeleton) to be combined through hydrogen bonding developing microfibrils. These fibrils then develop into a reticulated porous web bonded by inter- and intramolecular forces over the media surface. Reinforcement of new fibrils is made from the bottom, producing a cultivation is needed for scaffolds surface (21). To ramp up BC production, culture agitation is also employed, whereby strands of BC are produced by submerging them in culture media (2). BC production

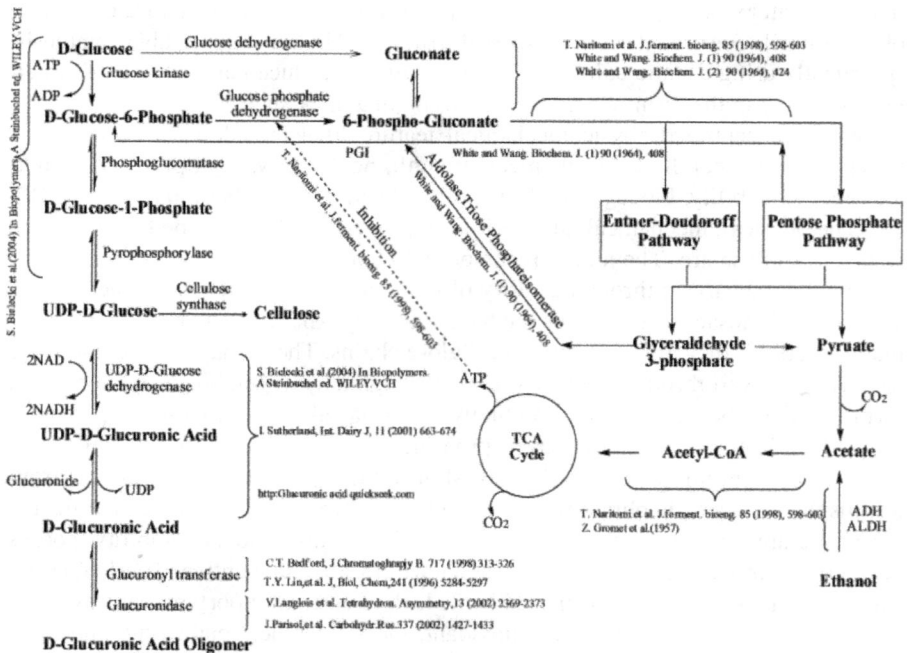

FIGURE 1.1 Bacterial cellulose synthesis pathway. Figure has been reproduced from (26) with permission from Elsevier.

Source: J.H. Ha, N. Shah, M. Ul-Islam, T. Khan, J.K. Park, Bacterial cellulose production from a single sugar α-linked glucuronic acid-based oligosaccharide, *Process Biochem.* 46 (2011) 1717–723. https://doi.org/10.1016/j.procbio.2011.05.024.

can be achieved through aerobic or anaerobic routes, also called biosynthesis or *in vitro* synthesis, differentiated by the use of either a cell or a cell-free environment. In biosynthesis, the enzymatic reaction is controlled by a specific enzyme, regulatory proteins, and various cofactors produced within the bacterial cell. Interestingly, these enzymatic reactions do not obstruct other anabolic activities going on in the cell (synthesis of nucleic acid/lipids/proteins, etc.). In anaerobic synthesis, however, a cell-free (24) environment for self-assembly of nanofibrils (22) is achieved. During the absence of a bacterial enzymatic reaction, specific cellulose-producing enzymes and cofactors are needed to keep up nanofibril production. A cell-free self-assembly of UDP-glucose has been studied extensively (24), despite the fact that complex polymerization from media to nanofibrils has still not been completely explored. The following section explores the production and challenges of BC in cell and cell-free environments.

1.4 MICROBIAL SYNTHESIS

UDP is the precursor for BC, which is produced by microbial cells in a four-step process.

- Activation of monosaccharides through glucose nucleotides.
- Polymerization of glucose units in cellulose chains.
- Addition of acyl group to individual glucose units.
- Production of β-(1$\rightarrow$4) glucan chains into extracellular environment (25).

Once the β-(1$\rightarrow$4) glucan chains are formed, polymerization starts to integrate thousands of individual chains into fibrils. These fibrils start forming small pellicles (membrane) at the air-media interface (9). When more pellicles are added, the membrane thickness increases toward the media, which entraps the microbial cells and results in depleting them of nutrients (or making them oxygen deficient) (2). Figure 1.1 illustrates the process in detail. It can be observed that glucose serves as the main carbon source and polymerizes to produce cellulose. However, various other carbon sources through multiple pathways can join the route of BC synthesis.

1.4.1 ADVANCEMENTS IN FERMENTATION APPROACHES

BC synthesis routes are differentiated in terms of application, production rate, capacitance, and economic feasibility and are classified through the production route (shaking, static, and agitated). Time- and effort-consuming static cultivation, cellulose-negative mutants in agitated cultivation, and lower yield, polymerization, mechanical strength, and crystallinity are the major challenges associated with different routes. To study and overcome these challenges, a comprehensive model of the BC production was proposed using bioreactions, carbon source, and oxygen transport as inputs. Other noticeable improvements include using a cylindrical silicone membrane vessel (27), rotating disk reactors (28), oxygen/glucose feeding reactors (29), and submerged fermentation through aerated/agitated cultivation, as illustrated in Figure 1.2.

1.4.2 STATIC FERMENTATIVE CULTIVATION

Static fermentation is the most commonly employed route for obtaining BC sheets/films/membranes through a cellulose-producing bacterial cell inoculated in the

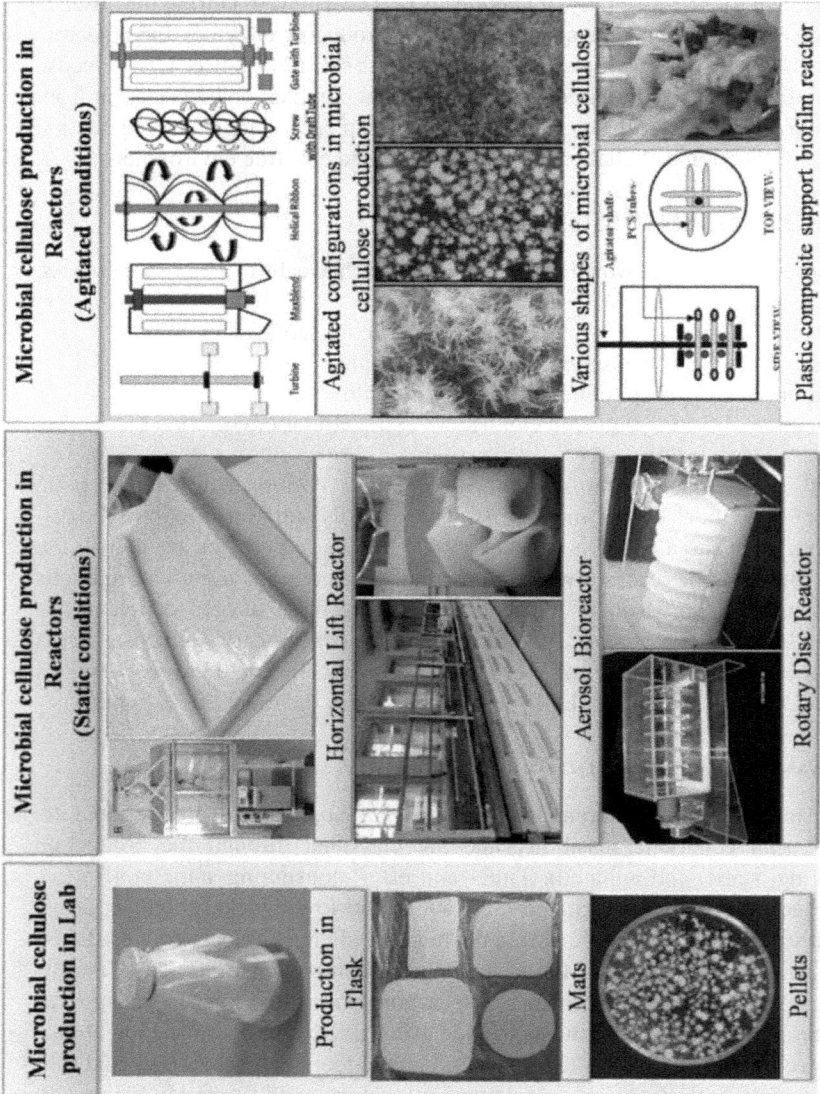

FIGURE 1.2 Illustration of various strategies for BC production. Figure has been reproduced from (33) with permission from Elsevier.

Source: Z. Shi, Y. Zhang, G.O. Phillips, G. Yang, Utilization of bacterial cellulose in food, *Food Hydrocoll.* 35 (2014) 539–545. https://doi.org/10.1016/j.foodhyd.2013.07.012.

culture media at optimum temperature/pH. Aerobic bacteria during static fermentation demand an oxygen supply for the incubation period of 7–10 days (this varies with bacteria strain) (3), during which bacterial cells proliferate and produce BC by partly utilizing available nutrients that results in lower BC yield. In static cultivation, BC is produced at the air-media interface and takes the shape of the container in which it is produced. As the new fibrils are formed, the thickness of BC increases, which slowly depletes the nutrients and makes the bacteria inactive by entrapping them (2). BC mostly finds application in the biomedical area, where a reticulated structure prepared through static cultivation is needed for scaffolds (30, 31). In spite of the demand, the lower production rate and yield limit its commercial application and serves as a major challenge that is handled though effective utilization of available nutrients (carbon source variation), giving optimum conditions while minimizing side product (3). Several studies reported high BC yield (from the static fermentation process) using a cell-free enzyme route, as no media is needed for cell growth (24) and all is consumed to generate BC. Jung et al. (32) reported intermittent bacterial feeding (in static cultivation) as an effective way of obtaining thicker (up to 30 mm) and differentiated BC.

1.4.3 SHAKING FERMENTATIVE CULTIVATION

As the name suggests, shaking fermentative cultivation produces BC in an incubator and uses a rotor or shaker to help inoculate microbial cells into the culture media. The distribution of microbial cells in the culture media and regulation aeration speed up the production process (2–3 days as opposed to 7–10 days for static fermentation). Shaking speed is usually kept between 100 and 250 rpm, which helps in forming different-size pellets. Besides speed, the shape of pellets also depends on the microbial strain type and the incubation period. For instance, 125 rpm produces sphere-like pellets that can be transformed into tail-like structures, and these structures integrate to form a large ball-like shape (34) with increased rotor speed. Studies reported that a shaking condition offers a 30% higher yield in *Komagataeibacter persimmonis* under similar growth/nutrient conditions (35).

1.4.4 AGITATION FERMENTATIVE CULTIVATION

Unlike pellicles in static fermentation, agitation fermentation in a controlled speed reactor produces small granules/pellicles and sphere-like particles (SCPs) (31, 36). BC produced from the agitated technique exhibits lower polymerization, mechanical strength, and crystallinity; however, these shortcomings are overlooked because of the short time, high yield, high cell density, and better oxygen contact needed for large-scale application. The yield in agitated fermentation can be further improved by tuning parameters such as oxygen transport, carbon source, microbial stain, and removal of cellulose fibrils. The faster rate compared to static and shaking cultivation is due to high microbial cell density (achieved due to controlled oxygen supply, pH, and nutrients) and foam prevention through antifoaming agents employed during incubation. Despite the promised incentive of high yield (achieved by a high energy supply to the fermenter), limitations of agitated fermentation include cellulose pellets instead of pellicles and growth of cellulose-negative strains lowering the BC yield (37).

1.5 BC STRUCTURE

Bacterial cellulose is a biopolymer produced by various strains of *Acetobacter* species such as *Achrobacter, Alcaligene, Aerobacter,* and *Azotobacter* (38). BC is a natural biomaterial comprising the purest form of cellulose produced by the Gram-negative bacterium *Acetobacter xylinum* (*Komagataeibacter xylinus*) utilizing glucose as the common substrate. BC possesses chemical resemblance to plant cellulose (i.e., $C_6H_{10}O_5)_n$, while differing from it in its nanoscaled fibrous morphology and physiological behavior. Bacterial cellulose is purer than plant cellulose as it is devoid of lignin, hemicellulose, and pectin. It comprises a three-dimensional fibrous network of nanofibers (ranged from 50 to 80 nm in width and 3 to 8 nm in thickness) arranged to form a hydrogel sheet, which gives it characteristic physicochemical and mechanical properties. It has high surface area, porosity, high water-holding capacity, excellent tensile strength, and biodegradability. During BC production, *A. xylinum* produces cellulose I (a ribbon-like polymer) and cellulose II (a thermodynamically stable polymer). The protofibrils of the glucose chain, secreted across the bacterial cell wall, aggregate to form cellulose ribbons that form a web-shaped network with a highly porous matrix. The presence of hydroxyl groups on the cellulosic surface provides the hydrophilicity and biodegradability of BC, thus offering a platform for chemical modifications (39, 40). These hydroxyl groups allow interactions with over 90% of water molecules, leading to high water retention capacity (41). Early studies on physical properties of BC revealed that the BC pellicle is strong, especially along the perpendicular direction of the fiber growth (42). The first-recorded Young's modulus was found to be 16–18 GPa across the surface of the BC plane and was further improved up to 30 GPa (43). Young's modulus of the BC sheets is increased by ribbon forms, which are easily oriented in one plane upon pressing (44). Wider ribbons increase Young's modulus since the effective cross-sectional area increases, creating an appropriate uniplanar orientation. It has high crystallinity (84%–89%) and high water-holding capacity (over 100 times its own weight) (45).

1.6 NEED AND TYPES OF BC STRUCTURE MODIFICATION

Bacterial cellulose has received considerable attention owing to its unique structure and admirable properties, but property optimization is required for its potential applications in the future. The synthesis of BC composite materials and their functionalization using different chemical modification techniques have been studied extensively. The synthesis of different BC composites and their investigations, functionalized BC, and even the regeneration of BC materials are some of the common aspects in structural modification of BC. Bacterial cellulose lacks appropriate functionalities for applications, which limits its application in certain fields. To gain control over its porosity and to avoid very slow degradation, BC can be modified by chemical (modification of chemical structure and functionalities) and physical methods (change in porosity, crystallinity, and fiber density) by applying versatile *in situ* and *ex situ* techniques. *In situ* modifications involve variation in the culture medium and carbon source and the addition of other supplements, while *ex situ* modifications are performed by chemical and physical treatment of synthesized BC.

1.7 CHEMICAL MODIFICATION

Bacterial cellulose polymer contains an unmodified C4-hydroxyl group at one end and a free C1-hydroxyl group at the opposite end, which provides the possible sites for chemical modifications of BC (46). Chemical modifications rely extensively on these hydroxyl groups not only in heterogeneous but also in homogeneous conditions.

BC can be chemically modified by *ex situ* techniques involving dipping the BC sheet into a certain polymer solution. However, these composites lack chemical bonds or any strong intermolecular forces between the BC fibers and the composite matrix; therefore, different chemical methods are employed to modify BC. One such modification method is treating the BC with an alkali (such as sodium hydroxide) (47). Surface modification of cellulosic materials, such as cellulose nanofibrils and cellulose nanocrystals, with other materials provides distinct properties that reflect multiple functionalities, thus opening up tremendous potential for more advanced applications (48). The fascinating 3D nanofibrillar network structure of BC allows its modifications, since it has a high number of surface hydroxyl groups and porous morphology. Therefore, enhancements in the physical, mechanical, and surface features can be obtained through alterations. This provides the basis for additional applications of BC through the blending of bioactive materials and alteration in porosity, biodegradability, and crystallinity (49). Many studies and investigations have been carried out to employ BC for environmental uses.

1.7.1 ACID HYDROLYSIS

A bacterial cellulose pellicle consists of bundles of nanofibers that can be converted to cellulose nanocrystals via acid hydrolysis. Cleavage of glycosidic bonds is gained when hydrogen ions penetrate the amorphous cellulose; as a result BC is transformed into a suspension of cellulose whiskers (cellulose nanocrystals). This helps in providing better properties and functionalities of BC, which can then be utilized for environmental applications such as biosensors for pollution detection. Furthermore, different hydrolysis conditions such as acid type, acid concentration, and reaction time and temperature give rise to different morphological features and properties of BC.

HCl and H_2SO_4 are strong inorganic acids used for the acid hydrolysis of cellulose. HCl generates limited nanocrystal dispersibility and low-density surface charges on the cellulose nanocrystals, while a very stable colloidal suspension is gained when H_2SO_4 is used. The sulfonation of the cellulose nanocrystals surface creates a high negative surface charge along with a decrease in the thermostability of cellulose nanocrystals caused by the presence of the sulfate groups $(-OSO_{3-})$ (50). The effect of hydrolysis time using sulfuric acid on the obtained cellulose nanocrystals was examined in a study done by Martínez-Sanz et al. The researchers found that as time increases the length of particles decreases, which is important when it comes to obtaining a noticeable increase in the crystallinity. Although the sulfuric acid treatment leads to a remarkable increase in the crystallinity, it negatively decreases the thermal stability of the produced nanocrystals, which makes it unfit for certain melticompoundable polymer-based nanocomposite applications (51). Corrêa et al. suggested that a stable and thermally resistant cellulose nanocrystal suspension is possible to obtain with the concomitant use of HCl and H_2SO_4 (52).

In addition, both the crystalline and amorphous regions of cellulose can be hydrolyzed using high acid concentrations. Elevated temperature causes crystallinity reduction and modifies the physical properties along with morphological characteristics of the nanocrystals (53). For cellulose nanocrystal production from cellulose, multiple plant sources were used, including banana pseudostems, coconut fibers, and cotton (54–57). However, the high cost of the purification process and energy consumption shifted the research focus to green alternative low-cost bacterial cellulose. Unlike plant cellulose, the bacterial cellulose is chemically pure, containing no hemicellulose, pectin, or lignin components. Its simple production and purification process makes it an attractive biopolymer for the production of cellulose nanocrystals.

Cellulose nanocrystals open up a wide range of applications such as the synthesis of antimicrobial and medical materials, enzyme immobilization, green catalysis, synthesis of drug carriers in therapeutic and diagnostic medicine, and biosensing. They possess unique properties, including hydrophilicity, biocompatibility, considerable tensile strength and stiffness, liquid crystalline behavior, and small particle size with high surface-to-volume ratios. In addition to that, their large number of hydroxyl groups facilitates the incorporation many chemical functionalities to their surface, furnishing superior dispersion in any solvent or polymer. Cellulose nanocrystals are receiving noticeable interest for biomedical, pharmaceutical, and environmental applications due to their high surface area and open pore structure (58). They are used in pH sensors, gas barrier films, and nanopaper, as a reinforcing agent for the fabrication of nanocomposites, and for the stabilization of oil/water interfaces (59).

After carbon dioxide, the second most toxic gas in the environment is hydrogen sulfide. Early detection of the gas would help in avoiding environmental pollution, enabling proper treatment. Sukhavattanakul and Manuspiya have successfully converted bacterial cellulose into cellulose nanocrystals via sulphuric acid hydrolysis that was further loaded with alginate-molybdenum trioxide nanoparticles and silver nanoparticles. The fabricated composite demonstrated good sensing capability for hydrogen sulfide gas (60).

Conventional packaging materials can be replaced with eco-friendly packaging materials possessing renewability, biodegradability, and large-scale availability. George et al. found that a nanocomposite made of hydroxypropyl methyl cellulose (HPMC), bacterial cellulose nanocrystals (BCNCs), and silver nanoparticles (AgNPs) reflect superior properties for food packaging materials. Also, HPMC and BCNCs are flexible, transparent, and good film-forming materials. Along with this, HPMC resists fats and oil and has been approved for making an edible film (61). George et al. acid hydrolyzed bacterial cellulose to make cellulose nanocrystals, yielding particles with a diameter of 20±5 nm and length of 290±130 nm. A gelatin-based edible nanocomposite film was synthesized. Appreciably, the BCNC combination added strength to the gelatin matrix and reduced the moisture affinity and moisture barrier properties of gelatin as a result of BCNCs interacting with hydrophilic amino acids sites present within gelatin. Furthermore, the segmental mobility of gelatin chains was also affected due to the formation of strong hydrogen bonds between BCNCs and gelatin, thus increasing glass transition temperature, degradation temperature, and the dynamic mechanical properties of gelatin, which is very important for edible packaging applications. The use of BCNCs assisted in making biodegradable, edible films at relatively low cost (62). BCNCs enhanced modulus and tensile strength but reduced elongation properties. The addition of AgNPs helped to restore the loss of elongation without disturbing other mechanical properties. This amalgamation of polymer technology and nanotechnology led to the fabrication of eco-friendly materials suitable for a variety of applications (62).

1.7.2 Enzymatic Hydrolysis

Acid hydrolysis is an environmentally hazardous process that causes a decrease in the mechanical performance of nanocrystals (63). Enzymatic hydrolysis delivers not only higher selectivity but also cost-effective nanocrystals in higher yields at milder operating conditions than chemical processes do. However, certain parameters such as enzyme efficiency, operating temperature, and selectivity still need to be addressed for the full exploitation of this technique.

Rovera et al. synthesized bacterial cellulose nanocrystals through enzyme hydrolysis with the aim that these nanoparticles would be efficient at making high oxygen barrier coatings for food packaging applications. Cellulase and endo-1,4-β-glucanases (EGs) were the two enzymes used for the treatment. Morphology and nanoparticle size was examined using dynamic light scattering (DLS), transmission electron microscopy (TEM), and atomic force microscopy (AFM). Evidently, cellulase enzyme treatment resulted in slightly bigger particle size than did EGs of around 250 nm, and their piled whiskers were found to be round in shape. Furthermore, the fabrication of pullulan/BCNC nanocomposite coatings enhanced the performance of the oxygen barrier of synthesized BCNCs utilized in the food packaging industry (64).

1.7.3 Modification with Polymers

1.7.3.1 BC-Chitosan

Chitosan, poly(1,4-β-d-glucopyranosamine), is a polysaccharide derived from natural chitin. It is a biocompatible, biodegradable, nontoxic, and antibacterial polymer. Bacterial cellulose has been modified by chitosan to combine the properties in the composites. BC-chitosan composites are used in the treatment of burns, skin ulcers, bedsores, and wounds. Romero et al. fabricated bacterial cellulose-chitosan (BC-Ch) nanocomposites. *Ex situ* and *in situ* methods were used to obtain the BC-Ch composite. In the *ex situ* technique, a BC sheet was immersed in an aqueous solution of chitosan to get the desired composite. The *in situ* technique involved the impregnation of chitosan during the formation of BC fibers as the BC production broth was mixed with both chitosan and the culture media of BC. The first type has smaller pores and a denser fiber network. The BC-Ch nanocomposite exhibited antioxidant and antimicrobial properties and greater mechanical characteristics. Especially, the composites displayed growth inhibition against bacterial strains such as *Pseudomonas aeruginosa* and *Staphylococcus aureus* and fungal strain *Candida albicans*. The accomplished BC-Ch composite surface prevents the formation of a biofilm. The antioxidant activity of BC increased as a result of chitosan incorporation (65).

In another study by Yin et al., the fabrication of BC-Ch composite and BC-chitooligosaccharide composite (BC-COS) was successfully achieved. Scanning electron microscopy (SEM) images revealed that the BC substrate host chitosan and chitooligosaccharide inside its matrix made the composite network denser with fewer pores as compared to the pure BC. Both membranes were able to exhibit antibacterial effect toward *Escherichia coli* and *Staphylococcus aureus* with a noticeable inhibition ratio of 99.99 ± 0.01% (*E. coli*) and 99.99 ± 0.01% (*S. aureus*) for BC-Ch, and 90.56 ± 0.06% (*E. coli*) and 99.64 ± 0.18% (*S. aureus*) for BC-COS composites. Impregnation of chitooligosaccharide into BC delivered composites with greater mechanical properties and better water-holding capacity in comparison to ones with chitosan. Furthermore, BC-COS displayed a good antioxidant activity. This blending has been designed to provide a variety of applications in the medical fields like antibacterial and wound dressing, and it can also be considered a promising candidate for food industry applications (66).

Nano-sized ZIF-67 crystals are reported to have great chemical and thermal stability. They are stable in boiling methanol at least for 5 days and up to 350 °C. Several studies mentioned that it can be synthesized by facile and environmentally friendly organic synthesis and with controlled experimental conditions; ZIF-67 particle size can be adjusted accordingly (67). Taking these findings into consideration, it is obvious that by merging the advantages of ZIF-67 with various polymers such as bacterial cellulose, the resulting composite will have a potentially better performance than that of pure ZIF-67. Li et al. reported that ZIF-67/bacterial cellulose/chitosan composite aerogel was employed to remove organic dyes and heavy metal ions from aqueous solutions. X-ray photoelectron spectroscopy (XPS) was used to investigate the adsorption process. Synthesized ZIF-67/BC/CH aerogel is capable of adsorbing 152.1 mg g^{-1} of Cr^{6+} and 200.6 mg g^{-1} of Cu^{2+}. Besides, the removal rate toward active red X-3B dye was found to be 100%. ZIF-67/BC/CH aerogel can be considered a promising high-efficient aerogel-based adsorbing material for wastewater treatment (68).

Urbina et al. have investigated the potentials of synthesized bacterial cellulose and chitosan composite (BC-Ch) as filtering membranes for copper in wastewaters. The fabrication of BC composite was carried out using *in situ* and *ex situ* methods. The morphological characterization using SEM analysis revealed that *in situ* techniques yielded a better incorporation of the chitosan into the BC matrix. BC filamentous network with substantial void volume was incorporated with Ch, which acted as an active catching site for the elimination of copper. Approximately 50% copper removal capacity of *in situ* BC-Ch filtering membrane was observed with the initial concentrations of 50 and 250 mg L^{-1} of copper. The synthesized BC-Ch composite can be employed in myriad applications for wastewater treatment (69).

1.7.3.2 BC-Polyvinyl Alcohol

Polyvinyl alcohol (PVA) is a hydrophilic water-soluble synthetic polymer used for biomedical applications. It is nontoxic and has excellent thermal stability, transparency, and high mechanical strength. PVA has gained considerable attention and is used in a wide range of applications owing to its excellent physical and optical properties.

In a recent study, voids between BC fibrils were filled with several ratios of PVA and glycerol (GLY) by the use of casting and immersion techniques. GLY worked as a plasticizer to promote plasticity and reduce brittleness, while PVA acted as the reinforcing agent. The synthesized film displayed 49.89% as the maximum value of elongation and resisted 13.78 MPa prior to rupture. The water vapor permeability (WVP) was found to be reduced upon incorporation of PVA within the porous network of BC, which was rectified with the use of GLY. This interaction among the polymers helps to increase the transparency and reduce the opacity values. The combination properties of the developed BC-PVA-GLY composite film can be used in food applications to prevent oxidation of vitamins, proteins, and lipids and also to avoid degradation of antioxidants in foods (70). The produced BC composites with glycerol and PVA are shown in Figure 1.3.

A bacterial cellulose, polyvinyl alcohol, biochar-nanosilver composite membrane (BC-PVA-C-Ag) was prepared by Zhang et al. with remarkable antibacterial activity toward *E. coli*. Membrane network structure was analyzed by SEM, X-ray diffractometry, Fourier transform infrared spectroscopy (FT-IR), and thermogravimetry-differential scanning calorimetry. The experiments showed the composite membrane to have long-lasting bacteriostatic ability, excellent durability, and the ability to resist silver loss that is required for water treatment and filtration application. Antibacterial activity revealed an average antibacterial zone of 16.1 ± 0.3 mm in diameter for C-Ag content of 0.05 g/L (71).

1.7.3.3 BC-Polydopamine

Polydopamine (PDA) has emerged as an important biopolymer material in the past few years and has found its application in several biomedical applications. It possesses various interesting properties such as self-assembly, universal adhesion, and ability to form coordination bonds with several metal ions, which are further reduced to metal nanoparticles (NPs).

Yang et al. have designed BC-based photocatalytic adsorbent nanocomposite film coated with polydopamine containing immobilized titanium dioxide nanoparticles on the modified bacterial cellulose surface for the adsorption and UV irradiated photocatalytic degradation of dyes. The synthetic process for $BC/PDA/TiO_2$ film (Figure 1.4) is of low-cost and eco-friendly facile operation. Chemical features and availability of appropriate functional groups provide the active sites for the removal of the organic pollutants and offer high photo-degradation performance toward dye molecules such as methyl orange, methylene blue, and Rhodamine B. Also, reusability and stability were excellent since the photo-degradation efficiency was reduced by only 5.5% after five cyclic tests. $BC/PDA/TiO_2$ composite film exhibited a unique structure and properties that can be successfully exploited in potential applications, including wastewater treatment and photocatalysis (72).

1.7.3.4 BC-Polypyrrole

Polypyrrole (PPy), a polymer obtained via oxidative polymerization of pyrrole, is a solid having intrinsically conducting properties used in electrical, optical, and medical fields. Synthesis of BC composites containing polypyrrole/titanium dioxide-silver nanoparticles ($BC/PPy/TiO_2$-Ag) was reported by Ghasemi and colleagues. The composite was used as a sensor to detect the presence of pathogenic bacteria. The dimensions of designed diagnostic film were 2×1 cm with 0.5 mm in thickness, and both

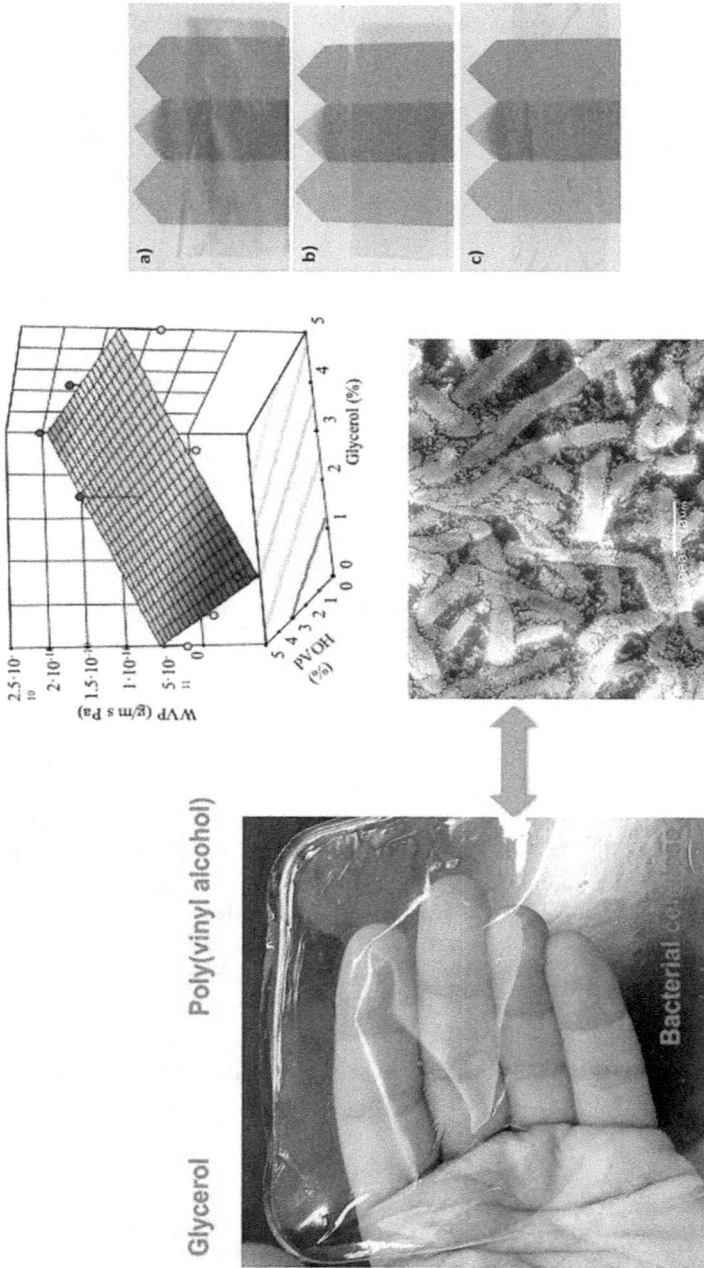

FIGURE 1.3 Pictorial representation of bacterial cellulose, glycerol, and polyvinyl alcohol based composites. Figure has been reproduced from (70) with permission from Elsevier.

Source: P. Cazón, G. Velazquez, M. Vázquez, Characterization of mechanical and barrier properties of bacterial cellulose, glycerol and polyvinyl alcohol (PVOH) composite films with eco-friendly UV-protective properties, *Food Hydrocoll.* 99 (2020) 105323.

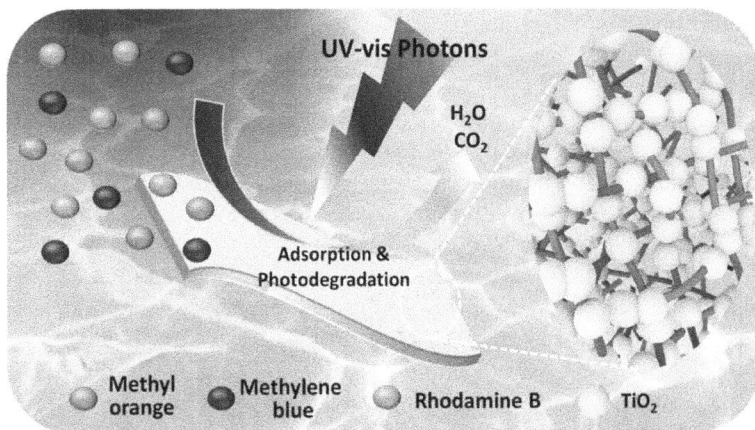

FIGURE 1.4 Illustration of BC/polydopamine/TiO_2 composite (72). Figure has been reproduced with permission from Elsevier.

Source: L. Yang, C. Chen, Y. Hu, F. Wei, J. Cui, Y. Zhao, X. Xu, X. Chen, D. Sun, Three-dimensional bacterial cellulose/polydopamine/TiO2 nanocomposite membrane with enhanced adsorption and photocatalytic degradation for dyes under ultraviolet-visible irradiation, *J. Colloid Interface Sci.* 562 (2020) 21–28.

ends of the sensor were linked to the multimeter of copper wires. Polypyrrole particles were like a thin shell that covered and blocked bacterial cellulose fibers. Bacterial cellulose bonded with polypyrrole with hydrogen bond interactions between the hydroxyl group of bacterial cellulose and the amine groups of polypyrrole, which boosts the tensile strength of the fabricated nanocomposite film. Then, TiO_2-Ag nanoparticles of irregular spherical shape with a range of 50 to 90 nm in size were uniformly distributed on the cellulose surface. *E. coli* PTCC 1330, *S. aureus* PTCC 1112, *S. aureus* ATCC25904, *A. hydrophyla* ATCC 35654, and *S. epidermidis* ATCC 700576 bacterial strains were examined for the bacterial growth detection efficiency of the fabricated film. The sensation ability toward Gram-negative bacteria was found to better than Gram-positive owing to positively charged nanoparticles used in making BC/PPy/TiO_2-Ag sensors. These kind of biosensors were easily portable and user friendly; they worked with a short response time for the target analysis due to their high sensitivity and can be considered a perfect diagnostic tool for ensuring biological safety (73). Using a simple hydrothermal method, a (carbon nanofiber@nitrogen-doped porous carbon) CNF@NPC core-sheath structured composite with excellent electrochemical performance were fabricated by Zhang et al. for rechargeable sodium-ion batteries (SIBs). CNF@NPC composite was obtained through bacterial cellulose/polypyrrole composites. The three-dimensional BC structure worked as a template to imbibe polypyrrole on its surface. For over 100 cycles, reversible specific capacity of CNF@NPC was found to be 240 mAh g^{-1} at 100 mA g^{-1}; for over 400 cycles, the cycling ability was 148.8 mAh g^{-1} at 500 mA g^{-1} and displayed superior rate performance of 146.5 mAh g^{-1} at 1,000 mA g^{-1}. CNF@NPC could prove to be a beneficial, environmentally friendly anode material, opening up a new application for SIBs (74).

1.7.3.5 BC-Polylactic Acid

Polylactic acid (PLA), a thermoplastic polyester, is a bioplastic obtained from lactic acid and utilized in the food packaging industry. Wardhono and Kanani developed a PLA biodegradable film reinforced with bacterial cellulose nanocrystals. BCNCs were obtained, first, through catalytic hydrolysis of bacterial cellulose using metal chloride salt facilitated by ultrasonic irradiation, and second, by using a solvent casting method. The crystals of rod-like and needle structure was used to strengthen the PLA matrix. The mechanical properties increased due to the loading of BCNCs in the PLA. It was obvious from the experiment that by increasing filler concentration of BCNCs the tensile strength of the composites increased, which indicated good interfacial reactions formed between PLA and BCNCs as well as equal dispersion of BCNCs in PLA. PLA-BCNC biocomposites displayed greater tensile strength than pure PLA,3.6 MPa for the composite film holding 5% w/w of BCNCs. This designed biocomposite film can potentially be applied for edible packaging materials (75).

1.7.3.6 BC-Polysaccharides Carboxymethyl Cellulose

Carboxymethyl cellulose (CMC) is a water-soluble polysaccharide widely used as a thickening and stabilizing agent in foods, cosmetics, and dyes. Shu et al. have successfully prepared an eco-friendly low-cost hierarchical porous carbon (HPC) composite material made from bacterial cellulose/polysaccharides carboxymethyl cellulose/citric acid, which performed perfectly as a supercapacitor. HPC materials were obtained by a one-step synthetic strategy of carbonization and activation process. The porous nanosheet morphology of HPC possesses 7.3% oxygen content, high specific surface area of 2490 m^2 g^{-1}, a current density of 0.5 A g^{-1} with a high specific capacitance of 350 F g^{-1}. After a 10,000-cycle test of HPC, it shows 96% capacitance retention. HPC opens up a next-generation energy storage system for supercapacitor applications (76).

1.7.4 MODIFICATION WITH ADSORBENTS

Bacterial cellulose could be modified with adsorbents to impart special characteristics and for its use in various fields. A novel BC membrane loaded with sodium nickel hexacyanoferrates (BC/NiHCF) (Figure 1.5) was synthesized for efficient adsorption

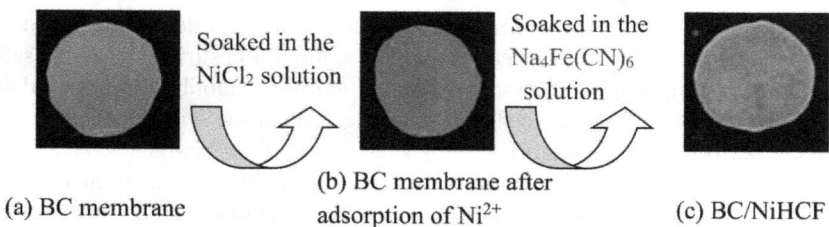

(a) BC membrane   Soaked in the NiCl$_2$ solution   (b) BC membrane after adsorption of Ni^{2+} Soaked in the Na$_4$Fe(CN)$_6$ solution   (c) BC/NiHCF

FIGURE 1.5 Schematic representation of the formation of BC/NiHCF membranes (5). Figure has been reproduced with permission from Elsevier.

Source: S. Zhuang, J. Wang, Removal of cesium ions using nickel hexacyanoferrates-loaded bacterial cellulose membrane as an effective adsorbent, *J. Mol. Liq.* 294 (2019) 111682.

and simultaneous removal of cesium ions from radioactive wastewater. Metal hexacyanoferrates (MHCF) are efficient adsorbents for cesium ion removal, and their BC composites displayed an ion exchange mechanism between the monovalent cations present in the lattice of crystal NiHCF and Cs ions from the wastewater (5).

1.7.5 MODIFICATION WITH NANOPARTICLES

Bacterial cellulose has a unique ultrafine three-dimensional (3D) network with a micro/nanoporous structure with well-separated nano- and microfibrils creating an extensive surface area. Surface area of BC contains hydroxyl and ether groups that provide effective reactive sites to uphold metallic ions on to its surface (Li et al., 2009). This high surface area can act as a template for the penetration of metal ions into its matrix. Moreover, the nanoporous structure of bacterial cellulose containing nanofibers provides surplus amount of sub-micron pores for hosting metal nanoparticles (77). Nanoparticles can be embedded into a BC porous network by either *in situ* or *ex situ* techniques.

1.7.5.1 *In situ* Formation of BC-Nanocomposites

In situ composite formation of BC represents an important modification technique in which nanoparticles can be *in situ* incorporated into BC fibrils while retaining the unique 3D nanoporous network of BC. This technique yields a novel nanocomposite having the properties of altered supramolecular structured BC along with the incorporated nanoparticle efficacies (78). *In situ* methodologies offer a control over size and morphologies of the synthesized nanoparticles via regulation and monitoring of BC template and *in situ* reaction conditions. BC fibrils prevent agglomeration of nanoparticles, ensuring dispersed nanoparticle formation inside the BC matrix. *In situ* generation of nanocomposites is simple and delivers nanoparticles with narrow size distribution. A variety of nanoparticles can be formed utilizing oxidation-reduction, precipitation, and sol-gel reactions.

1.7.5.2 *In situ* Formation of Nanostructures through Reduction

Several reducing agents have been reported for the *in situ* synthesis of nanocomposites. BC is used as a soft template, and reducing agents are used to control the size and morphology of the formed nanoparticles. Sodium borohydride ($NaBH_4$), triethanolamine, hydrazine (NH_2NH_2), hydroxylamine (NH_2OH), ascorbic acid, and polyethylenimine (PEI) are some of the reducing agents that have been reported for the *in situ* synthesis (79–81). These reducing agents can prevent agglomeration of nanoparticles through their stabilization. Some studies have also reported the use of BC as a reductant material to synthesize metal nanoparticles, which offers a cleaner and safer methodology for nanoparticle synthesis that is greatly effective in medical and catalytic fields (82).

1.7.5.3 *In situ* Formation of Nanostructures through Precipitation

Bacterial cellulose containing hydroxyl groups offers active sites for the adsorption of metal ions, while its nanoporous structure having sub-micron pores hosts the synthesized nanoparticles. In this method, BC is dipped in a metal ion solution followed by dipping in precipitating agent, which allows precipitation of metal nanoparticles followed by rinsing. Size and distribution of nanoparticles can be controlled by monitoring the

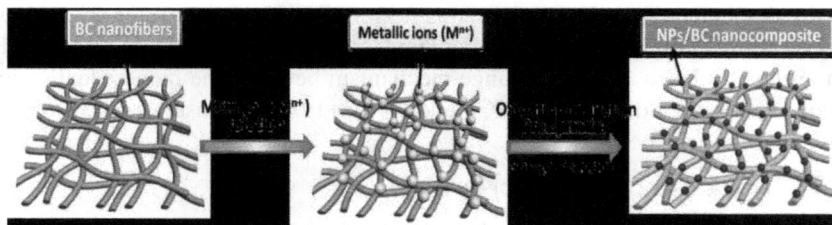

FIGURE 1.6 Schematic diagram of the *in situ* preparation of nanoparticles/BC nanocomposites (78). Figure has been reproduced with permission from Elsevier.

Source: W. Hu, S. Chen, J. Yang, Z. Li, H. Wang, Functionalized bacterial cellulose derivatives and nanocomposites, *Carbohydr. Polym.* 101 (2014) 1043–1060.

concentration of metal ions. At high concentration BC fibrils are ineffective in dispersing the nanoparticles resulting in agglomeration, while a lower concentration degrades the performance quality of the synthesized products. BC possesses sub-micron pores that readily stabilize the precipitated nanoparticles, forming a well-dispersed nanoparticle BC composite. Therefore, bacterial cellulose has proved to be a promising template for the synthesis of nanoparticles by *in situ* precipitation techniques.

Various studies have reported synthesis of metal nanoparticles by this technique such as BC-iron oxide (83), BC-magnetic nanoparticle (84), BC-ferrite (MFe_2O_4, M = Mn, Co, Ni, Cu) nanocomposites (85), BC-CdS and CdSe nanoparticles (86), BC-AgCl nanoparticles (77), and BC-Fe_3O_4 nanoparticles (87).

1.7.5.4 *In situ* Formation of Nanostructures through Sol-Gel Reaction

The ultrafine nanoporous structure of BC has been utilized by various researchers for the *in situ* generation of metal oxides through the sol-gel reaction. Inorganic ions are immobilized on to the surface of BC by immersing the matrix inside the precursor solution. Hydrolysis and condensation solidifies it into gel form, which upon heating leads to the formation of the desired oxides. These oxide nanoparticles are stabilized and dispersed over the BC template, interacting through van der Waals forces and hydrogen bonding interactions. Studies have reported several metal oxide-BC composites via sol-gel reactions such as BC-TiO_2 nanoparticles (88, 89), BC-titania networks (90), BC-silica nanocomposite (108), BC-titanium/vanadium oxide nanoparticles (88).

1.7.5.5 *Ex situ* Introduction of Components

Ex situ technique for the synthesis of functionalized BC composite involves the impregnation of solution through the *ex situ* introduction of different components. BC matrix containing polar hydroxyl groups bonds strongly with adsorbed metal ions. *Ex situ* modification of bacterial cellulose through adsorption of metal nanoparticles is much simpler and versatile than the *in situ* techniques. Various nanocomposites BC-MgO nanohybrids (91), BC-CuO nanohybrids (92), BC-silica nanoparticles (93), BC-poly(L-lactic acid) (PLLA) nanocomposites (94), BC–chitosan nanofibrils (95), and BC-montmorillonite nanocomposites have been synthesized through this methodology (96).

1.8 SELECTED ADVANCED APPLICATIONS OF MODIFIED BC

BC structural modification and combination with multiple materials have led to advanced applications in multiple fields, including environmental, medical, food, and other industrial sectors. Several shortcomings of pristine BC have been overcome through physical and chemical structural modification and composite preparations. Herein, we exemplify a few advanced applications of BC in multiple sectors based on BC structural modification and nature of combining partner.

Considering prospective environmental applications, modified BC has been employed for water and air filtration, absorption, and degradation and as pollutant sensors. Rohrbach et al. modified the cellulose-based filters with nanofibrillated cellulose (NFC), which exhibited impressive oil/water separation ability (97). In another study, the BC surface was modified with carboxymethylene to fabricate carboxymethylated BC composite (CM-BC) for effective lead and copper ions adsorption, and thus can be used for applications to purify water polluted with heavy metals (98). Furthermore, efforts were made to develop BC-based filters for air filtration. In this context, the air filter developed from cellulose fibers corrugated with non-phenolic resin provided effective absorption for small dust particles of 0.2 μm with 99.9% efficacy (99). Sukhavattanakul and Manuspiya reported the fabrication of metal nanoparticle-loaded thin BCNC film for detection of hydrogen sulfide (H_2S). The hybrid film displayed significant sensitivity for detecting H_2S at very low concentrations (60).

In medical fields, the key problem associated with clinical applications of pristine BC is its restricted surface loading and lack of functional groups to attach the bioactive molecules with therapeutic potentials. BC is insoluble in water, and common solvents restrict its surface functionalization with active chemical groups, thus hindering its use in various bioactive compounds, including drugs and proteins. Therefore, through multiple techniques the BC surface has been modified or biological features have been added to produce composite materials for achieving such goals. BC composites with hydroxyapatite, benzalkonium chloride, aloe vera chitosan, gelatin, and so on thus have been reported for effective wound dressings (2, 3). BC patches loaded with ionic liquids based on phenolic acids were developed that exhibited high compatibility, anti-inflammatory, antioxidant, and sustained drug release properties (100). In another study, BC membranes composited with chitosan and alginate layers were oxidized to achieve controlled release of epidermal growth factors under normal and infected conditions.

Besides medical and environmental uses, modified BC has been employed in other diverse applications. For example, the color, odor, shape, and flavor of BC have been modified for applications in the food industry. The water-holding and -retention features of BC were escalated by treatment in a basic medium. The treated BC was added to food, as it exhibits features of dietary sources and fat replacement (101). BC treated with red methyl was applied for developing sensors to evaluate the freshness of chicken. Likewise, BC-curcumin and BC-bromophenol-based sensors were developed for estimating the freshness of fish and guava, respectively. Furthermore, BC modified with acrylic resins, platinum, and silver have been utilized for optical, conductive, and catalytic applications, respectively (102–104).

1.9 CONCLUSIONS

BC has remained a hot research area for the last several decades. Efforts have been devoted to enhancing its production and productivity, exploring cheap raw sources, and subjecting BC to multiple applications. Considering the certain limitation associated with its physicochemical properties, strategies have been designed to modify its structure with multiple additives. The modified BC or its composites with different materials have coped impressively with BC's existing limitations and resulted in additional features and applications. Structural modifications of its composites have been developed using both pre- and post-synthetic approaches. BC has been modified with chemicals, polymers, nanomaterials, and natural materials, through enzymatic processes, and more. Enhancing its porosity, thermal and mechanical features, biocompatibility, biodegradability, and drug-carrying capabilities through the aforementioned structural modification techniques is still in progress and provides new fields of applications with each passing day.

1.10 ACKNOWLEDGMENT

This research was supported by the "The Research Council (TRC)" Oman through Block Research Funding Program (BFP/RGP/EBR/20/261) and (BFP/RGP/EBR/18/106).

REFERENCES

(1) M. Niaounakis, *Biopolymers: Applications and trends*, Oxford: William Andrew, 2015.
(2) N. Shah, M. Ul-Islam, W.A. Khattak, J.K. Park, Overview of bacterial cellulose composites: A multipurpose advanced material, *Carbohydr. Polym.* 98 (2013) 1585–1598.
(3) M.U. Islam, M.W. Ullah, S. Khan, N. Shah, J.K. Park, Strategies for cost-effective and enhanced production of bacterial cellulose, *Int. J. Biol. Macromol.* 102 (2017) 1166–1173.
(4) O.M.L. Alharbi, R.A. Khattab, I. Ali, Health and environmental effects of persistent organic pollutants, *J. Mol. Liq.* 263 (2018) 442–453.
(5) S. Zhuang, J. Wang, Removal of cesium ions using nickel hexacyanoferrates-loaded bacterial cellulose membrane as an effective adsorbent, *J. Mol. Liq.* 294 (2019) 111682.
(6) M. Ul-Islam, S. Khan, M.W. Ullah, J.K. Park, Bacterial cellulose composites: Synthetic strategies and multiple applications in bio-medical and electro-conductive fields, *Biotechnol. J.* 10 (2015) 1847–1861.
(7) M. Ul-Islam, S. Ul-Islam, S. Yasir, A. Fatima, M. Ahmed, Y.S. Lee, S. Manan, M. Wajid Ullah, Potential applications of bacterial cellulose in environmental and pharmaceutical sectors, *Curr. Pharm. Des.* 6 (2020) 5793–5806.
(8) M.W. Ullah, M. Ul-Islam, S. Khan, Y. Kim, J.H. Jang, J.K. Park, In situ synthesis of a bio-cellulose/titanium dioxide nanocomposite by using a cell-free system, *RSC Adv.* 6 (2016) 22424–22435.
(9) M.W. Ullah, M. Ul-Islam, S. Khan, Y. Kim, J.K. Park, Structural and physico-mechanical characterization of bio-cellulose produced by a cell-free system, *Carbohydr. Polym.* 136 (2016) 908–916.
(10) M.W. Ullah, S. Manan, S.J. Kiprono, M. Ul-Islam, G. Yang, Synthesis, structure, and properties of bacterial cellulose, 2019. https://doi.org/10.1002/9783527807437.ch4.

(11) M. Ul-Islam, T. Khan, W.A. Khattak, J.K. Park, Bacterial cellulose-MMTs nanoreinforced composite films: Novel wound dressing material with antibacterial properties, *Cellulose.* 20 (2013) 589–596. https://doi.org/10.1007/s10570-012-9849-3.

(12) A. Khalid, R. Khan, M. Ul-Islam, T. Khan, F. Wahid, Bacterial cellulose-zinc oxide nanocomposites as a novel dressing system for burn wounds, *Carbohydr. Polym.* 164 (2017) 214–221.

(13) A. Khalid, H. Ullah, M. Ul-Islam, R. Khan, S. Khan, F. Ahmad, T. Khan, F. Wahid, bacterial cellulose—TiO 2 nanocomposites promote healing and tissue regeneration in burn mice model, *RSC Adv.* 7 (2017) 47662–47668.

(14) M. Iguchi, S. Yamanaka, A. Budhiono, Bacterial cellulose—a masterpiece of nature's arts, *J. Mater. Sci.* 35 (2000) 261–270. https://doi.org/10.1023/A:1004775229149.

(15) M. Ul-Islam, N. Shah, J.H. Ha, J.K. Park, Effect of chitosan penetration on physicochemical and mechanical properties of bacterial cellulose, *Korean J. Chem. Eng.* 28 (2011) 1736–1743. https://doi.org/10.1007/s11814-011-0042-4.

(16) S. Khan, M. Ul-Islam, M.W. Ullah, M. Israr, J.H. Jang, J.K. Park, Nano-gold assisted highly conducting and biocompatible bacterial cellulose-PEDOT: PSS films for biology-device interface applications, *Int. J. Biol. Macromol.* 107 (2018) 865–873.

(17) S. Khan, M. Ul-Islam, M. Ikram, S.U. Islam, M.W. Ullah, M. Israr, J.H. Jang, S. Yoon, J.K. Park, Preparation and structural characterization of surface modified microporous bacterial cellulose scaffolds: A potential material for skin regeneration applications in vitro and in vivo, *Int. J. Biol. Macromol.* 117 (2018) 1200–1210.

(18) M. Ul-Islam, F. Subhan, S.U. Islam, S. Khan, N. Shah, S. Manan, M.W. Ullah, G. Yang, Development of three-dimensional bacterial cellulose/chitosan scaffolds: Analysis of cell-scaffold interaction for potential application in the diagnosis of ovarian cancer, *Int. J. Biol. Macromol.* 137 (2019) 1050–1059.

(19) T. Kamal, I. Ahmad, S.B. Khan, M. Ul-Islam, A.M. Asiri, Microwave assisted synthesis and carboxymethyl cellulose stabilized copper nanoparticles on bacterial cellulose nanofibers support for pollutants degradation, *J. Polym. Environ.* 27 (2019) 2867–2877.

(20) C. Seo, H.W. Lee, A. Suresh, J.W. Yang, J.K. Jung, Y.-C. Kim, Improvement of fermentative production of exopolysaccharides from Aureobasidium pullulans under various conditions, *Korean J. Chem. Eng.* 31 (2014) 1433–1437.

(21) M.W. Ullah, W.A. Khattak, M. Ul-Islam, S. Khan, J.K. Park, Metabolic engineering of synthetic cell-free systems: Strategies and applications, *Biochem. Eng. J.* 105 (2016) 391–405.

(22) Y. Kim, M.W. Ullah, M. Ul-Islam, S. Khan, J.H. Jang, J.K. Park, Self-assembly of biocellulose nanofibrils through intermediate phase in a cell-free enzyme system, Biochem. *Eng. J.* 142 (2019) 135–144.

(23) M. Ul-Islam, M.W. Ullah, S. Khan, J.K. Park, Production of bacterial cellulose from alternative cheap and waste resources: A step for cost reduction with positive environmental aspects, *Korean J. Chem. Eng.* 37 (2020) 925–937.

(24) M.W. Ullah, M. Ul-Islam, S. Khan, Y. Kim, J.K. Park, Innovative production of biocellulose using a cell-free system derived from a single cell line, *Carbohydr. Polym.* 132 (2015) 286–294.

(25) I.W. Sutherland, Microbial polysaccharides from Gram-negative bacteria, *Int. Dairy J.* 11 (2001) 663–674.

(26) J.H. Ha, N. Shah, M. Ul-Islam, T. Khan, J.K. Park, Bacterial cellulose production from a single sugar α-linked glucuronic acid-based oligosaccharide, *Process Biochem.* 46 (2011) 1717–1723. https://doi.org/10.1016/j.procbio.2011.05.024.

(27) M. Onodera, I. Harashima, K. Toda, T. Asakura, Silicone rubber membrane bioreactors for bacterial cellulose production, *Biotechnol. Bioprocess Eng.* 7 (2002) 289–294.

(28) K.-C. Cheng, J.M. Catchmark, A. Demirci, Effects of CMC addition on bacterial cellulose production in a biofilm reactor and its paper sheets analysis, *Biomacromolecules.* 12 (2011) 730–736.

(29) M. Hornung, M. Ludwig, H.P. Schmauder, Optimizing the production of bacterial cellulose in surface culture: A novel aerosol bioreactor working on a fed batch principle (Part 3), *Eng. Life Sci.* 7 (2007) 35–41.

(30) O. Shezad, S. Khan, T. Khan, J.K. Park, Physicochemical and mechanical characterization of bacterial cellulose produced with an excellent productivity in static conditions using a simple fed-batch cultivation strategy, *Carbohydr. Polym.* 82 (2010) 173–180.

(31) S. Tanskul, K. Amornthatree, N. Jaturonlak, A new cellulose-producing bacterium, Rhodococcus sp. MI 2: Screening and optimization of culture conditions, *Carbohydr. Polym.* 92 (2013) 421–428.

(32) J.Y. Jung, T. Khan, J.K. Park, H.N. Chang, Production of bacterial cellulose by Gluconacetobacter hansenii using a novel bioreactor equipped with a spin filter, *Korean J. Chem. Eng.* 24 (2007) 265–271.

(33) Z. Shi, Y. Zhang, G.O. Phillips, G. Yang, Utilization of bacterial cellulose in food, *Food Hydrocoll.* 35 (2014) 539–545. https://doi.org/10.1016/j.foodhyd.2013.07.012.

(34) Y. Hu, J.M. Catchmark, Formation and characterization of spherelike bacterial cellulose particles produced by Acetobacter xylinum JCM 9730 strain, *Biomacromolecules.* 11 (2010) 1727–1734.

(35) B. Hungund, S. Prabhu, C. Shetty, S. Acharya, V. Prabhu, S.G. Gupta, Production of bacterial cellulose from Gluconacetobacter persimmonis GH-2 using dual and cheaper carbon sources, *J Microb Biochem Technol.* 5 (2013) 31–33.

(36) D.R. Ruka, G.P. Simon, K.M. Dean, Bacterial cellulose and its use in renewable composites, *Nanocellulose Polym. Nanocomposites Fundam. Appl.* 32 (2014) 89.

(37) Y.-J. Kim, J.-N. Kim, Y.-J. Wee, D.-H. Park, H.-W. Ryu, Bacterial cellulose production by Gluconacetobacter sp. PKY5 in a rotary biofilm contactor, *Appl. Biochem. Biotechnol.* 137 (2007) 529–537.

(38) S. Khan, M. Ul-Islam, M.W. Ullah, M. Ikram, F. Subhan, Y. Kim, J.H. Jang, S. Yoon, J.K. Park, Engineered regenerated bacterial cellulose scaffolds for application in in vitro tissue regeneration, *RSC Adv.* 5 (2015) 84565–84573. https://doi.org/10.1039/c5ra16985b.

(39) Z. Hussain, W. Sajjad, T. Khan, F. Wahid, Production of bacterial cellulose from industrial wastes: A review, *Cellulose.* 26 (2019) 2895–2911.

(40) A.A. Alves, W.E. Silva, M.F. Belian, L.S.G. Lins, A. Galembeck, Bacterial cellulose membranes for environmental water remediation and industrial wastewater treatment, *Int. J. Environ. Sci. Technol.* 17 (2020) 3997–4008.

(41) K. Gelin, A. Bodin, P. Gatenholm, A. Mihranyan, K. Edwards, M. Strømme, Characterization of water in bacterial cellulose using dielectric spectroscopy and electron microscopy, *Polymer (Guildf).* 48 (2007) 7623–7631.

(42) A.J. Brown, XLIII.—On an acetic ferment which forms cellulose, *J. Chem. Soc. Trans.* 49 (1886) 432–439.

(43) A.N. Nakagaito, S. Iwamoto, H. Yano, Bacterial cellulose: The ultimate nano-scalar cellulose morphology for the production of high-strength composites, *Appl. Phys. A.* 80 (2005) 93–97.

(44) S. Yamanaka, K. Watanabe, N. Kitamura, M. Iguchi, S. Mitsuhashi, Y. Nishi, M. Uryu, The structure and mechanical properties of sheets prepared from bacterial cellulose, *J. Mater. Sci.* 24 (1989) 3141–3145.

(45) S. Gorgieva, J. Trček, Bacterial cellulose: Production, modification and perspectives in biomedical applications, *Nanomaterials.* 9 (2019) 1352.

(46) J.T. McNamara, J.L.W. Morgan, J. Zimmer, A molecular description of cellulose biosynthesis, *Annu. Rev. Biochem.* 84 (2015) 895–921.

(47) H. Jung, H.G. Yoon, W. Park, C. Choi, D.B. Wilson, D.H. Shin, Y.J. Kim, Effect of sodium hydroxide treatment of bacterial cellulose on cellulase activity, *Cellulose.* 15 (2008) 465–471.

(48) J.P. Silva, F.K. Andrade, F.M. Gama, Bacterial cellulose surface modifications, *Bact. Nanocellulose A Sophistic. Multifunct. Mater.* (2013) 11–91.

(49) M. Badshah, H. Ullah, A.R. Khan, S. Khan, J.K. Park, T. Khan, Surface modification and evaluation of bacterial cellulose for drug delivery, *Int. J. Biol. Macromol.* 113 (2018) 526–533.

(50) N.F. Vasconcelos, J.P.A. Feitosa, F.M.P. da Gama, J.P.S. Morais, F.K. Andrade, M. de S.M. de Souza, M. de Freitas Rosa, Bacterial cellulose nanocrystals produced under different hydrolysis conditions: Properties and morphological features, *Carbohydr. Polym.* 155 (2017) 425–431.

(51) M. Martínez-Sanz, A. Lopez-Rubio, J.M. Lagaron, Optimization of the nanofabrication by acid hydrolysis of bacterial cellulose nanowhiskers, *Carbohydr. Polym.* 85 (2011) 228–236.

(52) A.C. Corrêa, E. de Morais Teixeira, L.A. Pessan, L.H.C. Mattoso, Cellulose nanofibers from curaua fibers, *Cellulose.* 17 (2010) 1183–1192.

(53) M.C.B. de Figueirêdo, M. de Freitas Rosa, C.M.L. Ugaya, M. de S.M. de Souza, A.C.C. da Silva Braid, L.F.L. de Melo, Life cycle assessment of cellulose nanowhiskers, *J. Clean. Prod.* 35 (2012) 130–139.

(54) J.P.S. Morais, M. de Freitas Rosa, L.D. Nascimento, D.M. do Nascimento, A.R. Cassales, Extraction and characterization of nanocellulose structures from raw cotton linter, *Carbohydr. Polym.* 91 (2013) 229–235.

(55) H.-M. Ng, L.T. Sin, T.-T. Tee, S.-T. Bee, D. Hui, C.-Y. Low, A.R. Rahmat, Extraction of cellulose nanocrystals from plant sources for application as reinforcing agent in polymers, *Compos. Part B Eng.* 75 (2015) 176–200.

(56) A.L.S. Pereira, D.M. do Nascimento, J.P.S. Morais, N.F. Vasconcelos, J.P.A. Feitosa, A.I.S. Brígida, M. de F. Rosa, Improvement of polyvinyl alcohol properties by adding nanocrystalline cellulose isolated from banana pseudostems, *Carbohydr. Polym.* 112 (2014) 165–172.

(57) M.F. Rosa, E.S. Medeiros, J.A. Malmonge, K.S. Gregorski, D.F. Wood, L.H.C. Mattoso, G. Glenn, W.J. Orts, S.H. Imam, Cellulose nanowhiskers from coconut husk fibers: Effect of preparation conditions on their thermal and morphological behavior, *Carbohydr. Polym.* 81 (2010) 83–92.

(58) J. George, S.N. Sabapathi, Cellulose nanocrystals: Synthesis, functional properties, and applications, *Nanotechnol. Sci. Appl.* 8 (2015) 45.

(59) S. Xie, X. Zhang, M.P. Walcott, H. Lin, Cellulose nanocrystals (CNCs) applications: A review, *Eng Sci.* 2 (2018) 4–16.

(60) P. Sukhavattanakul, H. Manuspiya, Fabrication of hybrid thin film based on bacterial cellulose nanocrystals and metal nanoparticles with hydrogen sulfide gas sensor ability, *Carbohydr. Polym.* 230 (2020) 115566.

(61) K. JM, Mulder-Johnston CD. Edile and biodegradable polymer films, *Food Technol.* 51 (1997) 61–74.

(62) J. George, R. Kumar, V.A. Sajeevkumar, K.V. Ramana, R. Rajamanickam, V. Abhishek, S. Nadanasabapathy, Hybrid HPMC nanocomposites containing bacterial cellulose nanocrystals and silver nanoparticles, *Carbohydr. Polym.* 105 (2014) 285–292.

(63) S. Ahola, X. Turon, M. Osterberg, J. Laine, O.J. Rojas, Enzymatic hydrolysis of native cellulose nanofibrils and other cellulose model films: Effect of surface structure, *Langmuir.* 24 (2008) 11592–11599.

(64) C. Rovera, F. Fiori, S. Trabattoni, D. Romano, S. Farris, Enzymatic hydrolysis of bacterial cellulose for the production of nanocrystals for the food packaging industry, *Nanomaterials.* 10 (2020) 735.

(65) L.V. Cabañas-Romero, C. Valls, S. V Valenzuela, M.B. Roncero, F.I.J. Pastor, P. Diaz, J. Martínez, Bacterial cellulose—chitosan paper with antimicrobial and antioxidant activities, *Biomacromolecules.* 21 (2020) 1568–1577.

(66) N. Yin, R. Du, F. Zhao, Y. Han, Z. Zhou, Characterization of antibacterial bacterial cellulose composite membranes modified with chitosan or chitooligosaccharide, Carbohydr. *Polym.* 229 (2020) 115520.

(67) J. Qian, F. Sun, L. Qin, Hydrothermal synthesis of zeolitic imidazolate framework-67 (ZIF-67) nanocrystals, *Mater. Lett.* 82 (2012) 220–223.

(68) D. Li, X. Tian, Z. Wang, Z. Guan, X. Li, H. Qiao, H. Ke, L. Luo, Q. Wei, Multifunctional adsorbent based on metal-organic framework modified bacterial cellulose/chitosan composite aerogel for high efficient removal of heavy metal ion and organic pollutant, *Chem. Eng. J.* 383 (2020) 123127.

(69) L. Urbina, O. Guaresti, J. Requies, N. Gabilondo, A. Eceiza, M.A. Corcuera, A. Retegi, Design of reusable novel membranes based on bacterial cellulose and chitosan for the filtration of copper in wastewaters, *Carbohydr. Polym.* 193 (2018) 362–372.

(70) P. Cazón, G. Velazquez, M. Vázquez, Characterization of mechanical and barrier properties of bacterial cellulose, glycerol and polyvinyl alcohol (PVOH) composite films with eco-friendly UV-protective properties, *Food Hydrocoll.* 99 (2020) 105323.

(71) L. Zhang, S. Zheng, Z. Hu, L. Zhong, Y. Wang, X. Zhang, J. Xue, Preparation of polyvinyl alcohol/bacterial-cellulose-coated biochar—nanosilver antibacterial composite membranes, *Appl. Sci.* 10 (2020) 752.

(72) L. Yang, C. Chen, Y. Hu, F. Wei, J. Cui, Y. Zhao, X. Xu, X. Chen, D. Sun, Three-dimensional bacterial cellulose/polydopamine/TiO2 nanocomposite membrane with enhanced adsorption and photocatalytic degradation for dyes under ultraviolet-visible irradiation, *J. Colloid Interface Sci.* 562 (2020) 21–28.

(73) S. Ghasemi, M.R. Bari, S. Pirsa, S. Amiri, Use of bacterial cellulose film modified by polypyrrole/TiO2-Ag nanocomposite for detecting and measuring the growth of pathogenic bacteria, *Carbohydr. Polym.* 232 (2020) 115801.

(74) Z. Zhang, J. Zhang, X. Zhao, F. Yang, Core-sheath structured porous carbon nanofiber composite anode material derived from bacterial cellulose/polypyrrole as an anode for sodium-ion batteries, *Carbon N. Y.* 95 (2015) 552–559.

(75) E.Y. Wardhono, N. Kanani, Development of polylactic acid (PLA) bio-composite films reinforced with bacterial cellulose nanocrystals (BCNC) without any surface modification, *J. Dispers. Sci. Technol.* 43 (2019) 1488–1495.

(76) Y. Shu, Q. Bai, G. Fu, Q. Xiong, C. Li, H. Ding, Y. Shen, H. Uyama, Hierarchical porous carbons from polysaccharides carboxymethyl cellulose, bacterial cellulose, and citric acid for supercapacitor, *Carbohydr. Polym.* 227 (2020) 115346.

(77) W. Hu, S. Chen, X. Li, S. Shi, W. Shen, X. Zhang, H. Wang, In situ synthesis of silver chloride nanoparticles into bacterial cellulose membranes, *Mater. Sci. Eng. C.* 29 (2009) 1216–1219.

(78) W. Hu, S. Chen, J. Yang, Z. Li, H. Wang, Functionalized bacterial cellulose derivatives and nanocomposites, *Carbohydr. Polym.* 101 (2014) 1043–1060.

(79) H.S. Barud, C. Barrios, T. Regiani, R.F.C. Marques, M. Verelst, J. Dexpert-Ghys, Y. Messaddeq, S.J.L. Ribeiro, Self-supported silver nanoparticles containing bacterial cellulose membranes, *Mater. Sci. Eng. C.* 28 (2008) 515–518.

(80) G. Yang, J. Xie, F. Hong, Z. Cao, X. Yang, Antimicrobial activity of silver nanoparticle impregnated bacterial cellulose membrane: Effect of fermentation carbon sources of bacterial cellulose, *Carbohydr. Polym.* 87 (2012) 839–845.

(81) T. Zhang, W. Wang, D. Zhang, X. Zhang, Y. Ma, Y. Zhou, L. Qi, Biotemplated synthesis of gold nanoparticle—bacteria cellulose nanofiber nanocomposites and their application in biosensing, *Adv. Funct. Mater.* 20 (2010) 1152–1160.

(82) G. Yang, J. Xie, Y. Deng, Y. Bian, F. Hong, Hydrothermal synthesis of bacterial cellulose/AgNPs composite: A "green" route for antibacterial application, *Carbohydr. Polym.* 87 (2012) 2482–2487.

(83)   R.H. Marchessault, S. Ricard, P. Rioux, In situ synthesis of ferrites in lignocellulosics, *Carbohydr. Res.* 224 (1992) 133–139.

(84)   C. Katepetch, R. Rujiravanit, Synthesis of magnetic nanoparticle into bacterial cellulose matrix by ammonia gas-enhancing in situ co-precipitation method, *Carbohydr. Polym.* 86 (2011) 162–170.

(85)   N. Sriplai, W. Mongkolthanaruk, S. Pinitsoontorn, Synthesis and magnetic properties of bacterial cellulose—ferrite (MFe2O4, M= Mn, Co, Ni, Cu) nanocomposites prepared by co-precipitation method, *Adv. Nat. Sci. Nanosci. Nanotechnol.* 8 (2017) 35005.

(86)   X. Li, S. Chen, W. Hu, S. Shi, W. Shen, X. Zhang, H. Wang, In situ synthesis of CdS nanoparticles on bacterial cellulose nanofibers, *Carbohydr. Polym.* 76 (2009) 509–512.

(87)   W. Zhang, S. Chen, W. Hu, B. Zhou, Z. Yang, N. Yin, H. Wang, Facile fabrication of flexible magnetic nanohybrid membrane with amphiphobic surface based on bacterial cellulose, *Carbohydr. Polym.* 86 (2011) 1760–1767.

(88)   J. Gutierrez, S.C.M. Fernandes, I. Mondragon, A. Tercjak, Multifunctional hybrid nanopapers based on bacterial cellulose and sol—gel synthesized titanium/vanadium oxide nanoparticles, *Cellulose.* 20 (2013) 1301–1311.

(89)   R. Brandes, L. de Souza, V. Vargas, E. Oliveira, A. Mikowski, C. Carminatti, H. Al-Qureshi, D. Recouvreux, Preparation and characterization of bacterial cellulose/ TiO2 hydrogel nanocomposite, *J. Nano Res.* (2016) 73–80.

(90)   D. Zhang, L. Qi, Synthesis of mesoporous titania networks consisting of anatase nanowires by templating of bacterial cellulose membranes, *Chem. Commun.* (2005) 2735–2737.

(91)   S.S. Mirtalebi, H. Almasi, M.A. Khaledabad, Physical, morphological, antimicrobial and release properties of novel MgO-bacterial cellulose nanohybrids prepared by in-situ and ex-situ methods, *Int. J. Biol. Macromol.* 128 (2019) 848–857.

(92)   H. Almasi, L. Mehryar, A. Ghadertaj, Characterization of CuO-bacterial cellulose nanohybrids fabricated by in-situ and ex-situ impregnation methods, *Carbohydr. Polym.* 222 (2019) 114995.

(93)   A. Ashori, S. Sheykhnazari, T. Tabarsa, A. Shakeri, M. Golalipour, Bacterial cellulose/ silica nanocomposites: Preparation and characterization, *Carbohydr. Polym.* 90 (2012) 413–418.

(94)   Y. Kim, R. Jung, H.-S. Kim, H.-J. Jin, Transparent nanocomposites prepared by incorporating microbial nanofibrils into poly (L-lactic acid), *Curr. Appl. Phys.* 9 (2009) S69–S71.

(95)   S.C.M. Fernandes, L. Oliveira, C.S.R. Freire, A.J.D. Silvestre, C.P. Neto, A. Gandini, J. Desbriéres, Novel transparent nanocomposite films based on chitosan and bacterial cellulose, *Green Chem.* 11 (2009) 2023–2029.

(96)   W. Sajjad, T. Khan, M. Ul-Islam, R. Khan, Z. Hussain, A. Khalid, F. Wahid, Development of modified montmorillonite-bacterial cellulose nanocomposites as a novel substitute for burn skin and tissue regeneration, *Carbohydr. Polym.* 206 (2019) 548–556.

(97)   K. Rohrbach, Y. Li, H. Zhu, Z. Liu, J. Dai, J. Andreasen, L. Hu, A cellulose based hydrophilic, oleophobic hydrated filter for water/oil separation, *Chem. Commun.* 50 (2014) 13296–13299.

(98)   S. Chen, Y. Zou, Z. Yan, W. Shen, S. Shi, X. Zhang, H. Wang, Carboxymethylated-bacterial cellulose for copper and lead ion removal, *J. Hazard. Mater.* 161 (2009) 1355–1359.

(99)   S. Ma, M. Zhang, J. Nie, B. Yang, S. Song, P. Lu, Multifunctional cellulose-based air filters with high loadings of metal—organic frameworks prepared by in situ growth method for gas adsorption and antibacterial applications, *Cellulose.* 25 (2018) 5999–6010.

(100) Y. Choi, Y. Ahn, M. Kang, H. Jun, I.S. Kim, S. Moon, Preparation and characterization of acrylic acid-treated bacterial cellulose cation-exchange membrane, *J. Chem. Technol. Biotechnol. Int. Res. Process. Environ. Clean Technol.* 79 (2004) 79–84.

(101) S. LIN, L. Chen, H. Chen, Physical characteristics of surimi and bacterial cellulose composite gel, *J. Food Process Eng.* 34 (2011) 1363–1379.

(102) H. Yano, J. Sugiyama, A.N. Nakagaito, M. Nogi, T. Matsuura, M. Hikita, K. Handa, Optically transparent composites reinforced with networks of bacterial nanofibers, *Adv. Mater.* 17 (2005) 153–155.

(103) B.R. Evans, H.M. O'Neill, V.P. Malyvanh, I. Lee, J. Woodward, Palladium-bacterial cellulose membranes for fuel cells, *Biosens. Bioelectron.* 18 (2003) 917–923.

(104) J. Yang, D. Sun, J. Li, X. Yang, J. Yu, Q. Hao, W. Liu, J. Liu, Z. Zou, J. Gu, In situ deposition of platinum nanoparticles on bacterial cellulose membranes and evaluation of PEM fuel cell performance, *Electrochim. Acta.* 54 (2009) 6300–6305.

(105) S. Khan, M. Ul-Islam, M.W. Ullah, Y. Kim, J.K. Park. Synthesis and characterization of a novel bacterial cellulose–poly(3,4-ethylenedioxythiophene)–poly(styrene sulfonate) composite for use in biomedical applications, *Cellulose.* 22 (2015) 2141–2148.

(106) M.W. Ullah, W.A. Khattak, M. Ul-Islam, S. Khan, J.K. Park. Encapsulated yeast cell-free system: A strategy for cost-effective and sustainable production of bio-ethanol in consecutive batches, Biotechnol. Bioprocess Eng. 20 (2015) 561–575.

(107) W.A. Khattak, T. Khan, M. Ul-Islam, M.W. Ullah, S. Khan, F. Wahid, J.K. Park. Production, characterization and biological features of bacterial cellulose from scum obtained during preparation of sugarcane jaggery (gur), *J. Food Sci. Technol.* 52 (2015) 8343–8349.

(108) W. Soemphol, P. Charee, S. Audtarat, S. Sompech, P. Hongsachart, T. Dasri. Characterization of a bacterial cellulose-silica nanocomposite prepared from agricultural waste products, *Mater. Res. Express.* 7 (2020) 015085.

2 Cell-Free Nanocellulose Synthesis
Mechanism, Characterization, and Applications

*Muhammad Wajid Ullah, Mazhar Ul-Islam,
Ajmal Shahzad, Waleed Ahmad Khattak,
Shaukat Khan, Sehrish Manan, and Guang Yang*

CONTENTS

DOI: 10.1201/9781003118756-2

2.1 INTRODUCTION TO CELL-FREE SYSTEMS

The conventional microbial fermentation processes are often overshadowed by several limitations such as issues with microbial growth and pattern, cell viability, thermal instability, low yield, by-product formation, ineffective substrate utilization, and high downstream processing costs (1–3). Such issues can be overcome largely by using cell-free systems for the production of various fermentation products. A cell-free system represents a state-of-the-art conversion of a substrate into a product through a series of enzymatic reactions, with or without being catalyzed by specific cofactors (4). It is typically comprised of a transcription/translation machinery. The regulatory transcript/translation machinery contains different genetic tools (i.e., ribosomes, mRNA, tRNA, rRNA, and amino acids pool) and different regulatory elements, which collectively regulate the protein synthesis process in a cell-free environment (Figure 2.1). The concept of the cell-free system was first introduced by Buchner, who reported that biological processes could be carried out *in vitro* and demonstrated bioethanol production by a yeast-based cell-free system (5). The system relies on multienzyme reactions and aims to provide desired production formation, fewer unit operations, smaller reactor volume, higher volumetric yields, and reduced cycle duration (6). The development of new metabolic pathways or deletion/suppression of unrelated pathways could further offer high yield and productivity and enhance efficiency by funneling the energy consumption into protein synthesis. Moreover, the computational modeling of various biobricks could help improve robustness and utility and expand their applications to other fields. The cell-free approach could provide gateways to study the complex metabolic pathways in different organisms, which are otherwise difficult to study *in vivo*, and thus could provide insights to regulate and modulate such pathways. The cell-free systems offer several advantages such as tolerance to inhibitors, optimal enzyme-substrate interaction, limited enzyme loss, extended and continuous production, controlled variables, and preventing the accumulation of intermediate metabolites (2, 7). Further, the cell-free systems increase the selectivity, stability, durability, and efficiency of biochemical processes; thus, these can be used for low-cost production in a wide range of practical biotechnological applications.

After the first report by Buchner, the cell-free systems did not receive considerable attention for several decades, probably due to the lacking of potential for large-scale production of bioethanol, until the work of Welch and Scopes, who developed a reconstituted cell-free system by combining enzymes from different sources and provided desired concentrations of adenosine triphosphate (ATP). The reconstituted cell-free system effectively produced bioethanol with a much-improved yield (8). However, the addition of external ATP and enzyme purification significantly increased the production, while the enzymes were vulnerable to denaturation at high temperatures. Further, the system encountered several other limitations, such as feedback inhibition, degeneration and exhaustion of cofactors, and loss of reaction intermediates (9). Some of these issues have been addressed through the development of encapsulated yeast cell-free systems by Park and colleagues. The developed systems showed low-cost and enhanced production of bioethanol at elevated temperature in batch operation (10–14). Besides bioethanol, different cell-free systems

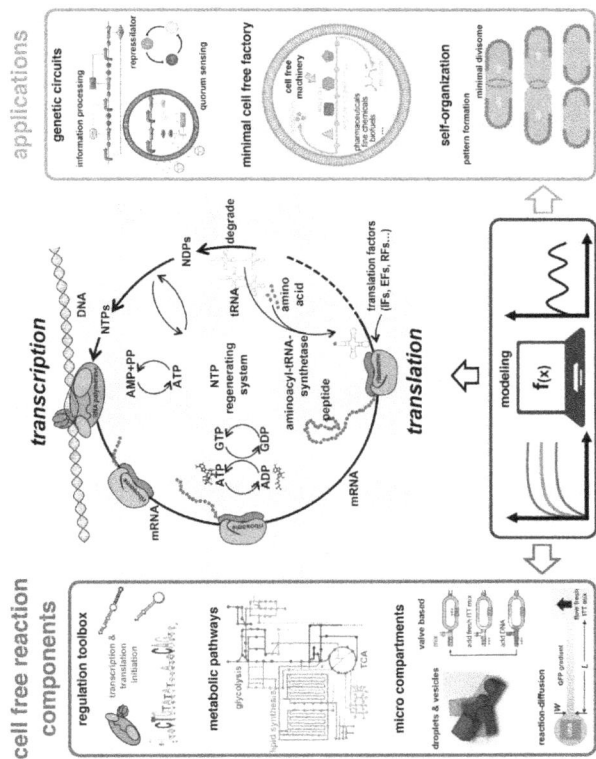

FIGURE 2.1 Illustration of cell-free protein synthesis, involved components, and application of this system in minimal cell models. A typical cell-free system contains a core transcription/translation machinery that interfaces with a large range of genetic tools and regulatory elements to control and program the protein synthesis process. Using the cell-free approach, the newly developed metabolic pathways can potentially increase the yield and productivity of the desired product, thereby mimicking the protein synthesis efficiency of a cellular system. Besides, the deletion or suppression of unrelated metabolic pathways can further enhance the efficiency by funneling the energy consumption into protein expression. Generating micro-compartments with microfluidic devices has pushed the cell-free systems toward high-throughput applications. Accurate computational modeling of various functional biobricks could improve their robustness and utility and assist in their integration. Together, these interdisciplinary approaches could be used to construct the model systems to investigate the depth of genetic circuits or fundamental aspects of self-organization, which are otherwise difficult to investigate *in vivo*, and ultimately facilitate the engineering of artificial cell factories. Figure reproduced from (26).

Source: Jia H, Heymann M, Bernhard F, et al. (2017) Cell-free protein synthesis in micro compartments: Building a minimal cell from biobricks. *N Biotechnol* 1–7. https://doi.org/10.1016/j.nbt.2017.06.014.

have been developed for production of bio-hydrogen (15–18), protein scaffolds (19), synthetic amino acids (20), and cell-free proteins including complex proteins (21, 22), toxic proteins (23), membrane proteins (24, 25), functional antibodies (25), and novel proteins with unusual amino acids (27, 28).

The use of synthetic biology principles, such as DNA synthesis through PCR, RNA synthesis through *in vitro* transcription, and production of polypeptides by *in vitro* transcription/translation (16), has greatly revolutionized the concept of cell-free systems. Synthetic biology focuses on constructing and implementing synthetic cell-free metabolic systems based on genetic and metabolic engineering-driven approaches. The lack of comprehensive information on their durability, the complexity of cellular processes, and the lack of control between *in vitro* and *in vivo* systems prompted metabolic engineers to reprogram the existing systems and design novel biological systems (18, 29). Moreover, the engineering of a cell is often complex, laborious, and expensive and has several limitations such as maintaining cellular viability and physiology and the presence of an external barrier in the form of a cell wall or membrane (30). Such issues have been greatly resolved by using advanced strategies such as compartmentalization, metabolic channeling, and co-immobilization (6, 25). Although the potential use of the synthetic cell-free system is largely unexplored, its potential benefits are quite clear.

The current chapter provides an overview of the concept and a brief history of cell-free technology and the latest developments. It mainly describes the synthesis of bacterial nanocellulose by using the cell-free system with a focus on its synthesis chemistry, molecular regulation, *in vitro* assembly and *de novo* synthesis of cellulose nanofibrils, and its molecular regulation. It also provides a comparison between the whole microbial cell system and the cell-free system for the biosynthesis of cellulose, as well as the difference between cellulose produced by both systems in terms of its structural, physicochemical, thermal, mechanical, and biological properties. The cell-free approach can be effectively used as a unique approach for one-pot *in situ* synthesis of cellulose-based nanomaterials, especially with bactericidal elements.

2.2 BACTERIAL NANOCELLULOSE

Bacterial nanocellulose (BNC), also known as bacterial cellulose (BC), is a natural polymer produced by several bacterial genera, including *Acetobacter*, *Rhizobium*, *Agrobacterium*, *Aerobacter*, *Achromobacter*, *Azotobacter*, *Salmonella*, *Escherichia*, and *Sarcina* (31, 32), a few algae such as *Valonia* and *Chaetamorpha* spp., fungi (33–35), and cell-free systems (36, 37) by utilizing glucose and other carbon sources. It is produced within the microbial cells in the form of β-1,4-glucan chains, which are excreted as protofibrils through the terminal complexes (TCs) located in the external barrier of microbial cells. These β-1,4-glucan chains crystallize and form the ribbon-shaped microfibrils, which ultimately form pellicles comprised of bundles (37, 38). In the microbial cell system, it is produced aerobically at the air-medium interface in the form of a hydrogel or a membrane (39). This unique synthesis method and crystallization provide BNC with unique structural, physicochemical, thermal, mechanical, and biological properties as compared to plant cellulose (40). Its unique properties include ultra-purity, high crystallinity (60%–90%), a high degree of polymerization

(DP), three-dimensional (3D) fibrous structure, moldability into different shapes, high hydrophilicity, and a high water-retaining capacity (40, 41).

Currently, BNC research is mainly focused on improving the yield and productivity, enhancing the existing and imparting novel features, minimizing the production cost, producing on an industrial scale, and expanding its applications to different fields. To this end, extensive efforts have been carried out by identifying novel microbial strains and genetic engineering of known strains (42), exploring low-cost substrates (43, 44), developing advanced reactors such as rotating disc reactors, rotary biofilm contactors, spin filter bioreactors, and silicon membrane reactors (43), and functionalizing BNC by developing its composites with different materials (45, 46). It is receiving immense consideration for applications in different areas such as medicine (47–49), bioprinting (50–52), food (53), energy (54), optoelectronics (55, 56), sensors (57, 58), cosmetics (59, 60), and the environment (61–63). Specifically, it is a preferable choice for biomedical applications such as wound healing, tissue engineering, and drug delivery systems (64–68), owing to its specialized properties for such applications.

2.2.1 SYNTHESIS CHEMISTRY

Synthesis of cellulose from glucose is a biochemical process in which each step is mediated by a specific enzyme, with or without the help of a cofactor. In general, cellulose synthesis from glucose by microbial cells is a four-step process (69). In the first step, monosaccharides are activated through the formation of sugar nucleotides. Herein, the sugar nucleotides provide various monosaccharides by means of interconversion through epimerization, dehydrogenation, and decarboxylation reactions, each catalyzed by a specific enzyme. The second step involves the assembly of repeating units through the sequential addition of sugar units via polymerization. During this step, the cellulose backbone is formed by the sequential addition of d-glucose-1-phosphate and UDP-glucose, catalyzed by the UDP-glucose pyrophosphorylase and cellulose synthase, respectively. In the third step, the acyl groups (if present) are simultaneously added to the glucan units. The specific sugar transferases, such as 1-acyl-sn-glycerol-3-phosphate acyltransferase, carry out the transfer of an acyl group and form the acetyl Co-A (initiator of the TCA cycle). The sequential transfer of different monosaccharides and acyl groups from their respective donors by highly specific sugar transferases determines the structure of the polysaccharide. In the final step, the produced glucan chains are excreted to the external environment across the external barrier, that is, a cell wall or membrane. It is an energy-dependent step that utilizes cellular ATP molecules. It is a complex process that is readily affected by several mutations in the gene regulating the enzymes involved, which usually results in the internalization of cellulose to the periplasm with inevitable lethal effects.

Ullah et al. confirmed the presence of several enzymes for carbohydrate metabolism in the crude extract of *G. hansenii*-based cell-free system via LC-MS/MS LTQ Orbitrap analysis (Table 2.1). The enzymes glucokinase, phosphoglucomutase, UDP-glucose pyrophosphorylase, and cellulose synthase are directly involved in cellulose synthesis from D-glucose (Figure 2.2). Glucose-6-phosphate (G6P) produced by glucokinase serves as a substrate not only for the principal cellulose synthesis pathway

TABLE 2.1

Illustration of the *G. hansenii* PJK Enzymes in the Cell-Free Extract Involved in the Biocellulose Synthesis. The enzymes were analyzed by LC-MS/MS LTQ Orbitrap using the Mascot algorithm.

Accession no.	GI no.	Description	Taxonomy	Mass (Da)	PI
EFG85049.1	gi|295978312	Glucokinase	*G. hansenii*	34282	6.51
EFG84192.1	gi|295977434	Phosphoglucomutase	*G. hansenii*	59762	6.03
EFG85649.1	gi|295978924	UDP-glucose pyrophosphorylase	*G. hansenii*	22755	4.36
EFG83224.1	gi|295976446	Cellulose synthase catalytic subunit (UDP-forming)	*G. hansenii*	175801	6.94
EFG84542.1	gi|295977790	UDP-D-glucose dehydrogenase	*G. hansenii*	47825	5.61
EFG83101.1	gi|295976316	1-acyl-sn-glycerol-3-phosphate acyltransferase	*G. hansenii*	41281	11.15
EFG83324.1	gi|295976549	Diguanylate cyclase/ phosphodiesterase	*G. hansenii*	84871	6.11
EFG83841.1	gi|295977078	Glucose dehydrogenase	*G. hansenii*	84604	5.73
WP003620002.1	gi|489715879	Aldolase	*G. hansenii*	29824	6.42
EFG84043.1	gi|295977283	Triosephosphate isomerase	*G. hansenii*	25791	4.96
EFG82935.1	gi|295976147	Pyruvate dehydrogenase (acetyl-transferring)	*G. hansenii*	34832	5.35
EFG85628.1	gi|295978903	Glucose-6-phosphate dehydrogenase	*G. hansenii*	57043	6.04

Source: Reproduced from Khattak WA, Ullah MW, Ul-Islam M, et al. (2014b). (36)

but also for the pentose phosphate (PP) and tricarboxylic acid (TCA) pathways. The flux of G6P through the two routes determines the level of cellulose synthesis. Similarly, UDP-glucose serves as a common substrate for UDP-glucose dehydrogenase and cellulose synthase; thus, its flux through the two pathways determines the level of cellulose synthesis. UDP-glucose dehydrogenase requires NAD as a cofactor; thus, a low concentration of NAD may partially inhibit its activity, thus could potentially lead to high cellulose production. In addition, the presence of diguanylic cyclase and two phosphodiesterases (PDE-A and PDE-B) may also serve to regulate the cellulose synthase activity through a specific activator—bis-(3→5)-cyclic diguanylic acid. The presence of PDE-A and PDE-B aids in accelerating the polymerization of UDP-glucose into cellulose. In addition to the principal cellulose synthesis pathway, glucose can potentially enter other pathways such as Entner-Doudoroff (ED) and PP pathways, which produce triose sugar that is subsequently converted

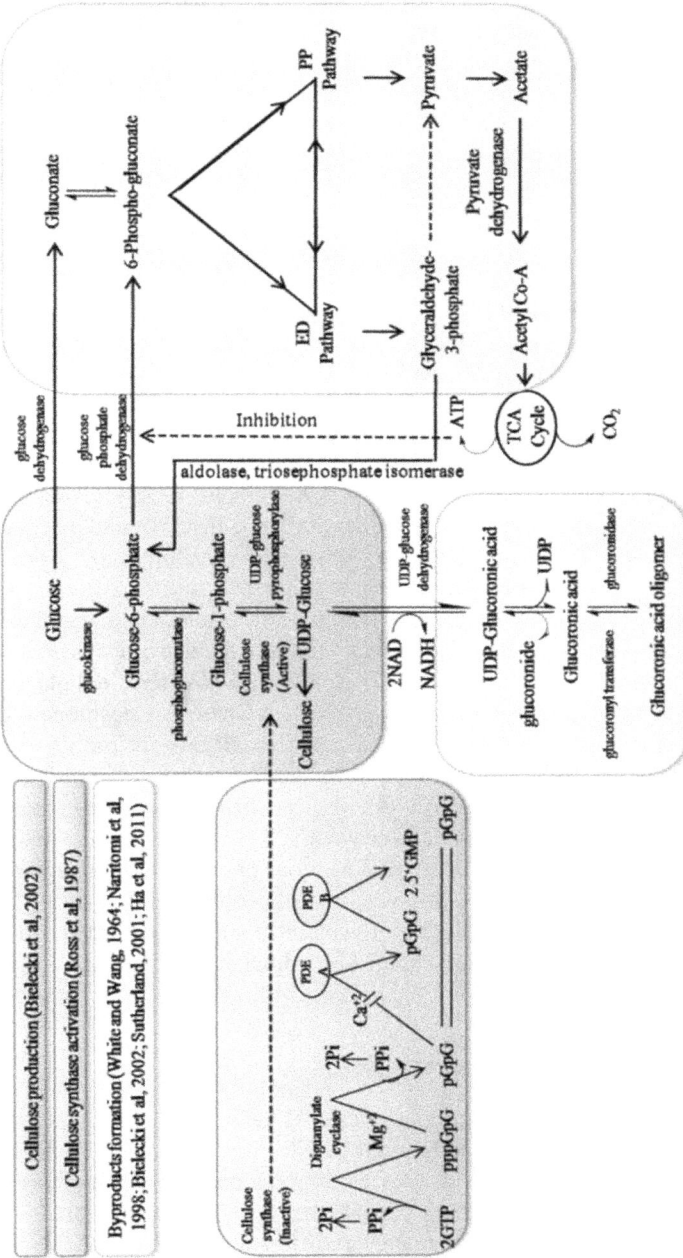

FIGURE 2.2 Schematic representation of biocellulose production by the cell-free system through the principal glucose pathway and other pathways interconnected through the activation of cellulose synthase. The scheme was developed based on the literature review and the results of LC-MS/MS LTQ Orbitrap analysis. Figure reproduced from (36).

Source: Khattak WA, Ullah MW, Ul-Islam M, et al. (2014b) Developmental strategies and regulation of cell-free enzyme system for ethanol production: A molecular prospective. *Appl Microbiol Biotechnol* 98. https://doi.org/10.1007/s00253-014-6154-0.

into pyruvate. The pyruvate is converted into acetyl-Co-A by pyruvate dehydrogenase, which serves to produce bulk ATPs to enhance the efficacy of the cell-free system for cellulose production. It is worth mentioning here that a high concentration of ATP inhibits G6P dehydrogenase and reduces its activity by several folds; thus, it may retard the PP pathway and favor cellulose synthesis through the principal pathway in the cell-free system. Overall, each step in the cellulose synthesis by a cell-free system is regulated by the availability of specific enzymes, substrate, and/or cofactor, and the absence of any of these components retard or abolishes the cellulose synthesis pathway.

2.2.2 MOLECULAR REGULATION OF NANOCELLULOSE SYNTHESIS

At least four different genes regulate BNC synthesis by microbial cells named *axcess A*, *axcess B*, *axcess C*, and *axcess D* that are arranged in the form of an operon and encode proteins AxCESA, AxCESB, AxCESC, and AxCESD, respectively (70). Usually, a point mutation (i.e., single base change) results in the frameshift of the operon (71). The AxCESA and AxCESB proteins catalyze and regulate the polymerization of individual β-1,4-glucan chains, where the former contains the amino acid motifs suggestive of processive β-glycosyltransferase and bind to UDP-glucose (72). The AxCESB protein reversibly binds to large quantities of c-di-GMP and activates the AxCESB, where c-di-GMP autoinhibit the enzyme by breaking a salt bridge, which otherwise tethers a conserved gating loop that controls and coordinates the substrate at the active site (71, 73). A disruption of the salt bridge via mutagenesis usually generates a constitutively active cellulose synthase (CS). The c-di-GMP-activated BcsA-B complex contains a nascent cellulose polymer whose terminal glucose unit resides at a novel location above the BcsA's active site where it is positioned for catalysis (73). The other two proteins—AxCESC and AxCESD—are proposed to mediate the extrusion of β-1,4-glucan chains and their subsequent crystallization into the cellulose subelementary fibrils (SEF) (63, 74). The AxCESD is a cylinder-shaped protein formed by an octamer as a functional unit where all N-termini of the octamer are positioned inside the AxCESD cylinder and form four-passageways for separate extrusion of four β-1,4-glucan chains along with the dimer interface in a twisted manner to the external medium (75). These four protein subunits and perhaps other components form the cellulose-producing complexes, which can be viewed as pores (i.e., TCs) at the microbial cell surface.

2.2.3 IN VITRO ASSEMBLY OF CELLULOSE NANOFIBRILS

The β-1,4-glucan chains produced inside the microbial cells are excreted into the medium through TCs, which aggregate and form the BNC hydrogel at the air-medium interface (Figure 2.3). TCs are located as a series of pores on the outer membrane of the microbial cells and are observed as particle arrays by freeze-fracture electron microscopy and adopt varying arrangements in different organisms (37, 38, 70). These are present as a single linear row in *A. xylinum* and brown and red algae; as a solitary rosette in green algae and land plants; as multiple rows in glaucophycean algae, some red and green algae, slime mold, and tunicates; and as diagonal rows of

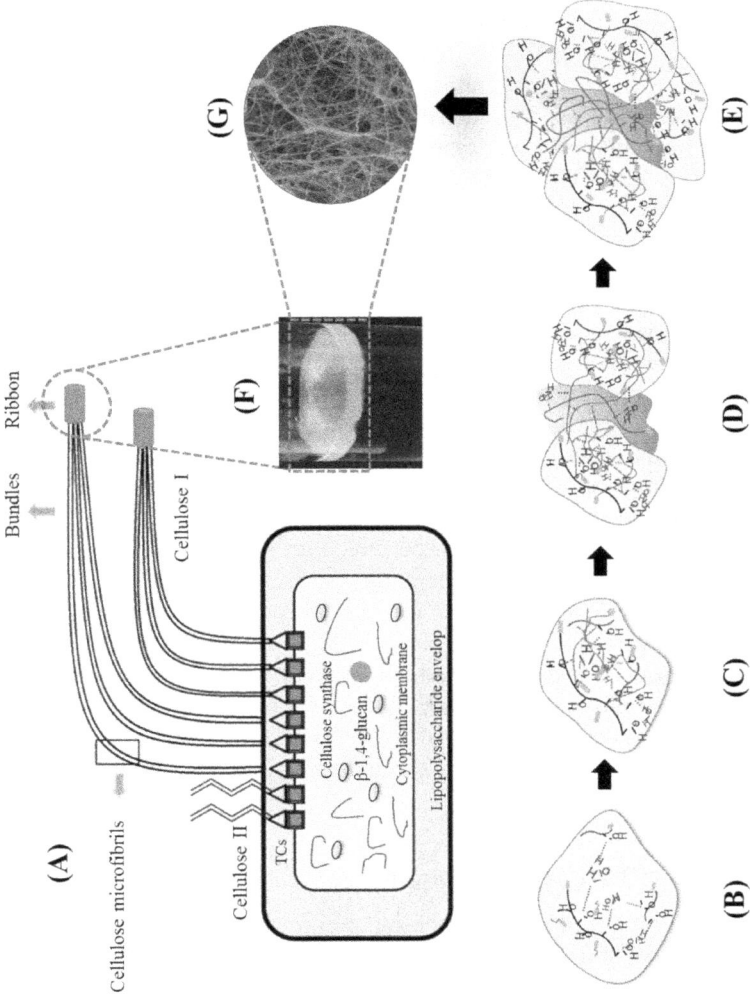

FIGURE 2.3 Schematic illustration of (A) synthesis of β-1,4-glucan chains and their excretion from the bacterial cells across the cell wall through TCs, involving the (B) synthesis and aggregation of fibrils, (C) formation of pellicles, (D) movement of pellicles toward the air-medium interface due to density gradient (cell-free system), and (E) and formation of BC sheet at the air-medium interface in the form of (F) a hydrogel, which is seen as (G) the reticulated fibrous structure forming a network of cellulose fibers. Figure adapted from (37, 48).

Source: Kim Y, Ullah MW, Ul-Islam M, et al. (2019) Self-assembly of bio-cellulose nanofibrils through intermediate phase in a cell-free enzyme system. *Biochem Eng J* 142:135–144. https://doi.org/10.1016/j.bej.2018.11.017.

Lin FC, Brown RM, Drake RR, Haley BE (1990) Identification of the uridine 5'??-diphosphoglucose (UDP-Glc) binding subunit of cellulose synthase in Acetobacter xylinum using the photoaffinity probe 5-Azido-UDP-Glc. *J Biol Chem* 265:4782–4784.

particles in xanthophycean algae (76). TCs are made up of four subunits: AxCESA, AxCESB, AxCESC, and AxCESD (70). Among these, the AxCESC and AxCESD are proposed to mediate the extrusion of β-1,4-glucan chains and their subsequent crystallization during the cellulose assembly (72); however, their complete action mechanism is yet to be explored. The extruded β-1,4-glucan chains aggregate and form the randomly and loosely arranged twisted subelementary fibrils and form the premature cellulose. These loosely arranged fibrils subsequently become compact through crystallization and form the well-arranged ribbons and bundles (39) and ultimately a BNC hydrogel or a membrane. The thickness of hydrogel increases with time when more β-1,4-glucan chains are secreted, extruded through TCs, crystallized as bundles and ribbons, and ultimately added to the hydrogel (77). The hydrogel grows downward until all microbial cells are trapped inside and become inactive or die due to the depletion of oxygen (39, 78). The described molecular configuration of BNC promotes its unique structural, physicochemical, thermal, mechanical, and biological features, which are rated better than the plant cellulose (40).

Unlike in microbial cell systems where BC is produced aerobically, the cell-free fermentation operates under anaerobic conditions; thus, it minimizes the risk of contamination by other microorganisms. In a cell-free system, the enzymes produce β-1,4-glucan chains inside the culture medium, which self-aggregate via hydrogen bonding and form the elementary microfibrils (77). Thereafter, a higher order of cellulose organization is achieved through the self-aggregation of multiple elementary fibrils through hydrogen bonding and forms the macrofibrils (39, 79). The thickness of macrofibrils increases when more β-1,4-glucan chains are synthesized and added to the preexisting ones, resulting in the formation of bundles that finally turn into ribbons. Compared to the microbial cell system, the cell-free system produces visibly noticeable cellulose nanofibrils within the culture medium, which form a few loosely arranged clumps; however, initially at a much slower rate. The density and thickness of such clumps increase with time and form small-sized pellicles, which move toward the surface due to low density than the medium and form a thinner sheet at the air-medium interface. The thickness of the sheet keeps on increasing with time as more pellicles are added. In a cell-free system, three different forms of cellulose morphologies, such as premature cellulose, pellicles, and sheets, can be observed (67, 68).

2.2.4 DE NOVO SYNTHESIS OF CELLULOSE NANOFIBRILS

As described earlier, BNC is produced in the cytoplasm of microbial cells, whereas final crystallization of cellulose fibrils takes place in or near the outer membrane; this process is largely dependent on the ordered nature of TCs (80). Compared to the microbial cell system, the cell-free systems offer *de novo* synthesis of β-1,4-glucan chains *in vitro*, which crystallize and form the premature cellulose and cellulose pellicles within the culture medium (37, 38). The *de novo* synthesis of various bioproducts by using cell-free systems has been well established (81, 82). Kim et al. reported that the cellulose synthesis by the cell-free system was indeed a *de novo* synthesis process due to the following facts: (1) nature of preculture used in the development of the cell-free system, (2) successful repeated batch operation, and (3) decreased glucose concentration during the incubation. The synthesis of nascent β-1,4-glucan chains in each batch and their accumulation over time inside the culture medium and the amount of decreased glucose concentration indicates their *de novo* synthesis.

2.3 COMPARISON OF MICROBIAL AND CELL-FREE BIOSYNTHESIS OF CELLULOSE

BNC synthesis by the cell-free system differs from the microbial cell system in several aspects, as discussed later and demonstrated in Figure 2.4. Further, a comparative analysis of the production and features of cellulose produced by the microbial cell and the cell-free systems is given in Table 2.2.

2.3.1 ANAEROBIC CULTIVATION

BNC synthesis by microbial cell system is an aerobic synthesis. The cellulose-producing bacterial cells are mostly facultative or strict aerobe; thus, a continuous supply of oxygen is required for normal growth. However, this process of providing oxygen increases the risk of contamination by other microbial cells (83–85). In contrast, the cell-free system does not require the presence of oxygen due to the absence of microbial cells in the system, thus bypassing the risk of any contamination.

2.3.2 MEDIUM COMPOSITION

BNC synthesis by microbial cell system requires an expensive and complex medium that essentially contains carbon and nitrogen sources as well as other minerals. The carbon source in the medium is partly utilized for the growth and proliferation of microbial cells while the rest is converted into cellulose, thus significantly reducing the overall yield in terms of cellulose production at the account of the consumed carbon source. In contrast, the provided carbon source is solely converted into cellulose by the enzymes in the cell-free system, representing a high yield.

2.3.3 ENERGY REQUIREMENT

The extrusion of β-1,4-glucan chains from microbial cells to the external environment is a less-known phenomenon during which the bacterial signaling molecule c-di-GMP activates the BNC synthesis, which is then translocated across the TCs by a complex of cellulose synthase as described previously. It is an energy-dependent process and utilizes cellular ATP molecules (69). Thus, a well-organized system is required to ensure efficient extrusion of β-1,4-glucan chains across the cell membrane into the external environment. In contrast, a cell-free system is an open system and bypasses the complicated energy-dependent extrusion process. The β-1,4-glucan chains are directly synthesized within the culture medium, which float as uniformly distributed microfibrils throughout the culture medium and are accumulated over time and form a hydrogel or membrane at the air-medium interface as described earlier. Further, it bypasses the internalization of cellulose, which sometimes occurs in

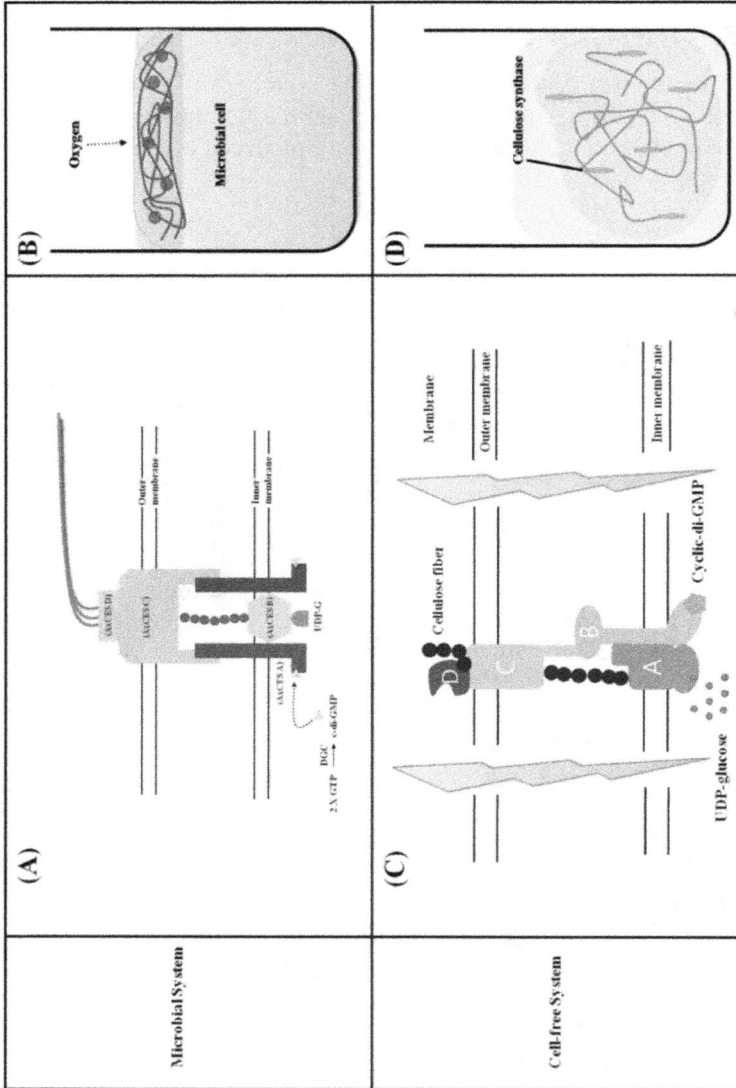

FIGURE 2.4 Illustration of schematic models of (A) cellulose-synthesizing machinery in the microbial system, (B) production of cellulose in the microbial system, (C) cellulose-synthesizing machinery in a cell-free system, and (D) synthesis of biocellulose in the cell-free system. Figure reproduced from (37).

Source: Kim Y, Ullah MW, Ul-Islam M, et al. (2019) Self-assembly of bio-cellulose nanofibrils through intermediate phase in a cell-free enzyme system. *Biochem Eng J* 142:135–144. https://doi.org/10.1016/j.bej.2018.11.017.

TABLE 2.2
A Comparison of Production and Features of Cellulose Produced by the Microbial Cell and the Cell-Free Systems.

Parameters	Microbial cell systems	Cell-free system
Cultivation method	Aerobic	Anaerobic
Medium composition	Complex	Simple
Yield	Relatively low	Relatively high
Energy requirement	High	Low
Cellulose type	Type-I	Type-II
Crystallinity	Relatively high	Relatively low
Water-holding capacity	Relatively low	Relatively high
Water release rate	Rapid	Prolonged
Thermal stability	Relatively low	Relatively high
Mechanical strength	Relatively low	Relatively high
Fiber diameter	Relatively thick	Relatively thin
Porosity	Relatively high	Relatively high
Cost	High	Low

the microbial cell system due to a mutation in the genes regulating this process and causes inevitable lethal effects.

2.3.4 CELLULOSE PRODUCTION EFFICIENCY

As described earlier, the utilization of glucose by microbial cells for growth and proliferation reduces the overall yield; a comparison of yield in terms of consumed glucose by microbial and cell-free systems was reported by (70). The comparative efficacy of both systems for cellulose production was evaluated after 3, 5, 10, and 15 days under the same experimental conditions. Initially (up to Day 5), the amount of cellulose produced by the microbial cell system was much more than the cell-free system; however, the yield was higher for the cell-free system. The low cellulose yield by the microbial system was due to the parallel glucose consumption for growth and proliferation along with cellulose production. During the 5–15 days, the rate of cellulose production was superseded by the cell-free system and showed a 57.68% yield as compared to the 39.62% yield by the microbial cell system. Overall, the study showed that both systems produced a comparable amount of cellulose; however, the cell-free system produced a significantly higher yield than the microbial cell system.

2.4 CHARACTERIZATION OF CELL-FREE NANOCELLULOSE

The structural properties of nanocellulose vary according to its synthesis method, which in turn decides its specific applications in different fields.

2.4.1 STRUCTURAL MORPHOLOGY

Scanning electron microscopic observation of cellulose produced by microbial cells (Figure 2.5) and cell-free systems (Figure 2.6) reveals a three-dimensional arrangement of web-shaped nanofibrils, where the fibrils are interconnected through inter- and intramolecular hydrogen bonds that stabilize the reticulate structure (72). The fibrils are loosely arranged with plenty of empty spaces, forming the pores, and increase the surface area (86–88). The nanocellulose produced by both microbial and cell-free systems varies in size and density of nanofibrils on top and bottom surfaces and fiber diameter. Kim et al. reported that the fiber diameter of premature

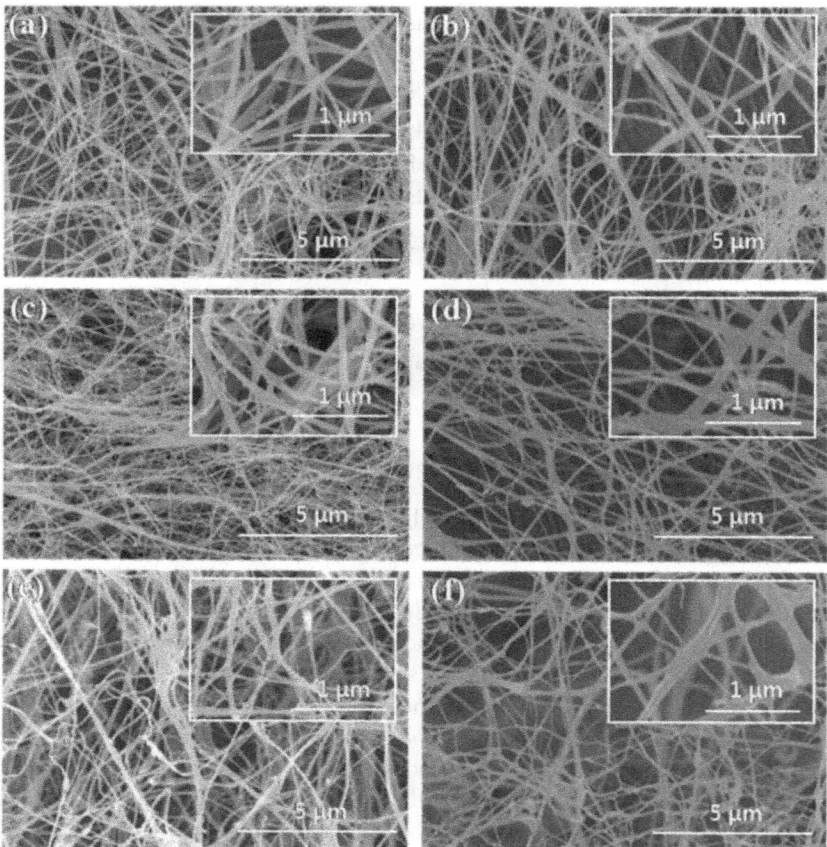

FIGURE 2.5 Field-emission scanning electron microscopy analysis of (a) top and (b) bottom surfaces of 5-day BC, (c) top and (d) bottom surfaces of 10-day BC, and (e) top and (f) bottom surfaces of 15-day BNC produced under static conditions at 30 °C and pH 5.0. Figure reproduced from (67).

Source: Schrecker ST, Gostomski PA (2005) Determining the water holding capacity of microbial cellulose. *Biotechnol Lett* 27:1435–1438. https://doi.org/10.1007/s10529-005-1465-y.

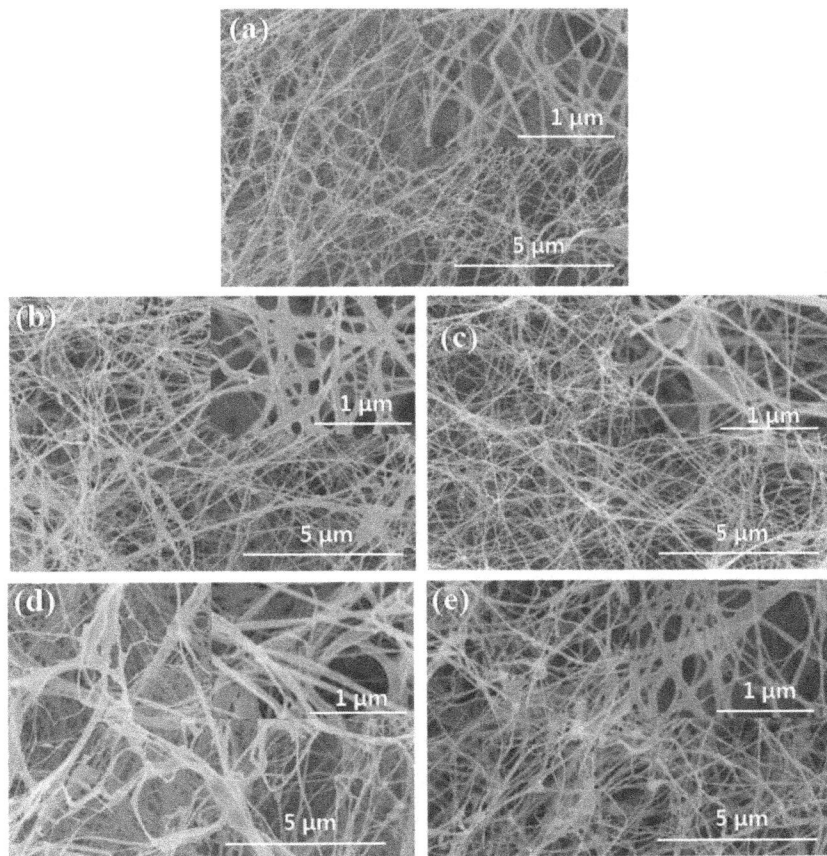

FIGURE 2.6 Field-emission scanning electron microscopy analysis of (a) top surface of 5-day premature biocellulose, (b) top and (c) bottom surfaces of the 10-day biocellulose pellicle, and (d) top and (e) bottom surfaces of 15-day biocellulose sheet produced under static conditions at 30 °C and pH 5.0. Figure reproduced from (37).

Source: Kim Y, Ullah MW, Ul-Islam M, et al. (2019) Self-assembly of bio-cellulose nanofibrils through intermediate phase in a cell-free enzyme system. *Biochem Eng J* 142:135–144. https://doi.org/10.1016/j.bej.2018.11.017.

cellulose (59.5±6.4 nm) was lower compared to both the top (68.1±12.6 nm) and bottom (95.3±6.5 nm) surfaces of BNC (67) due to the slow synthesis of cellulose during the early phase of incubation in the cell-free environment. In contrast, the bottom diameter of the 5-day BNC was higher than the top surface due to the addition of newly synthesized fibrils at the top surface by microbial cells (89). Similarly, the bottom diameter of cellulose produced by the cell-free system remained lower than BNC after 10 days due to its incomplete growth of pellicles. After 15 days, the top and bottom diameters of cell-free nanocellulose varied significantly due to the addition of more pellicles from the bottom side to the pre-existing pellicles at the top. In contrast, no significant difference in top and bottom diameters of BNC was observed, which could be because of the halting of cellulose production by microbial cells due to depletion of oxygen, deficiency of nutrients, or death of microbial cells (39, 78). Overall, this study demonstrates a completely different nanofibril diameter trend at the top and bottom surfaces in microbial and cell-free systems. This variation could be attributed to the different synthesis mechanisms, consumption of oxygen during the early phase in the culture medium of microbial cells, and different *in vitro* assembly of cellulose nanofibrils (37, 77, 90). Overall, a marked difference occurs between the microbial and cell-free systems in that the newly synthesized cellulose nanofibrils are located at the top surface in BNC while the older ones are at the bottom surface, and vice versa in the cell-free systems, indicating the independence of the cell-free system from the availability of oxygen.

2.4.2 POLYMORPHIC STRUCTURE

X-ray diffraction analysis of nanocellulose produced by the cell-free system demonstrates cellulose II polymorphic structure as compared to the cellulose I structure by the BNC produced by the microbial cell system (Figure 2.7) (38). In the microbial cell

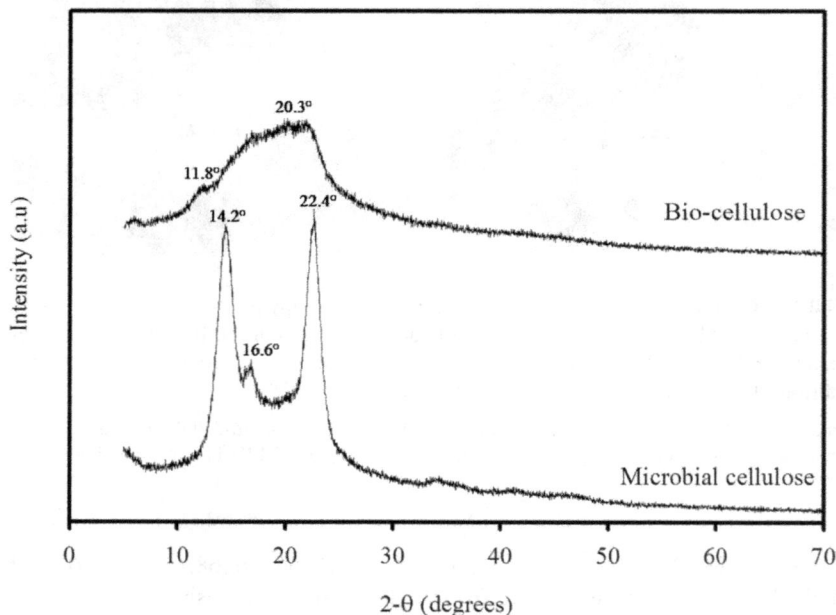

FIGURE 2.7 X-ray diffraction patterns of cellulose produced by the bacterial cell and cell-free systems. Both cellulose types were produced statically at 30 °C and pH 5.0 and freeze-dried before analysis. Figure reproduced from (68).

Source: Seo C, Lee HW, Suresh A, et al. (2014) Improvement of fermentative production of exopolysaccharides from Aureobasidium pullulans under various conditions. *Korean J Chem Eng* 31:1433–1437. https://doi.org/10.1007/s11814-014-0064-9.

system, the newly synthesized β-1,4-glucan chains align with each other due to the movement of bacterial cells and lock into a specific crystalline arrangement to form the cellulose I structure (76). In contrast, the β-1,4-glucan chains are randomly synthesized within the culture medium, which fold into thermodynamically stable cellulose II structure or remain as non-crystalline cellulose. The cellulose I structure can also be obtained *in vitro* under specific conditions such as by using detergent, as reported for several plant species such as cotton fibers (91), mung bean (92), and blackberry (93).

2.4.3 Physiological Properties

The empty spaces (i.e., pores) present in the morphological structure of cellulose allow the absorption of liquids (e.g., water and polymer solutions) and solid particles (e.g., medium components and other particles), which result in its swelling. Several studies have reported that BNC can accommodate 100–200 times its dry weight in water (94, 95). This ability to accommodate liquid and solid molecules by nanocellulose is important in making its composites with other materials, and the moist environment is beneficial during the wound healing process where it facilitates the penetration of active substances, enables easy and painless recovery, accelerates wound healing, and prevents damage to newly formed skin (90). The water molecules reside in the porous matrix through physical adsorption and through hydrogen bonding (96). In addition to its porous nature, BNC's 3D fibril arrangement, high surface area per unit mass, and hydrophilic nature endow it with high water-holding capacity (WHC) and slow water release rate (WRR) (86, 94, 95). In addition to high WHC, cellulose follows a slow and steady water release due to its well-organized and uniformly distributed nanofibrils. These features are highly dependent on the activity of producing microbial cells and the synthesis method, chemical composition of the production medium, culture time and method, amount of inoculum, and the carbon source (97, 98). A comparison of WHC and WRR from cellulose produced by the microbial cells and the cell-free system is shown in Figure 2.8, which demonstrates a high WHC and slow WRR by the cellulose produced by the cell-free system as compared to the microbial cell system (38).

2.4.4 Thermal Properties

The thermal properties of nanocellulose depend on its source, crystallinity, and flexibility. Thermal degradation of nanocellulose takes place in three steps: dehydration (i.e., removal of water molecules), depolymerization (i.e., the breakage of glycosidic linkages among the glucose units), and decomposition (i.e., degradation of glycosyl units), which ultimately results in the formation of charred residue (99). Ullah et al. reported that cellulose produced by the cell-free system was thermally more stable even than BNC owing to its more compact and well-arranged fibril structure. They reported that the onset and endset temperatures of BNC and cell-free nanocellulose were 246 °C and 297 °C, and 327 °C and 352.7 °C, during which both cellulose types lost 87% and 84.7% of their original weights, respectively (Figure 2.9). The weight loss in both celluloses was almost comparable at the end of the experiment. Compared to BNC, the cell-free nanocellulose showed better thermal stability and was comparable to the regenerated cellulose (cellulose II) (100).

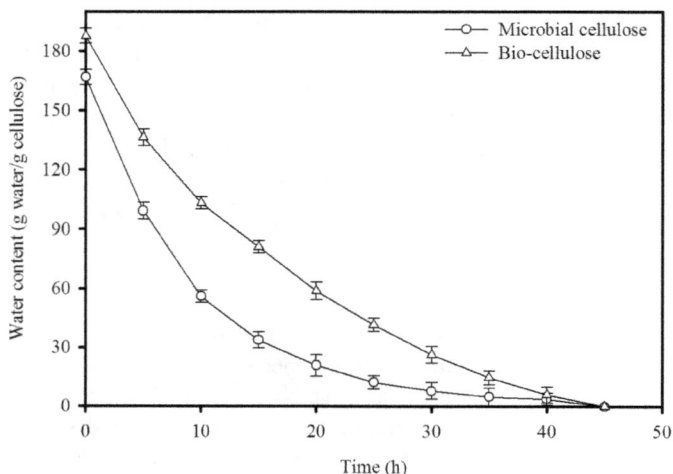

FIGURE 2.8 The water-holding capacity and release rate of bacterial nanocellulose and cell-free cellulose, produced by microbial cells and the cell-free system, respectively. Both cellulose types were produced statically at 30 °C and pH 5.0. Figure reproduced from (38).

Source: Klemm D, Schumann D, Udhardt U, Marsch S (2001) Bacterial synthesized cellulose—Artificial blood vessels for microsurgery. *Prog Polym Sci* 26:1561–1603. https://doi.org/10.1016/S0079-6700(01)00021-1.

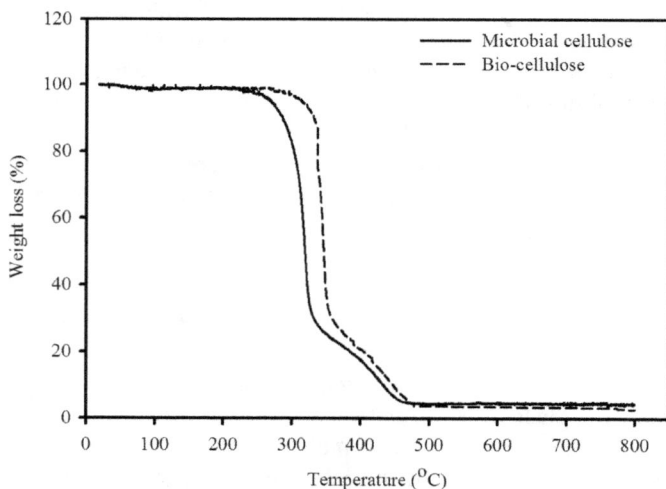

FIGURE 2.9 Thermal gravimetric analysis curves of bacterial nanocellulose and cell-free cellulose samples, produced by bacterial cells and the cell-free system, respectively. Both cellulose types were produced statically at 30 °C and pH 5.0 and freeze-dried before analysis. Figure reproduced from (38).

Source: Klemm D, Schumann D, Udhardt U, Marsch S (2001) Bacterial synthesized cellulose—Artificial blood vessels for microsurgery. *Prog Polym Sci* 26:1561–1603. https://doi.org/10.1016/S0079-6700(01)00021-1.

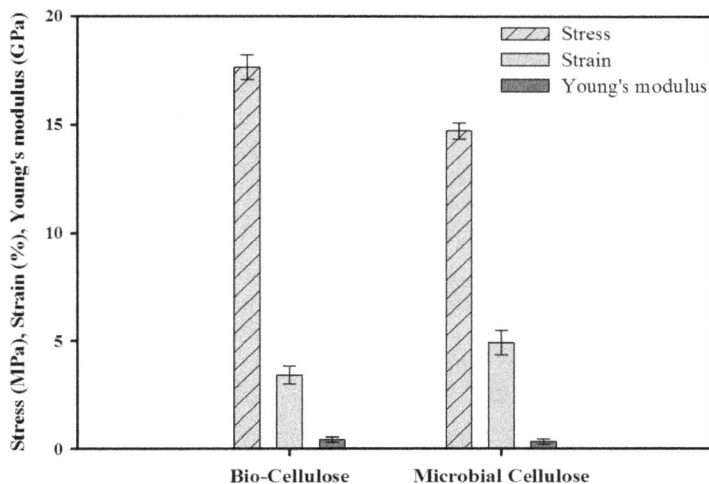

FIGURE 2.10 Mechanical strength of bacterial nanocellulose and cell-free cellulose samples. Tensile strength at breaking point, percent strain, and Young's modulus were determined using the universal testing machine. Both cellulose types were produced statically at 30 °C and pH 5.0 and freeze-dried before analysis. Figure reproduced from (38).

Source: Klemm D, Schumann D, Udhardt U, Marsch S (2001) Bacterial synthesized cellulose—Artificial blood vessels for microsurgery. *Prog Polym Sci* 26:1561–1603. https://doi.org/10.1016/S0079-6700(01)00021-1.

2.4.5 Mechanical Properties

The mechanical properties of nanocellulose produced by the microbial and cell-free system are highly dependent on their crystallinity and degree of polymerization, which are in turn dependent on the arrangement of the fibrils (101). The alignment of nanocellulose by applying external stimuli such as electric and magnetic fields, microfluidic channels, and templates is greatly improved, thus enhances its mechanical strength. In addition to alignment, the compact arrangement and the strength of individual fibers greatly improve the mechanical strength of BNC. The relatively aligned nanofibrils in BNC due to bacterial movement are expected to possess better mechanical strength, and the dense and well-arranged arranged fibers in cell-free nanocellulose demonstrate better mechanical properties (i.e., tensile strength, elongation strain, and Young's modulus) (Figure 2.10) (38). The extended incubation in a cell-free system favored improved fiber density and strength. An improved fiber density and strength can also be achieved for BNC by extending the BNC production by supplementing additional carbon sources or extending its availability. The compact and well-distributed fibrils favor a uniform response to the applied force (102, 103).

2.5 APPLICATIONS: ONE-POT SYNTHESIS OF NANOMATERIALS

As BNC lacks antibacterial, antifungal, magnetic, and antioxidant properties as well as possesses limited biocompatibility and optical transparency, such issues are overcome by developing its composites with other material via *in situ* or *ex situ* impregnation or through regeneration methods, with the aim not only to improve its existing properties but also to endow it with additional properties (102). For instance, the use of BNC in biomedical applications such as wound healing and tissue regeneration requires antibacterial and antifungal properties and high biocompatibility. Therefore, to endow BNC with antimicrobial properties, its composites are developed with different nanomaterials, including nanoparticles such as silver (15), titanium dioxide (104, 105), and zinc (106), as well as different polymers such as chitosan (107), which possess antibacterial properties. However, the conventional *in situ*, *ex situ*, and regeneration methods have various limitations. For example, nanomaterials with antibacterial properties cannot be added to the culture medium of bacterial cells for *in situ* impregnation into the BNC matrix due to their bactericidal properties. Further, the nanomaterials settle down during the static incubation while BNC gel or membrane is produced at the air-medium interface; thus, the nanoparticles' uptake efficiency of static incubation is very low. In contrast, although the nanoparticles' uptake efficiency during the *ex situ* impregnation *via* blending under shaking is high, this method does not ensure the uniform distribution of nanoparticles in the BNC matrix as well does not allow the impregnation of large-size particles. Although the regeneration method overcomes the aforementioned limitations of *in situ* and *ex situ* impregnation methods, the dissolution of cellulose fibrils results in loss of fibrous structure, which in turn negatively affects other properties of BNC that are important for its applications from a biomedical perspective. To overcome the issues of conventional composite synthesis approaches, Ullah et al. developed a one-pot composite synthesis strategy by using the *G. hansenii*-based cell-free system (43) (Figure 2.11). During the nanocomposite synthesis, the suspended titanium dioxide nanoparticles interacted with the growing cellulose fibrils and were trapped within the matrix through physical interaction and hydrogen bonding. This strategy offered two main advantages: it bypassed the toxicity of nanoparticles toward the bacterial cells and ensured the efficient uptake and homogenous distribution of nanoparticles within the cellulose/TiO$_2$ nanocomposite. Further, this one-pot composite development strategy

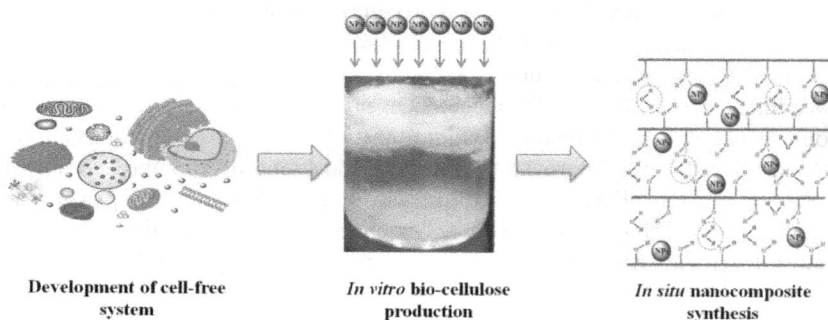

Development of cell-free   *In vitro* bio-cellulose   *In situ* nanocomposite
system   production   synthesis

FIGURE 2.11   One-pot *in situ* development of cellulose/TiO$_2$ nanocomposites in a cell-free system.

of using a cell-free system could be employed with a broad range of nanostructures of different shapes and sizes as well as polymers for biomedical and other applications.

2.6 CONCLUSIONS AND PROSPECTS

Bacterial nanocellulose has received immense consideration in recent years, especially for its application in the biomedical field, owing to its unique structural, physicochemical, mechanical, thermal, and biological features, both alone and in the form of composites with other materials. Compared to the microbial production, the cellulose synthesis by cell-free system offers several advantages such as simple and control synthesis, *de novo* synthesis, high-yield, low-cost production, and novel composite development strategies as well as produces cellulose with better features. In a cell-free system, cellulose is synthesized *in vitro* without involving a complex extrusion system that self-aggregates in three stages: formation of premature cellulose, cellulose pellicles, and cellulose sheet. The submerged fiber production in the cell-free environment and later agglomeration provides insight into incorporating different micro- and nano-sized materials between the growing nanofibrils. Further, the anaerobic synthesis of cellulose in the cell-free system bypasses the risk of contamination associated with BNC production by microbial cells.

The cell-free approach for production of different biocommodities has been receiving immense consideration over the last few decades due to its unique advantages; nevertheless, it still faces several challenges such as a lack of stable enzymes, the high cost of cofactors, the imbalance of cofactors and their regeneration system, and the unavailability of low-cost substrates. Additionally, the cell-free enzymes must remain stable (i.e., thermal and proteolytic stability), intact, and catalytically active for several weeks in a cell- and membrane-free environment. Extensive efforts, such as establishing novel and advanced algorithms, using multiscale modeling approaches, designing via computation, fabricating advanced nanoscale biomaterials, discovering novel scaffolds, developing minimal cells and simple and efficient strategies for implementing the cell-free system, and exploring less expensive energy sources for cofactor regeneration, would ultimately broaden the application of the cell-free system to different fields.

2.7 ACKNOWLEDGMENT

This work was supported by the National Natural Science Foundation of China (21774039, 51973076), BRICS STI Framework Programme 3rd call 2019 (2018YFE0123700), China Postdoctoral Science Foundation (2016M602291), and the Fundamental Research Funds for Central Universities, Open Research Fund of State Key Laboratory of Polymer Physics and Chemistry, and Changchun Institute of Applied Chemistry, Chinese Academy of Sciences.

REFERENCES

(1) Ahrem H, Pretzel D, Endres M, et al. (2014) Laser-structured bacterial nanocellulose hydrogels support ingrowth and differentiation of chondrocytes and show potential as cartilage implants. *Acta Biomater* 10:1341–1353. https://doi.org/10.1016/j.actbio.2013.12.004

(2) Aljohani W, Ullah MW, Zhang X, Yang G (2018) Bioprinting and its applications in tissue engineering and regenerative medicine. *Int J Biol Macromol* 107:261–275. https://doi.org/10.1016/j.ijbiomac.2017.08.171

(3) Bai FW, Anderson WA, Moo-Young M (2008) Ethanol fermentation technologies from sugar and starch feedstocks. *Biotechnol. Adv.* 26:89–105

(4) Benziman M, Haigler CH, Brown RM, et al. (1980) Cellulose biogenesis: Polymerization and crystallization are coupled processes in Acetobacter xylinum. *Proc Natl Acad Sci U S A* 77:6678–6682. https://doi.org/10.1073/pnas.77.11.6678

(5) Buchner E (1897) Alkoholische Gärung ohne Hefezellen. *Berichte der Dtsch Chem Gesellschaft* 30:1110–1113. https://doi.org/10.1002/cber.189703001215

(6) Cannon RE, Anderson SM (1991) Biogenesis of Bacterial Cellulose. *Crit Rev Microbiol* 17:435–447. https://doi.org/10.3109/10408419109115207

(7) Casteleijn MG, Urtti A, Sarkhel S (2013) Expression without boundaries: Cell-free protein synthesis in pharmaceutical research. *Int J Pharm* 440:39–47. https://doi.org/10.1016/j.ijpharm.2012.04.005

(8) Czaja W, Krystynowicz A, Bielecki S, Brown RM (2006) Microbial cellulose—The natural power to heal wounds. *Biomaterials* 27:145–151

(9) Dahman Y (2009) Nanostructured biomaterials and biocomposites from bacterial cellulose nanofibers. *J Nanosci Nanotechnol.* https://doi.org/10.1166/jnn.2009.1466

(10) Di Z, Shi Z, Ullah MW, et al. (2017) A transparent wound dressing based on bacterial cellulose whisker and poly(2-hydroxyethyl methacrylate). *Int J Biol Macromol* 105:638–644. https://doi.org/10.1016/j.ijbiomac.2017.07.075

(11) Ding SY, Liu YS, Zeng Y, et al. (2012) How does plant cell wall nanoscale architecture correlate with enzymatic digestibility? *Science* 80 (338):1055–1060. https://doi.org/10.1126/science.1227491

(12) Dondapati SK, Wüstenhagen DA, Kubick S (2018) Functional analysis of membrane proteins produced by cell-free translation. *Methods Mol Biol* 1685:171–185. https://doi.org/10.1007/978-1-4939-7366-8_10

(13) Endler A, Sánchez-Rodríguez C, Persson S (2010) Cellulose squeezes through. *Nat Chem Biol* 6:883–884. https://doi.org/10.1038/nchembio.480

(14) Farooq U, Ullah MW, Yang Q, et al. (2020) High-density phage particles immobilization in surface-modified bacterial cellulose for ultra-sensitive and selective electrochemical detection of Staphylococcus aureus. *Biosens Bioelectron* 157:112163. https://doi.org/10.1016/j.bios.2020.112163

(15) Fatima A, Yasir S, Khan MS, et al. (2021) Plant extract-loaded bacterial cellulose composite membrane for potential biomedical applications. *J Bioresour Bioprod* 6:26–32. https://doi.org/10.1016/j.jobab.2020.11.002

(16) Forster AC, Church GM (2007) Synthetic biology projects in vitro. *Genome Res.* 17:1–6

(17) Fu L, Zhou P, Zhang S, Yang G (2013) Evaluation of bacterial nanocellulose-based uniform wound dressing for large area skin transplantation. *Mater Sci Eng C* 33:2995–3000. https://doi.org/10.1016/j.msec.2013.03.026

(18) Gao Q, Shen X, Lu X (2011) Regenerated bacterial cellulose fibers prepared by the NMMO·H 2O process. *Carbohydr Polym.* https://doi.org/10.1016/j.carbpol.2010.09.029

(19) Gelin K, Bodin A, Gatenholm P, et al. (2007) Characterization of water in bacterial cellulose using dielectric spectroscopy and electron microscopy. *Polymer (Guildf)* 48:7623–7631. https://doi.org/10.1016/j.polymer.2007.10.039

(20) George J, Ramana KV, Bawa AS, Siddaramaiah (2011) Bacterial cellulose nanocrystals exhibiting high thermal stability and their polymer nanocomposites. *Int J Biol Macromol* 48:50–57. https://doi.org/10.1016/j.ijbiomac.2010.09.013

(21) Hodgman CE, Jewett MC (2012) Cell-free synthetic biology: Thinking outside the cell. *Metab Eng* 14:261–269. https://doi.org/10.1016/j.ymben.2011.09.002

(22) Hu S-Q, Gao Y-G, Tajima K, et al. (2010) Structure of bacterial cellulose synthase subunit D octamer with four inner passageways. *Proc Natl Acad Sci* 107:17957–17961. https://doi.org/10.1073/pnas.1000601107

(23) Iguchi M, Yamanaka S, Budhiono A (2000) Bacterial cellulose—a masterpiece of nature's arts. *J Mater Sci* 35:261–270. https://doi.org/10.1023/A:1004775229149

(24) Jandt U, You C, Zhang YHP, Zeng AP (2013) Compartmentalization and metabolic channeling for multienzymatic biosynthesis: Practical strategies and modeling approaches. *Adv. Biochem. Eng. Biotechnol.* 137:41–65

(25) Jasim A, Ullah MW, Shi Z, et al. (2017) Fabrication of bacterial cellulose/polyaniline/single-walled carbon nanotubes membrane for potential application as biosensor. *Carbohydr Polym* 163:62–69. https://doi.org/10.1016/j.carbpol.2017.01.056

(26) Jia H, Heymann M, Bernhard F, et al. (2017) Cell-free protein synthesis in micro compartments: Building a minimal cell from biobricks. *N Biotechnol* 1–7. https://doi.org/10.1016/j.nbt.2017.06.014

(27) Jiang Y, Zhang Y, Wu H, et al. (2008) Protamine-templated biomimetic hybrid capsules: Efficient and stable carrier for enzyme encapsulation. *Chem Mater.* https://doi.org/10.1021/cm701959e

(28) Jung JY, Khan T, Park JK, Chang HN (2007) Production of bacterial cellulose by Gluconacetobacter hansenii using a novel bioreactor equipped with a spin filter. *Korean J Chem Eng* 24:265–271. https://doi.org/10.1007/s11814-007-5058-4

(29) Kaewnopparat S, Sansernluk K, Faroongsarng D (2008) Behavior of freezable bound water in the bacterial cellulose produced by Acetobacter xylinum: An approach using thermoporosimetry. *AAPS PharmSciTech* 9:701–707. https://doi.org/DOI 10.1208/s12249-008-9104-2

(30) Karim AS, Jewett MC (2016) A cell-free framework for rapid biosynthetic pathway prototyping and enzyme discovery. *Metab Eng* 36:116–126. https://doi.org/10.1016/j.ymben.2016.03.002

(31) Khan S, Ul-Islam M, Khattak WA, et al. (2015a) Bacterial cellulose-poly(3,4-ethylenedioxythiophene)-poly(styrenesulfonate) composites for optoelectronic applications. *Carbohydr Polym* 127:86–93. https://doi.org/10.1016/j.carbpol.2015.03.055

(32) Khan S, Ul-Islam M, Khattak WA, et al. (2015b) Bacterial cellulose-titanium dioxide nanocomposites: Nanostructural characteristics, antibacterial mechanism, and biocompatibility. *Cellulose.* https://doi.org/10.1007/s10570-014-0528-4

(33) Khan S, Ul-Islam M, Ullah MW, et al. (2015c) Engineered regenerated bacterial cellulose scaffolds for application in in vitro tissue regeneration. *RSC Adv* 5:84565–84573. https://doi.org/10.1039/c5ra16985b

(34) Khan T, Park JK (2008) The structure and physical properties of glucuronic acid oligomers produced by a Gluconacetobacter hansenii strain using the waste from beer fermentation broth. *Carbohydr Polym* 73:438–445. https://doi.org/10.1016/j.carbpol.2007.12.010

(35) Khattak WA, Ul-Islam M, Ullah MW, et al. (2014a) Yeast cell-free enzyme system for bio-ethanol production at elevated temperatures. *Process Biochem* 49. https://doi.org/10.1016/j.procbio.2013.12.019

(36) Khattak WA, Ullah MW, Ul-Islam M, et al. (2014b) Developmental strategies and regulation of cell-free enzyme system for ethanol production: A molecular prospective. *Appl Microbiol Biotechnol* 98. https://doi.org/10.1007/s00253-014-6154-0

(37) Kim Y, Ullah MW, Ul-Islam M, et al. (2019) Self-assembly of bio-cellulose nanofibrils through intermediate phase in a cell-free enzyme system. *Biochem Eng J* 142:135–144. https://doi.org/10.1016/j.bej.2018.11.017

(38) Klemm D, Schumann D, Udhardt U, Marsch S (2001) Bacterial synthesized cellulose—Artificial blood vessels for microsurgery. *Prog Polym Sci* 26:1561–1603. https://doi.org/10.1016/S0079-6700(01)00021-1

(39) Kubiak K, Kurzawa M, Jedrzejczak-Krzepkowska M, et al. (2014) Complete genome sequence of *Gluconacetobacter xylinus* E25 strain-Valuable and effective producer of bacterial nanocellulose. *J Biotechnol.* https://doi.org/10.1016/j.jbiotec.2014.02.006

(40) Kudlicka K, Brown RMJ (1997) Cellulose and callose biosynthesis in higher plants. *Plant Physiol* 115:643–656. https://doi.org/115/2/643 [pii]

(41) Kwok R (2010) Five hard truths for synthetic biology. *Nature* 463:288–290.

(42) Lai-Kee-Him J, Chanzy H, Müller M, et al. (2002) In vitro versus in vivo cellulose microfibrils from plant primary wall synthases: Structural differences. *J Biol Chem* 277:36931–36939. https://doi.org/10.1074/jbc.M203530200

(43) Le Bras D, Strømme M, Mihranyan A (2015) Characterization of dielectric properties of nanocellulose from wood and algae for electrical insulator applications. *J Phys Chem B.* https://doi.org/10.1021/acs.jpcb.5b00715

(44) Lee CM, Gu J, Kafle K, et al. (2015) Cellulose produced by *Gluconacetobacter xylinus* strains ATCC 53524 and ATCC 23768: Pellicle formation, post-synthesis aggregation and fiber density. *Carbohydr Polym* 133:270–276. https://doi.org/10.1016/j.carbpol.2015.06.091

(45) Li S, Huang D, Zhang B, et al. (2014) Flexible supercapacitors based on bacterial cellulose paper electrodes. *Adv Energy Mater* 4:1301655. https://doi.org/10.1002/aenm.201301655

(46) Li S, Jasim A, Zhao W, et al. (2018) Fabrication of pH-electroactive bacterial cellulose/polyaniline hydrogel for the development of a controlled drug release system. *ES Mater Manuf* 41–49. https://doi.org/10.30919/esmm5f120

(47) Lin S Bin, Hsu CP, Chen LC, Chen HH (2009) Adding enzymatically modified gelatin to enhance the rehydration abilities and mechanical properties of bacterial cellulose. *Food Hydrocoll* 23:2195–2203. https://doi.org/10.1016/j.foodhyd.2009.05.011

(48) Lin FC, Brown RM, Drake RR, Haley BE (1990) Identification of the uridine 5???-diphosphoglucose (UDP-Glc) binding subunit of cellulose synthase in Acetobacter xylinum using the photoaffinity probe 5-Azido-UDP-Glc. *J Biol Chem* 265:4782–4784

(49) Lu F, Smith PR, Mehta K, Swartz JR (2015) Development of a synthetic pathway to convert glucose to hydrogen using cell free extracts. *Int J Hydrogen Energy* 40:9113–9124. https://doi.org/10.1016/j.ijhydene.2015.05.121

(50) Lu Y (2017) Cell-free synthetic biology: Engineering in an open world. *Synth Syst Biotechnol* 2:23–27. https://doi.org/10.1016/j.synbio.2017.02.003

(51) Ludwicka K, Jedrzejczak-Krzepkowska M, Kubiak K, et al. (2016) Medical and cosmetic applications of bacterial nanocellulose. In: *Bacterial Nanocellulose: From Biotechnology to Bio-Economy.* Amsterdam: Elsevier.

(52) Mahmoudi K, Hosni K, Hamdi N, Srasra E (2014) Kinetics and equilibrium studies on removal of methylene blue and methyl orange by adsorption onto activated carbon prepared from date pits-A comparative study. *KOREAN J Chem Eng* 32:274–283. https://doi.org/10.1007/s11814-014-0216-y

(53) Mbituyimana B, Mao L, Hu S, et al. (2021) Bacterial cellulose/glycolic acid/glycerol composite membrane as a system to deliver glycolic acid for anti-aging treatment. *J Bioresour Bioprod.* https://doi.org/10.1016/j.jobab.2021.02.003

(54) McCarthy RR, Ullah MW, Booth P, et al. (2019a) The use of bacterial polysaccharides in bioprinting. *Biotechnol Adv* 37:107448. https://doi.org/10.1016/j.biotechadv.2019.107448

(55) McCarthy RR, Ullah MW, Pei E, Yang G (2019b) Antimicrobial Inks: The anti-infective applications of bioprinted bacterial polysaccharides. *Trends Biotechnol.* 37:1153–1155

(56) Meftahi A, Khajavi R, Rashidi A, et al. (2010) The effects of cotton gauze coating with microbial cellulose. *Cellulose* 17:199–204. https://doi.org/10.1007/s10570-009-9377-y

(57) Morgan JLW, McNamara JT, Zimmer J (2014) Mechanism of activation of bacterial cellulose synthase by cyclic di-GMP. *Nat Struct Mol Biol* 21:489–496. https://doi.org/10.1038/nsmb.2803

(58) Myung S, Rollin J, You C, et al. (2014) In vitro metabolic engineering of hydrogen production at theoretical yield from sucrose. *Metab Eng* 24:70–77. https://doi.org/10.1016/j.ymben.2014.05.006

(59) Ohashi H, Kanamori T, Shimizu Y, Ueda T (2010) A highly controllable reconstituted cell-free system--a breakthrough in protein synthesis research. *Curr Pharm Biotechnol* 11:267–271. https://doi.org/BSP/CPB/E-Pub/0047-11-3 [pii]

(60) Park J, Park Y, Jung J (2003) Production of bacterial cellulose byGluconacetobacter hansenii PJK isolated from rotten apple. *Biotechnol Bioprocess Eng* 8:83–88. https://doi.org/Doi 10.1007/Bf02940261

(61) Peng L, Kawagoe Y, Hogan P, Delmer D (2002) Sitosterol-beta-glucoside as primer for cellulose synthesis in plants. *Science* 295:147–50. https://doi.org/10.1126/science.1064281

(62) Percival Zhang YH (2010) Production of biocommodities and bioelectricity by cell-free synthetic enzymatic pathway biotransformations: Challenges and opportunities. *Biotechnol Bioeng* 105:663–667. https://doi.org/10.1002/bit.22630

(63) Rebuffet E, Frick A, Järvå M, Törnroth-Horsefield S (2017) Cell-free production and characterisation of human uncoupling protein 1–3. *Biochem Biophys Reports* 10: 276–281. https://doi.org/10.1016/j.bbrep.2017.04.003

(64) Sajjad W, He F, Ullah MW, et al. (2020) Fabrication of bacterial cellulose-curcumin nanocomposite as a novel dressing for partial thickness skin burn. *Front Bioeng Biotechnol* 8. https://doi.org/10.3389/fbioe.2020.553037

(65) Saxena IM, Brown RM (2005) Cellulose biosynthesis: Current views and evolving concepts. *Ann. Bot.* 96:9–21

(66) Saxena IM, Kudlicka K, Okuda K, Brown RM (1994) Characterization of genes in the cellulose-synthesizing operon (acs operon) of Acetobacter xylinum: Implications for cellulose crystallization. *J Bacteriol* 176:5735–5752

(67) Schrecker ST, Gostomski PA (2005) Determining the water holding capacity of microbial cellulose. *Biotechnol Lett* 27:1435–1438. https://doi.org/10.1007/s10529-005-1465-y

(68) Seo C, Lee HW, Suresh A, et al. (2014) Improvement of fermentative production of exopolysaccharides from Aureobasidium pullulans under various conditions. *Korean J Chem Eng* 31:1433–1437. https://doi.org/10.1007/s11814-014-0064-9

(69) Shah N, Ul-Islam M, Khattak WA, Park JK (2013) Overview of bacterial cellulose composites: A multipurpose advanced material. *Carbohydr. Polym.* 98:1585–1598

(70) Shezad O, Khan S, Khan T, Park JK (2010) Physicochemical and mechanical characterization of bacterial cellulose produced with an excellent productivity in static conditions using a simple fed-batch cultivation strategy. *Carbohydr Polym* 82:173–180. https://doi.org/10.1016/j.carbpol.2010.04.052

(71) Shi X, Shi Z, Wang D, et al. (2016) Microbial cells with a Fe3O4 doped hydrogel extracellular matrix: Manipulation of living cells by magnetic stimulus. *Macromol Biosci* 16:1506–1514. https://doi.org/10.1002/mabi.201600143

(72) Shi Z, Zhang Y, Phillips GO, Yang G (2014) Utilization of bacterial cellulose in food. *Food Hydrocoll.* 35:539–545

(73) Shoda M, Sugano Y (2005) Recent advances in bacterial cellulose production. *Biotechnol Bioprocess Eng* 10:1–8. https://doi.org/10.1007/bf02931175

(74) Shoukat A, Wahid F, Khan T, et al. (2019) Titanium oxide-bacterial cellulose bioadsorbent for the removal of lead ions from aqueous solution. *Int J Biol Macromol* 129: 965–971. https://doi.org/10.1016/j.ijbiomac.2019.02.032

(75) Sonnabend A, Spahn V, Stech M, et al. (2017) Production of G protein-coupled receptors in an insect-based cell-free system. *Biotechnol Bioeng* 114:2328–2338. https://doi.org/10.1002/bit.26346

(76) Stech M, Nikolaeva O, Thoring L, et al. (2017) Cell-free synthesis of functional antibodies using a coupled in vitro transcription-translation system based on CHO cell lysates. *Sci Rep* 7:12030. https://doi.org/10.1038/s41598-017-12364-w

(77) Sutherland IW (2001) Microbial polysaccharides from Gram-negative bacteria. *Int Dairy J*:663–674

(78) Talebnia F, Niklasson C, Taherzadeh MJ (2005) Ethanol production from glucose and dilute-acid hydrolyzates by encapsulated S. cerevisiae. *Biotechnol Bioeng*. https://doi.org/10.1002/bit.20432

(79) Tang W, Jia S, Jia Y, Yang H (2010) The influence of fermentation conditions and post-treatment methods on porosity of bacterial cellulose membrane. *World J Microbiol Biotechnol* 26:125–131. https://doi.org/10.1007/s11274-009-0151-y

(80) Taokaew S, Seetabhawang S, Siripong P, Phisalaphong M (2013) Biosynthesis and characterization of nanocellulose-gelatin films. *Materials (Basel)*. https://doi.org/10.3390/ma6030782

(81) Ul-Islam M, Ahmad F, Fatima A, et al. (2021) Ex situ synthesis and characterization of high strength multipurpose bacterial cellulose-aloe vera hydrogels. *Front Bioeng Biotechnol* 9. https://doi.org/10.3389/fbioe.2021.601988

(82) Ul-Islam M, Ha JH, Khan T, Park JK (2013) Effects of glucuronic acid oligomers on the production, structure and properties of bacterial cellulose. *Carbohydr Polym* 92:360–366. https://doi.org/10.1016/j.carbpol.2012.09.060

(83) Ul-Islam M, Khan T, Park JK (2012a) Water holding and release properties of bacterial cellulose obtained by in situ and ex situ modification. *Carbohydr Polym* 88:596–603. https://doi.org/10.1016/j.carbpol.2012.01.006

(84) Ul-Islam M, Khan T, Park JK (2012b) Nanoreinforced bacterial cellulose-montmorillonite composites for biomedical applications. *Carbohydr Polym* 89:1189–1197. https://doi.org/10.1016/j.carbpol.2012.03.093

(85) Ul-Islam M, Khan S, Ullah MW, Park JK (2015) Bacterial cellulose composites: Synthetic strategies and multiple applications in bio-medical and electro-conductive fields. *Biotechnol J* 10:1847–1861. https://doi.org/10.1002/biot.201500106

(86) Ul-Islam M, Khan S, Ullah MW, Park JK (2019) Comparative study of plant and bacterial cellulose pellicles regenerated from dissolved states. *Int J Biol Macromol* 137: 247–252. https://doi.org/10.1016/j.ijbiomac.2019.06.232

(87) Ullah MW, Khattak WA, Ul-Islam M, et al. (2014) Bio-ethanol production through simultaneous saccharification and fermentation using an encapsulated reconstituted cell-free enzyme system. *Biochem Eng J* 91:110–119. https://doi.org/10.1016/j.bej.2014.08.006

(88) Ullah MW, Khattak WA, Ul-Islam M, et al. (2015) Encapsulated yeast cell-free system: A strategy for cost-effective and sustainable production of bio-ethanol in consecutive batches. *Biotechnol Bioprocess Eng* 20:561–575. https://doi.org/10.1007/s12257-014-0855-1

(89) Ullah MW, Khattak WA, Ul-Islam M, et al. (2016) Metabolic engineering of synthetic cell-free systems: Strategies and applications. *Biochem Eng J* 105. https://doi.org/10.1016/j.bej.2015.10.023

(90) Ul-Islam M, Khattak WA, Ullah MW, et al. (2014) Synthesis of regenerated bacterial cellulose-zinc oxide nanocomposite films for biomedical applications. *Cellulose* 21. https://doi.org/10.1007/s10570-013-0109-y

(91) Ullah MW, Manan S, Kiprono SJ, Ul-Islam M Y (2019) Synthesis, structural, and properties of bacterial cellulose. In: *Nanocelluose: From Fundamentals to Advanced Materials*. Weinheim: Wiley-VCH Verlag GmbH & Co. KGaA., pp. 81–113

(92) Ul-Islam M, Shehzad A, Khan S, et al. (2014) Antimicrobial and biocompatible properties of nanomaterials. *J Nanosci Nanotechnol*. https://doi.org/10.1166/jnn.2014.8761

(93) Ul-Islam M, Subhan F, Islam SU, et al. (2019) Development of three-dimensional bacterial cellulose/chitosan scaffolds: Analysis of cell-scaffold interaction for potential application in the diagnosis of ovarian cancer. *Int J Biol Macromol* 137:1050–1059. https://doi.org/10.1016/j.ijbiomac.2019.07.050

(94) Ullah MW, Ul-Islam M, Khan S, et al. (2015) Innovative production of bio-cellulose using a cell-free system derived from a single cell line. *Carbohydr Polym* 132:286–294. https://doi.org/10.1016/j.carbpol.2015.06.037

(95) Ullah MW, Ul-Islam M, Khan S, et al. (2016a) Structural and physico-mechanical characterization of bio-cellulose produced by a cell-free system. *Carbohydr Polym* 136:908–916. https://doi.org/10.1016/j.carbpol.2015.10.010

(96) Ullah MW, Ul-Islam M, Khan S, et al. (2016b) In situ synthesis of a bio-cellulose/titanium dioxide nanocomposite by using a cell-free system. *RSC Adv* 6:22424–22435. https://doi.org/10.1039/c5ra26704h

(97) Ullah MW, Ul Islam M, Khan S, et al. (2017) Recent advancements in bioreactions of cellular and cell-free systems: A study of bacterial cellulose as a model. *Korean J Chem Eng* 34:1591–1599. https://doi.org/10.1007/s11814-017-0121-2

(98) Ul-Islam M, Ullah MW, Khan S, et al. (2016) Recent advancement in cellulose based nanocomposite for addressing environmental challenges. *Recent Pat Nanotechnol* 10. https://doi.org/10.2174/1872210510666160429144916

(99) Ul-Islam M, Ullah MW, Khan S, et al. (2017) Strategies for cost-effective and enhanced production of bacterial cellulose. *Int J Biol Macromol* 102:1166–1173. https://doi.org/10.1016/j.ijbiomac.2017.04.110

(100) Ul-Islam M, Ullah MW, Khan S, Park JK (2020) Production of bacterial cellulose from alternative cheap and waste resources: A step for cost reduction with positive environmental aspects. *Korean J Chem Eng* 37:925–937. https://doi.org/10.1007/s11814-020-0524-3

(101) Wang L, Hu S, Ullah MW, et al. (2020) Enhanced cell proliferation by electrical stimulation based on electroactive regenerated bacterial cellulose hydrogels. *Carbohydr Polym* 249:116829. https://doi.org/https://doi.org/10.1016/j.carbpol.2020.116829

(102) Wei P, Wang Q, Hang B, et al. (2017) High-level cell-free expression and functional characterization of a novel aquaporin from Photobacterium profundum SS9. *Process Biochem* 1. https://doi.org/10.1016/j.procbio.2017.05.014

(103) Welch P, Scopes RK (1985) Studies on cell-free metabolism: Ethanol production by a yeast glycolytic system reconstituted from purified enzymes. *J Biotechnol* 2:257–273. https://doi.org/10.1016/0168-1656(85)90029-X

(104) Ye X, Wang Y, Hopkins RC, et al. (2009) Spontaneous high-yield production of hydrogen from cellulosic materials and water catalyzed by enzyme cocktails. *ChemSusChem* 2:149–152. https://doi.org/10.1002/cssc.200900017

(105) You C, Zhang YHP (2013) Self-assembly of synthetic metabolons through synthetic protein scaffolds: One-step purification, co-immobilization, and substrate channeling. *ACS Synth Biol* 2:102–110. https://doi.org/10.1021/sb300068g

(106) Zaks A (2001) Industrial biocatalysis. *Curr. Opin. Chem. Biol* 5:130–136

(107) Zhang YHP, Evans BR, Mielenz JR, et al. (2007) High-yield hydrogen production from starch and water by a synthetic enzymatic pathway. *PLoS One* 2. https://doi.org/10.1371/journal.pone.0000456

3 Bacterial Cellulose and Its Composites for Biomedical and Industrial Applications

Shaukat Khan, Naveed Ur Rahman, Adnan Haider, Muhammad Wajid Ullah, and Mazhar Ul-Islam

CONTENTS

3.1 INTRODUCTION

Biodegradable materials are increasingly in demand by consumers due to the high environmental risks of using nonbiodegradable materials. Cellulose, the most abundant natural biopolymer (1), including wood, cotton, hemp, and bacterial cellulose (BC), has been in use for centuries due to its wide availability and biodegradability.

DOI: 10.1201/9781003118756-3

BC, the purest form of natural cellulose, is however newer in study and applications. Although the chemical structure is the same as that of plant cellulose, marked differences are observed between BC and plant cellulose (PC). Plant fibers contain impurities such as hemicellulose, pectin, and lignin and consist of only 40%–70% cellulose (2), whereas BC is the purest form of cellulose, containing no impurities, and possesses good mechanical properties. These unique properties have made BC a popular subject for extensive study and characterization in recent times (3).

BC is extracellulary synthesized by certain bacteria such as *Acetobacter, Gluconacetobacter, Azotobacter, Salmonella, Sarcina ventriculi,* and *Pseudomonas* (4). However, *Gluconacetobacter xylinum, Gluconacetobacter pasteurianus,* and *Gluconacetobacter hansenii* are the most bacteria used for BC production (5). Bacteria produces BC through the polymerization of glucose in the form of β-1,4-glucan chains and secretion of the produced cellulose chains through tiny pores in its cell membrane (6). The chains in turn assemble and crystallize to produce ribbons. This leads to a nanofibrous 3D network on the surface of the liquid medium in the form of a pellicle (7). BC bears a type I cellulose crystal structure (8), with a high degree of crystallinity (around 90%) (9). The highly hydrophilic character of BC is due to its surface hydroxyl groups (10). BC nanofibers have a diameter of around 20 to 100 nm and show a higher surface area than do those of plant cellulose (11). Grande et al. (12) determined its physicomechanical features—the Young's modulus was found to be 6.86 GPa, the tensile strength was 241.42 MPa, and the maximum elongation at break was 8.21%.

BC is used in various applications owing to its exceptional mechanical properties. The demand for BC-based products with superior properties to meet the requirements in various fields is increasingly growing. This leads to the production of BC composites with a variety of nanofillers with target characteristics. The nature of nanofiller used depends on the target applications, for example biopolymers are augmented with BC to increase their biocompatibility while antimicrobial nanomaterials are used to impart antimicrobial property to native BC, while conducting and magnetic materials are added to the BC matrix for target electronic and sensor applications. Besides the variety of nanofillers, the nanocomposite synthetic method also varies in accordance to the nature of nanofiller used and the target applications. In the *in situ* method, the nanofiller is added to the BC fermentation medium and the composite is synthesized during the BC production process, whereas in the *ex situ* method, the already produced BC pellicle is immersed in the nanofiller solution or suspension resulting in nanocomposite synthesis. These nanocomposites are widely used in biomedical and industrial applications (152, 153). In the following sections, we will discuss the synthesis of BC nanocomposites and their target applications.

3.2 BC COMPOSITES FOR ENHANCED BIOMEDICAL APPLICATIONS

3.2.1 COMPOSITES WITH BIOPOLYMERS

Collagen is a fiber-forming biopolymer found in numerous animal tissues. Collagen has high tensile strength, making it suitable for biomedical applications. Zhijiang and Guang (13) synthesized a BC/collagen composite by immersing BC pellicles in acetic acid solution of collagen followed by freeze drying. The composite was suggested as suitable material for wound dressing due to its high crystallinity, increased thermal stability, and tensile strength. Saska et al. (14) cross-linked BC with collagen through glycine esterification. The obtained composites showed enhanced flexibility increasing the ease of handling in surgery. Luo et al. (15) impregnated BC with collagen through the *in situ* method of adding collagen to BC culture media. The obtained BC/collagen composite showed a different crystallinity, thickness, and color from BC alone.

Chitosan, a biopolymer, possesses some exceptional properties such as absorption of wound exudates, wound healing, and antimicrobial properties. Ul-Islam et al. (16) reported BC/chitosan composites through immersion of BC pellicles in 1% chitosan solution in 1% acetic acid. The effects of solution temperature and immersion time were determined. An increase in water-holding capacity, water release rate, and mechanical properties was found for the composites synthesized in shaking incubation mode and at 50 °C temperature, elevating their applicability in the biomedical fields. Phisalaphong et al. (56) synthesized BC/chitosan composites through the *in situ* impregnation method by adding low molecular weight chitosan into BC production media. The obtained nanocomposites showed smaller pore size compared to BC, high surface area, high water-holding capacity, and better mechanical properties than BC in dry and wet state. Kim et al. (18) impregnated never-dried BC with chitosan solution in acetic acid. Chitosan penetrated the BC fiber network and formed a multilayer network with enhanced thermal stability (the degradation temperature increased from 263 °C to 366 °C with an increase in chitosan ratio from 1.2% to 4.5%) while reduction was noticed in the tensile strength and Young's modulus. Kim et al. (18) synthesized BC/chitosan composites through immersion of BC in chitosan solution followed by a freeze-drying step. The obtained nanocomposites showed almost similar mechanical properties to BC, but an enhancement in biocompatibility, thermal stability, and degree of crystallinity was observed. Lin et al. (19) also prepared BC composites with chitosan through immersion of BC pellicles into chitosan solution. The obtained nanocomposites showed higher water-holding capacity and antimicrobial properties. The nanocomposites were applied as wound dressings and showed quicker healing compared to pristine BC.

Other biopolymers such as gelatin and starch have also been used to prepare BC composites. Nakayama et al. (20) prepared BC/gelatin nanocomposites through the widely used *ex situ* impregnation method, immersing BC in gelatin aqueous solution. The nanocomposites showed high mechanical strength (3 MPa) and elastic modulus (23 MPa). Grande et al. (21) synthesized BC/starch composites through a self-assembly method. Starch was added to the BC growth media and autoclaved, leading to the gelatinization of starch. BC production in such media lead to the incorporation of starch into BC. Wan et al. (22) also synthesized such composites. Glycerin (30% w/w to starch) was mixed in starch followed by BC immersion into this solution. The obtained nanocomposites showed better tensile strength and Young's modulus but lower elongation at break than did pure starch.

3.2.2 Composites with Ceramics

The good biocompatibility and nontoxicity of certain minerals like hydroxyapatite, calcium carbonate, and silica have led to their vast biomedical applications. Nanoclays on the other hand are known for their mechanical reinforcement properties.

Hydroxyapatite (HAp), normally used in bone drug delivery (23) and bone grafting (24, 25), is basically a calcium phosphate–bearing phosphate mineral and displays chemical composition and structure similar to minerals of bone. Grande et al. (26) used the *in situ* method to incorporate HAp into BC-producing media leading to BC/HAp composites. The obtained nanocomposites were found to be suitable for biomedical applications. Wan et al. (27) phosphorylated BC to enhance its ability to induce HAp production. The results showed better HAp production with phosphorylated BC compared to pristine BC. Zimmermann et al. (28) adsorbed negatively charged carboxymethyl cellulose (CMC) on the BC surface to induce nucleation of calcium-deficient hydroxyapatite. Saska et al. (29) incubated BC pellicles in $CaCl_2$ solution and Na_2HPO_4 solution, and the obtained nanocomposites were found effective for defect regeneration in tibial bones in rats. Hutchens et al. (30) synthesized a BC/cdHAp nanocomposite mimicking natural bone biomineralization.

Silica, the major sand component, is obtained through sand mining and quartz purification. Yano et al. (31) utilized the *in situ* as well as *ex situ* impregnation method to produce BC/silica composites. Maeda et al. (32) synthesized a BC/silica nanocomposite through the *ex situ* method followed by hot pressing. The composite films showed enhanced tensile strength and Young's modulus of 185 MPa and 17 GPa, respectively. BC/silica nanocomposites have been proposed as adsorbents for water purification (33), aerogels (34), and light-emitting materials (34, 35).

Calcium carbonate ($CaCO_3$) is also found in mineralized tissues (36). BC matrix has been impregnated with $CaCO_3$ (37, 38). Stoica-Guzun et al. (36) studied the effect on BC pellicles through sonication in $CaCO_3$ using $NaCO_3$ and $CaCl_2$ as reactants. The results showed the ultrasound irradiation resulted in bigger crystals with different shapes compared to non-irradiated samples. Later, Stoica-Guzun et al. (39) investigated the effect of microwave irradiation using the same reactants. Huge morphological and polymorphism differences for irradiated $CaCO_3$ crystals were found suitable for medical and industrial applications.

Scientists have also shown that polymer nanoclay composites show enhanced biodegradability along with better mechanical and thermal stability and barrier properties compared to pure polymers (40, 41). Perotti et al. (42) synthesized BC/laponite clay composites bearing different polymer-to-clay ratios by dipping BC in a water dispersion of clay particles. The obtained nanocomposites were highly uniform and stable leading to improved tensile strength and Young's modulus. Ul-Islam et al. (43) impregnated BC with montmorillonite (MMT) at different clay concentrations. They investigated the BC/MMT composites against *Staphylococcus aureus* (*S. aureus*) and *Escherichia coli* (*E. coli*) and found that the composites showed better antimicrobial property at high MMT concentration.

3.2.3 Composites with Antimicrobial Agents

The eligibility of an antimicrobial agent for nanocomposite preparation basically depends on its efficacy and specificity against target pathogens and contaminants. An ideal antimicrobial agent should be inexpensive, fast acting, nanocarcinogenic, and nontoxic to animal tissue and not interfere in wound healing processes (44). Antiseptics are preferred over antibiotics due to their ability to retain their property even with increased bacterial resistance (45). The antimicrobial property has been introduced into BC through impregnation of antimicrobial organic and polymeric materials and also inorganic substances. The antiseptic used for making BC composites include bio- and synthetic polymers with antibacterial activity (46–48), cationic antiseptics (e.g., biguanides and quaternary ammonium compounds) (49, 50), antimicrobial peptides (48), antibiotics (51), and inorganic materials (e.g., metals and metal oxides and graphene nanoparticles) (52–54).

3.2.3.1 Biopolymers with Antimicrobial Activity

Some biopolymers possess antimicrobial activity, elevating their potential as wound healing materials. Synthesis of BC composites with these biopolymers will impart antimicrobial activity to BC. Chitosan, due to the presence of an amino group and its cationic nature, shows antimicrobial activity against Gram-positive and Gram-negative bacteria. Besides, it promotes type III collagen production in the wounds and also stimulates the proliferation of fibroblast cells which accelerates healing (47). BC/chitosan composites have demonstrated a noticeable inhibition of Gram-positive *S. aureus* and Gram-negative *E. coli* bacteria. When compared with commercial transparent film dressings and hydrocolloid dressing in Sprague Dawley rat wound healing, the results revealed that the BC/chitosan composites exhibited a better wound contraction ratio, showing their potential as wound dressing materials (19). Kingkaew et al. investigated the release rate, absorption capacities, and antimicrobial activity of BC/chitosan composites synthesized with chitosan possessing different molar weights (55). Although the lowest molar weight chitosan showed the highest adsorption capacity, chitosan penetrated into the BC membrane and filled the pores regardless of its molar weight. However, the release rate decreases with increasing molar weight. Composites with higher chitosan concentration (lower molar weight) illustrated higher antibacterial activity against *Aspergillus niger* (*A. niger*) and *S. aureus*. The *ex situ* method of BC/chitosan composite preparation is considered better compared to the *in situ* method involving chitosan incorporation during the BC production. Kingkaew et al. also reported *in situ* synthesis of BC/chitosan composites; however, the composites failed to show antimicrobial activity due to the very small amount of chitosan impregnated in the BC membrane (56).

3.2.3.2 Antimicrobial Peptides

Antimicrobial peptides are membrane-active, highly efficient antimicrobials effective against various resistant microbes, thus elevating their value in medical fields. The general pathway of their activity involves the disruption of the pathogen cell membrane. These peptides play important roles in innate and adaptive immune systems and show little trend of resistance development. Therefore, they represent a new class of antimicrobial agents (57). A new peptide was recorded by the symbiotic culture of polyhexanide biguanide and green tea. The resultant 3D BC/polyhexanide biguanide membrane showed high potential as chronic wound dressings (48). Some researchers have reported synthesis of plant cellulose composites with peptides (58) and their composites with chemically altered cellulose fibers (59). The unique properties of peptides elevate the need for further investigation in this area.

3.2.3.3 Cationic Antimicrobial Agents

Cationic antimicrobial agents like biguanides, quaternary ammonium salts, and bisbiguanides have been in use for decades as disinfectants and antiseptics in medical and everyday life. It is quite easy for these antibacterial agents to disrupt the single protective layer of Gram-positive bacteria. However, Gram-negative bacteria possess an additional outer layer made of porins and liposaccharides containing very fine restricting conduits that limit the movement of any foreign body and therefore hinder its interaction with the cell membrane (60). Therefore, each antimicrobial agent is required to be tested before impregnation into the BC membrane for wound dressing applications.

The most utilized biguanide polymer is poly (hexamethylene biguanide) hydrochloride (PHMB) (60). Its wide application as an antimicrobial agent in linen disinfection has led to the production of BC dressings incorporated with PHMB (50). BC composite dressings with PHMB have shown good antimicrobial activity against a number of bacteria, including *S. aureus* and the phytopathogenic *Xanthomonas campestris* (*X. campestris*), *Klebsiella pneumonia* (*K. pneumonia*), and *Pseudomonas syringae* (*P. syringae*), as well as yeast (49). In addition to PHMB, other cationic surfactants have also been introduced into the BC membrane to impart biocidal activity for wound dressing applications. Wei et al. (61) tested freeze-dried BC/benzalkonium chloride composites against Gram-positive *S. aureus* and Gram-negative *E. coli* and *Bacillus subtilis* (*B. subtilis*). The BC composites showed limited activity against the Gram-negative *E. coli* compared to its better activity against Gram-positive bacteria. These BC antimicrobial composites were suggested as prospective commercial films for severe wound treatments. Moritz et al. (62) impregnated octenidine dihydrochloride into the BC membrane through the *ex situ* method. The composite film showed a high antibacterial property against *S. aureus* even after 6 months of storage and showed only very little cytotoxic effect on human kertinocyte cells.

3.2.3.4 Antibiotics

The BC membrane can also be impregnated with some broad-spectrum antibiotic agents. Chloramphenicol in considered a broad-spectrum antibacterial. Its mode of action is different from those antimicrobial agents containing cationic groups. It blocks the fundamental functions of ribosomes through binding to the 50S ribosomal subunits (63). Chloramphenicol (CAP)-loaded dialdehyde (DABC) and nonoxidized BC have been tested for their antimicrobial activity against *S. aureus*, *E. coli*, and *Streptococcus pneumoniae* (*S. pneumoniae*). Although the drug-loading capacity of DABC was very low, both membranes exhibited effective activity against the tested bacteria. The fibroblast cell line L929 on the CAP/DABC showed better adhesion and proliferation than the nonoxidized BC did. This study revealed the potential of CAP/DABC as wound dressing materials (51).

3.2.3.5 Inorganic Compounds

Due to their large surface area and particle shapes, the inorganic nanomaterials show impressive antimicrobial properties (64). Nanoparticles (NPs) face some major disadvantages as they aggregate very easily and also possess uncontrollable release of ions and show cytotoxicity to animal cells. The recent trend of fabricating antimicrobial composites through a combination of polymers with potent antibacterial NPs is showing a greater research trend (65). Particularly, nanosilver has shown a broad-spectrum antimicrobial activity that is more effective than bulk silver. Silver has been in use as an antimicrobial agent for a long time. The antibacterial action of nanosilver is due to the interaction of released silver cations with the peptidoglycans of the bacterial cell wall, leading to protein unfolding, which ultimately blocks the activity of oxygen metabolizing enzymes and leads to bacterial death. Also, silver diffuses through the cell membrane interacting with thiol groups of enzymatic and respiratory proteins and also with the cell DNA. Due to their broad action mechanism, silver is highly active against both Gram-positive and Gram-negative bacteria (66).

Silver NPs have been impregnated into the BC matrix through various methods. Normally, previously synthesized NPs are impregnated into BC or BC is added to the synthetic mixture for *in situ* synthesis of Ag NPs on BC (67). The *ex situ* method is an easily controlled process, but the aggregates formation restricts its applicability (68). In comparison, the *in situ* method is more appropriate for the synthesis of uniform composites. Another method for introduction of slow silver releasing to the BC matrix is through the impregnation of such substances that contain silver such as silver sulfadiazine (SSD). SSD, due to its wide range of antifungal and antimicrobial activity, is widely used in wound therapy. The antimicrobial action mechanism involves the release of silver ions (69). SSD interacts not only with bacterial cells but also with host cells; however, this effect can be minimized by controlling the ion release (70). The impregnation of SSD material with BC will result in an effective nontoxic biomaterial with valuable antimicrobial and wound healing properties. BC/SSD membranes have shown good biocompatibility with epidermal and fibroblast cells while effectively hindering the growth of *Pseudomonas aeruginosa* (*P. aeruginosa*), *S. aureus*, and *E. coli* bacteria. Therefore, these BC/SSD composites are considered effective antimicrobial wound dressing materials for burn treatment (71).

BC/ZnO nanocomposites have also demonstrated good antibacterial activity against *E. coli* and *S. aureus*. The nanocomposite was synthesized through ultrasonic-assisted *in situ* synthesis of ZnO nanoparticles on BC pellicles leading to uniform composites (72). ZnO nanoparticles have also been incorporated in regenerated BC resulting in uniform composites with good antibacterial properties. The bacterial inhibition was found dependent on the NP concentration in the nanocomposite (54). Xiong et al. (73) demonstrated the effect of surface hydroxyl groups of BC on the morphology of *in situ* synthesized ZnO nanoparticles due to their strong interaction with Zn^{2+}. Flower-like aggregates are formed on the BC surface, whereas spindle-shaped NPs are formed in the absence of BC. Photoactive substances produce free radicals in the presence of UV light leading to photocatalytic disinfection. These radicals react with organic molecules in water and air and thus provide antimicrobial activity (74). BC composites with photoactive TiO_2 nanoparticles have been found very effective against Gram-positive and Gram-negative bacteria and also yeast (75, 76). The antibacterial action mechanism was found to be the combination of various mechanisms such as cell membrane damage through direct contact of ions, production of reactive oxygen species (ROS) such as oxygen free radicals, superoxide, and hydrogen peroxide causing the damage of surface phospholipid layer of bacterial cell membrane (76). Also, the ROS generated were shown to disrupt the life cycle of bacteria through oxidation of thiol groups on the amino acids of bacterial proteins (74, 76).

Besides metal and metal oxide NPs, other nanoparticles have also shown antimicrobial properties. Montmorillonite, a clay, has shown broad biomedical applications. MMT and its derivatives are biocompatible with animal cells and have shown a broad spectrum of antimicrobial activities against Gram-positive and Gram-negative bacteria, thus promoting fast wound healing (77). BC/MMT composites with pure MMT and Ca-, Na-, and Cu-modified MMT were fabricated, and their antimicrobial activity was determined to know their effectiveness and stability as wound dressing and artificial skin material. The composites demonstrated good antibacterial activity against *S. aureus* and *E. coli*; however, the activity was directly related to the amount of MMT present. The BC/Cu-MMT showed better inhibition of the bacteria compared to the other composites. These composites were recommended for application as wound dressing and skin regeneration materials (43). BC composites with graphene oxide nanosheets have also shown good antimicrobial activity against *S. aureus* and *E. coli*. (53).

3.3 APPLICATIONS OF BC COMPOSITES

3.3.1 BIOMEDICAL APPLICATIONS

The major problem for BC in broad clinical applications is its limited surface loading and absence of functional groups to anchor bioactive compounds with diagnostic or therapeutic applications. The insolubility of BC in water and common organic solvents hinders the functionalization of active surface chemical groups while keeping its 3D structure and biocompatibility intact (78). Therefore, the general entrapment of bioactive compounds like proteins, drugs, and polyelectrolytes that can regenerate tissues like proteins, drugs, and polyelectrolytes is barely achieved under normal conditions, presenting a major challenge for full adaptation of BC to its potential therapeutic applications (79). To overcome such limitations, a number of functionalization techniques have been developed to add biorecognition, charged interfacial groups, conductivity, or electrostatic potential to generate the desired material for widespread applications in regenerative medicine (80). To impart the desired biological, chemical, and physical features to BC, specially designed composites with two or more specific components are under constant investigation to develop novel biomaterials for tissue regeneration applications (81).

Although BC is inert toward pH variations and ionic strength, many studies have shown production of functional BC composites with broad range applications in diagnostic (82) and biomedical applications (7). BC microstructure is shown to undergo adhesion and adsorption with proteins and polysaccharides of different polarity, allowing the sequential functionalization of BC (83). Also, many conjugation processes have been reported with xyloglucan (83), alginate (84), chitosan (85), and gelatin (20). These modifications are related to improving certain properties of the material, including increasing hydration capacity or bonding efficiency. For example, the BC/pectin composites have shown a 20-fold increase in compression modulus and a greater resistance to compression and stress (86). Also, BC with cross-linked carboxymethyl cellulose showed better entrapment capacity toward ibuprofen compared to the unmodified BC, showing good potential as a drug delivery system (87).

Other modifications target the increased porosity of BC to increase drug diffusion and cell communication (80, 154). Therefore, the basic aim of BC composite production is to seek the development of novel biomaterials for bioengineering, adding the advantage of biologically active components to the BC microstructure. The general objective of nanocomposite synthesis includes better wound healing, prevention of bacterial growth, scaffolding, and diagnostic applications.

3.3.1.1 Wound Healing

BC is considered as an excellent wound dressing material as it absorbs wound exudates, reduces pain, and prevents infections (88). However, BC composites with different biomolecules are synthesized for specific target applications such as better tissue regeneration or improved cell adhesion. These specifically designed materials are considered potential candidates for the next generation of wound healing devices and regenerative medicine. In this context, the BC/Ag nanocomposite is considered the most important composite material as wound dressing due to its bactericidal and bacteriostatic effects. In such composites, BC acts as a stabilizing agent to control nucleation and thus prevent aggregation of nanoparticles, leading to uniform nanocomposites (89). The surface porosity and hydrophilicity of BC play an important role in the synthesis of and stabilization of Ag NPs. Although Ag NPs can cause cytotoxicity in some cases (90), the use of BC/Ag composites as wound dressing materials have shown numerous benefits, avoiding various kinds of bacterial infections including *S. aureus, E. coli, B. subtilis, K. pneumoniae,* and *P. aeruginosa* (91, 92). These properties suggest BC/Ag nanocomposites as an effective wound dressing material. BC can also be modified as a drug reservoir and administration system, benefiting from its macromolecular structure, flexibility, and transparency for tracking the wound regeneration process. These applications deal with the *in situ* functionalization of the 3D BC network with polyelectrolytes and drugs for control drug delivery. Different studies have reported the efficacy of hydroxyapatite (30), benzalkonium chloride (61), vaccarine (93), and aloe vera (94) as wound dressings.

Although many reports have shown the functionalization of BC interfacial fibers with positively charged (80) or negatively charged groups (95), only a few of them have described their application as a drug delivery system or biomedical device. For example, Picheth et al. (96) oxidized BC membrane covered with chitosan and alginate layers and were able to show controlled release of epidermal growth factors under normal conditions and also modulate their release in case of bacterial infection on the skin (96). Such products were only possible due to the presence of the negatively charged carboxyl groups present at the material interface allowing spray-assisted "layer by layer" coating. This methodology not only preserves the properties and integrity of BC but also allows the adsorption of interfacial layers for controlled drug delivery (97, 98).

Some reports also explored the potentials of BC as a drug reservoir for a range of clinical applications, from the treatment of superficial skin diseases to cancerous tissues. Many drugs including diclofenac (99), tetracycline (100), doxorubicin (101), and ibuprofen (102) have been successfully introduced into BC for local or transdermal delivery. In several studies, a functional biomaterial such as hydroxyethyl cellulose (103), sodium alginate (101), light irradiation of BC (104), or modifying the drug content (100) was introduced to maintain controlled drug release. The application of such BC composites as drug release and wound-healing materials have great potential in providing an adequate environment for epithelialization and tissue repair through combined drug release and healing processes.

3.3.1.2 Ophthalmic Scaffolds and Contact Lenses

An innovative application of BC and its composites involves the adhesion and proliferation of retinal pigment epithelium (RPE) and keratinocyte cells. Such BC composites can reduce the rejection rate of corneas and greatly improve the treatment of ocular diseases through increasing local neovascularization, decreasing side effects, and decreasing surgical recovery time. Therefore, numerous BC composites have been shown to adapt the material properties required for ocular therapy. For example, incorporation of polyvinyl alcohol has resulted in increased transmittance and UV absorbance of BC (105); Goncalves et al. (106) showed an increase in RPE proliferation through an increase in hydrophilicity of pristine BC achieved by incorporation of carboxymethyl cellulose and chitosan. Also, reports have shown that different functionalization resulted in a 3D structure that was more suitable for applications such as glaucoma relief and artificial cornea (81, 84). Therefore, BC composites can not only support the growth of corneal stroma cells but also preserve the complete view of the patient. These composites are therefore considered potential materials for use as ocular scaffolds and also to replace the less biocompatible hydroxyapatite or poly (methyl) methacrylate (PMMA) systems currently in clinical use (107).

The ability of BC to be molded into different shapes offers its potential for use in eye-related diseases, for example several press methodologies have led to BC-based convex-shaped stable contact lenses for correction of astigmatism, presbyopia, myopia, and hyperopia (108, 109). BC-based contact lenses can also act as drug delivery systems, maintaining a controlled drug release in treating eye infections. Cavicchioli et al. (110) have reported the incorporation of cyclodextrin/ciprofloxacin complexes into BC before pressing it into contact lenses. Therefore, BC-based contact lenses offer great potential for application as wound dressings after eye surgery, to improve recovery of ocular burns, and to replace the antibiotics in eye drops.

3.3.1.3 Bone, Cartilage, and Connective Tissue Repair

BC has also been tested in bone regeneration applications. Bone tissue consists of collagen supporting osteocytes, osteoblasts, and lining cells and also the solid matrix consisting of calcium phosphates such as hydroxyapatite and tricalcium phosphate. Patients with fractures and bone diseases need biocompatible grafts for filling the defective area and regenerating tissue. Normally the transplant of allogeneic or autologous bone presents the risk of rejection and transmission of pathogens and faces the limitation of limited sizes and shapes; its replacement with BC is a potential therapeutic option. Currently, pristine BC is used as an artificial replacement of dura mater in patients suffering from trauma or surgical procedures. Recent reports have shown BC membrane to show lower inflammatory response, prevent adhesion to brain tissue, and ultimately lead to repair in dura mater defect in rabbits. The results were better than commercially available dura mater substituent (NormalGEN®) (112).

Also, BC is easily stitched in neurosurgery and improves the post-surgery healing period (113). A number of BC composites are intended for better bone regeneration leading to improved osteogenic potential and faster healing. Particularly, surface-phosphorylated BC with high porosity is studied due to its ability to form complexes with calcium leading to better mineralization rates during the regeneration process and an increased proliferation of osteoprogenitor cells (114, 115).

The bone regeneration applications are based on the similarity of the 3D BC fiber network with collagen; thus, mineralization with hydroxyapatite is considered to mimic bone in various reconstitutes or fillings (116, 117). Therefore, these BC-based bone implants are utilized to fill defects and also provide long-lasting support in case of injury without replacement. However, recent studies have employed new strategies for therapeutic natural bone regeneration. For example, Hu et al. (118) synthesized BC-calcium phosphate composites loaded with cellulase enzymes; the composite is capable of local releasing calcium phosphate and maintaining a viable scaffold for mouse embryo pre-osteoblasts (MC3T3-E1) before biodegradation by the cellulase enzyme, thus acting as a temporary bone substitute (118).

3.3.1.4 Artificial Blood Vessels and Cardiovascular Implants

BC has been molded into different shapes during its synthesis for better cell adhesion and proliferation and for the development of several prostheses in the last two decades. For example, Klemm et al. (119) demonstrated that BC is an excellent substitute of anterosclerotic coronaries as artificial vessels, showing sufficient mechanical properties and molding ability to replace small dimeter vessels (<5 nm). The produced material (BASYC®) was successfully implanted in the carotid arteries of pigs and rats. The implant showed long-term stability and maintained the blood passage without any obstruction for a period as long as 3 months (120). Comparative studies with commonly used vascular grafts (PET and ePTFE), showed BC as an advantageous material with slow surface generation of thrombin (121). BC-derived composites are beneficial as replacement of atherosclerotic blood vessels due to their improved properties; for example, a BC composite with PVA showed low toxicity, higher resistance to traction, and better suture retention, which is an important property for maintaining long-term stability (122).

Other composites specially designed for improved cell adhesion and that complement activation prevention are synthesized by polyethylene glycol (PEG) grafting on BC nanofibers. These nanocomposites reduce the water contact angle by more than 50%, thus increasing the hydrophilicity of BC through PEG brush layer on BC nanofibers (123). A similar BC-PEG composite showed better biocompatibility with fibroblast 3T3 cells (124). Therefore, synthesis of more hydrophilic BC composites will present unmatched potential as implants of blood vessels covering cardiovascular stents or to be used as cardiac valves.

3.3.2 Industrial Applications of BC Composites

3.3.2.1 Food Applications and Food Packaging

BC is also used in the food industry because of its good rheological properties. It is easy to modify the BC production media to add color, flavor, shape, and texture to the produced BC for use in food. BC is basically a dietary fiber that offers valuable benefits for food digestion and decreases the risk of chronic diseases, including diabetes, obesity, constipation, and cardiovascular diseases (125). BC is also used as an additive for the production of low-calorie foods. Lin et al. (126) increased the water retention property of BC through treatment with an alkaline medium. The resulting compound was named alkaline treated cream and was added to surimi, as the treated BC exhibited properties suitable as fat replacement and a dietary source. Chau et al. (127) found that administering BC to hamsters resulted in a decrease of triglycerides and total cholesterol in serum and also decreased cholesterol in the liver; BC was considered a suitable alternative to gelatin and similar products. Okiyama et al. (128) studied the effect of BC on food rheology. They found that BC, unlike xanthan gum, prevents cocoa precipitation of chocolate drinks through entrapment in its fibrillar mesh, thus acting as an exceptional suspending agent. BC also improves the mechanical properties of food dressings through reduced adhesion and can replace the stabilizer gum in ice creams due to similar properties. BC is also used as body reinforcement in tofu. The main problem for tofu transport is its fragility, but its mechanical properties are significantly improved with a minimal percentage of BC.

BC has also found applications in active and smart food packaging. Edible BC antimicrobial membranes are developed using lactoferrin as an active component (129). These antimicrobial films hindered the viability of *S. aureus* and *E. coli*, elevating their application as a packaging material for highly perishable foods. Jipa et al. (130) incorporated sorbic acid (SA) as an antimicrobial agent into BC. A monolayer film was prepared using powder BC (BCP) and polyvinyl alcohol (PVA), which in turn was coated with BC to produce multilayer films. The release rate, water sensitivity, and antimicrobial capacity was influenced by the SA and BCP concentration. Also, the concentration and water solubility of SA influenced the antimicrobial activity of the composite films (130).

Kuswandi et al. (131) immobilized red methyl in BC film through the absorption method for the development of a new label sensor. The sensor color and chicken storage time showed a linear relationship, suggesting the possible application of the sensor for evaluation of the freshness of chicken cuts. Similarly, a BC/curcumin film responded as a color indicator to volatile amines that are released during fish spoilage (132), and BC/bromophenol blue responds through a color change to the freshness of guava fruit. Pourjavaher et al. (133) immersed BC into an anthocyanin solution extracted from *Brassica oleracea* for the production of a pH indicator. The composite films responded well to pH when exposed to pH 2–10.

3.3.2.2 Sensors

The BC-based sensors take advantage of BC's surface hydroxyl groups, its large surface area, and its high water retention and absorption capacity (134). BC nanofibers are applied as cover to quartz crystal microbalance (QCM) as a humidity sensor. An increase in BC weight was recorded with an increase in humidity in the test chamber. Therefore, it was suggested as a simple, low-cost, and sensitive humidity sensor based on BC fiber coating on QCM (134). Another similar approach led to a formaldehyde sensor using QCM and BC, covering BC with polyethylenimine (PEI) (135). The

BC surface hydroxyl groups and PEI surface amide groups led to hydrogen bonding. The BC/PEI nanocomposite fibers were successfully used as formaldehyde sensors where PEI acted as the detector. The sensor exhibited good reproducibility, reversibility, and a linear relationship with the concentration increase of formaldehyde (1 to 100 ppm at ambient temperature).

Wang et al. (136) utilized a different approach for the synthesis of a BC/Au sensor. The Au NPs were synthesized in the presence of a BC suspension. The heme proteins like hemoglobin (Hb), horseradish peroxide, and myoglobin (Mb) were immobilized on the BC/Au nanocomposite surface. The heme proteins showed electrocatalytic activity toward the H_2O_2 reduction in the presence of a mediating agent hydroquinone (HQ). The H_2O_2 biosensor showed biocatalytic activity with rapid response, high sensitivity, and low detection limit. The author followed the same approach for the development of a BC/Au biosensor for glucose detection (137).

3.3.2.3 Separation Membranes

The nanofibrous 3D structure of BC makes it a suitable material for element separation, dialysis, and filtration or as an ion exchange membrane. He et al. (138) synthesized BC/silica composites through the vacuum infiltration method and studied its adsorption capacity. The composite showed recyclable compressibility and super elasticity, high hydrophobicity, and ten times its weight in oil binding capacity. Dubey et al. (139) studied a BC/chitosan composite for ethanol separation from ethanol/azeotrope mixture in water (EtOH/H_2O). Compared to the PVA and chitosan mixture that dissolves at high water concentration, the BC/chitosan composite was stable. It also showed better dimensional and thermal stability and high mechanical properties. Choi et al. (140) explored BC/acrylic acid (BC/AAc) composite as ion exchange membranes.

3.3.2.4 Optical Materials

Optical materials need to have high transparency, flexibility, high stability, and be dispersion free. Nanostructures with components less than 1/10 wavelength size will not show dispersion (141). Therefore, BC composites with fiber diameter around 40–70 nm will exhibit excellent optical properties, dimensional stability, and good flexibility (142). Yano et al. (143) impregnated dry BC films with acrylic resins. The obtained nanocomposites showed a fiber content of 70%, a less than 10% loss of transparency, a very low coefficient of thermal expansion, and good mechanical properties compared to plastics. Nogi et al. (144) synthesized a dimethanol dimethacrylate BC/tricyclodecan composite with different BC percentage. They observed that 7.4 weight% BC caused a light transmission impairment of only 2.4% but reduced the thermal expansion coefficient of the acrylic resin from 86×10^{-6} to 38×10^{-6} K^{-1}.

BC nanofibers were also acetylated for better transparency of acrylic resins reinforced with BC nanofibers (145). Reduction in refractive index was found with acetylation and loss of regular transmission of the material with 63% of fiber content from 13% to 3.4%. Also, the coefficient of thermal expansion obtained was around 1×10^{-6} K^{-1}.

Fernandes et al. (146) synthesized BC/chitosan composites with a transmittance of 90% in the range of 400–700 nm. The composite of BC with uncured chitosan resulted in materials with high Young's modulus and low elongation at fracture in addition to good thermal stability and low O_2 permeability.

3.3.2.5 Energy Storage

BC conductive composites can be used in energy storage applications. BC's ability to catalyze palladium (Pd) precipitation produces structures with very high surface area. These BC/Pd materials are considered suitable materials for energy storage applications (147). BC has been found suitable for developing polyelectrolyte membranes and fuel cells. Yang et al. (148) developed BC/Pt composite through *in situ* reduction. The obtained nanocomposite showed high electrocatalytic activity. BC films doped with protonic acids showed high proton conductivity in the films, suggesting the applicability of BC as a fuel cell membrane.

For modern-day technological applications, the material applied for energy storage should possess flexibility along with mechanical strength. Therefore, the production of stretchable and conductive membranes is the need of the day. Pyrolyzed BC, having interwoven carbon nanofibers, can be an alternative mechanically robust 3D conducting network (149). The pyrolyzed BC/polydimethylsiloxane composites were prepared using BC precursor. These composite films showed electrical conductivity of 0.2–0.41 Scm^{-1} and well maintained their electromechanical properties under high tension (150). BC has also been applied in supercapacitors. For example, Chen et al. (151) developed a supercapacitor through MnO_2 coating on BC (BC/MnO$_2$) as a positive electrode and BC coated with nitrogen was used as negative electrode. The device showed charge/discharge ability with an energy density of 32.91 W h kg^{-1} and high cycling stability.

3.4 CONCLUSION AND PROSPECTS

This chapter is devoted to the biomedical and industrial applications of BC and its nanocomposites. BC composites are specially designed not only for their target biomedical applications such as wound healing, tissue regeneration, tissue implants, blood vessel and skeletal grafts but also for their value-added industrial applications, including biosensor, food industry, separation membrane, and optical and energy materials. The chapter provides a detailed review of the efforts made to devise BC composites with nanomaterials, biopolymers, and other nanofillers for the aforementioned applications. Furthermore, a comprehensive summary of the recent studies regarding the applications of such composites is provided. BC is a versatile and indispensable material for practical biomedical and industrial applications due to its ease of production, high crystallinity, better mechanical properties, high purity, nano woven morphology, and good biocompatibility and moldability.

Currently, the commercialization of BC and its value-added composites is hampered due to its high production cost and structural limitations. For example, the high cost of BC fermentation medium and the low yield by BC-producing bacterial strains leads to high production cost at the industrial level. Also, BC lacks surface adhesion sites for cells, limiting its biocompatibility. Furthermore, the lack of anti-microbial and antioxidant activity, electrical and magnetic conductivity, and optical transparency limits its practical applications in the biomedical and industrial fields. However, these limitations can be overcome through several strategies; for example, low cost and high yield can be achieved by utilizing low-cost carbon sources such as industrial wastes and through genetic engineering of the BC-producing bacterial strain, respectively. Also, development of novel effective surface modification and composite synthetic strategies will further enhance the biomedical and industrial applications of BC. Advanced approaches such as the development of a cell-free enzyme system for BC production and impairment of foreign genes for composite synthesis will overcome the shortcomings of in-use composite synthetic approaches. In conclusion, BC-centered research will mainly focus on lowering the production cost, improving properties, and discovering novel methods for the synthesis of value-added BC composites.

REFERENCES

(1) Coughlan MP. *Cellulose hydrolysis: The potential, the problems and relevant research at Galway.* Portland Press Limited; 1985.

(2) Madsen B, Gamstedt EK. Wood versus plant fibers: Similarities and differences in composite applications. *Advances in Materials Science and Engineering.* 2013; 2013: 564346.

(3) Iguchi M, Yamanaka S, Budhiono A. Bacterial cellulose—a masterpiece of nature's arts. *Journal of Materials Science.* 2000;35(2):261–70.

(4) Shoda M, Sugano Y. Recent advances in bacterial cellulose production. *Biotechnology and Bioprocess Engineering.* 2005;10(1):1.

(5) Schierbaum F. Book Review: Polysaccharides and polyamides in the food industry. properties, production, and patents. By A. Steinbüchel and SK Rhee (Editors). *Starch Stärke.* 2005;57(9):453.

(6) Czaja WK, Young DJ, Kawecki M, Brown RM. The future prospects of microbial cellulose in biomedical applications. *Biomacromolecules.* 2007;8(1):1–12.

(7) Koizumi S, Yue Z, Tomita Y, Kondo T, Iwase H, Yamaguchi D, et al. Bacterium organizes hierarchical amorphous structure in microbial cellulose. *The European Physical Journal E.* 2008;26(1–2):137–42.

(8) Atalla RH, Vanderhart DL. Native cellulose: A composite of two distinct crystalline forms. *Science.* 1984;223(4633):283–5.

(9) Lee KY, Buldum G, Mantalaris A, Bismarck A. More than meets the eye in bacterial cellulose: Biosynthesis, bioprocessing, and applications in advanced fiber composites. *Macromolecular Bioscience.* 2014;14(1):10–32.

(10) Gelin K, Bodin A, Gatenholm P, Mihranyan A, Edwards K, Strømme M. Characterization of water in bacterial cellulose using dielectric spectroscopy and electron microscopy. *Polymer.* 2007;48(26):7623–31.

(11) Guo J, Catchmark JM. Surface area and porosity of acid hydrolyzed cellulose nanowhiskers and cellulose produced by *Gluconacetobacter xylinus. Carbohydrate Polymers.* 2012;87(2):1026–37.

(12) Grande CJ, Torres FG, Gomez CM, Troncoso OP, Canet-Ferrer J, Martinez-Pastor J. Morphological characterisation of bacterial cellulose-starch nanocomposites. *Polymers and Polymer Composites.* 2008;16(3):181–5.

(13) Zhijiang C, Guang Y. Bacterial cellulose/collagen composite: Characterization and first evaluation of cytocompatibility. *Journal of Applied Polymer Science*. 2011;120(5): 2938–44.

(14) Saska S, Teixeira LN, de Oliveira PT, Gaspar AMM, Ribeiro SJL, Messaddeq Y, et al. Bacterial cellulose-collagen nanocomposite for bone tissue engineering. *Journal of Materials Chemistry*. 2012;22(41):22102–12.

(15) Luo H, Xiong G, Huang Y, He F, Wang Y, Wan Y. Preparation and characterization of a novel COL/BC composite for potential tissue engineering scaffolds. *Materials Chemistry and Physics*. 2008;110(2–3):193–6.

(16) Ul-Islam M, Shah N, Ha JH, Park JK. Effect of chitosan penetration on physico-chemical and mechanical properties of bacterial cellulose. *Korean Journal of Chemical Engineering*. 2011;28(8):1736.

(17) Torres F, Commeaux S, Troncoso O. Biocompatibility of bacterial cellulose based biomaterials. *Journal of Functional Biomaterials*. 2012;3(4):864–78.

(18) Kim J, Cai Z, Lee HS, Choi GS, Lee DH, Jo C. Preparation and characterization of a bacterial cellulose/chitosan composite for potential biomedical application. *Journal of Polymer Research*. 2011;18(4):739–44.

(19) Lin W-C, Lien C-C, Yeh H-J, Yu C-M, Hsu S-h. Bacterial cellulose and bacterial cellulose—chitosan membranes for wound dressing applications. *Carbohydrate Polymers*. 2013;94(1):603–11.

(20) Nakayama A, Kakugo A, Gong JP, Osada Y, Takai M, Erata T, et al. High mechanical strength double-network hydrogel with bacterial cellulose. *Advanced Functional Materials*. 2004;14(11):1124–8.

(21) Grande CJ, Torres FG, Gomez CM, Troncoso OP, Canet-Ferrer J, Martínez-Pastor J. Development of self-assembled bacterial cellulose—starch nanocomposites. *Materials Science and Engineering: C*. 2009;29(4):1098–104.

(22) Wan Y, Luo H, He F, Liang H, Huang Y, Li X. Mechanical, moisture absorption, and biodegradation behaviours of bacterial cellulose fibre-reinforced starch biocomposites. *Composites Science and Technology*. 2009;69(7–8):1212–7.

(23) Shinto Y, Uchida A, Korkusuz F, Araki N, Ono K. Calcium hydroxyapatite ceramic used as a delivery system for antibiotics. *The Journal of Bone and Joint Surgery British Volume*. 1992;74(4):600–4.

(24) Aoki H. Medical applications of hydroxyapatite. *Ishiyaku Euro America*. 1994:13–74.

(25) Goto T, Kojima T, Iijima T, Yokokura S, Kawano H, Yamamoto A, et al. Resorption of synthetic porous hydroxyapatite and replacement by newly formed bone. *Journal of Orthopaedic Science*. 2001;6(5):444–7.

(26) Grande CJ, Torres FG, Gomez CM, Bañó MC. Nanocomposites of bacterial cellulose/hydroxyapatite for biomedical applications. *Acta Biomaterialia*. 2009;5(5):1605–15.

(27) Wan Y, Huang Y, Yuan C, Raman S, Zhu Y, Jiang H, et al. Biomimetic synthesis of hydroxyapatite/bacterial cellulose nanocomposites for biomedical applications. *Materials Science and Engineering: C*. 2007;27(4):855–64.

(28) Zimmermann KA, LeBlanc JM, Sheets KT, Fox RW, Gatenholm P. Biomimetic design of a bacterial cellulose/hydroxyapatite nanocomposite for bone healing applications. *Materials Science and Engineering: C*. 2011;31(1):43–9.

(29) Saska S, Barud H, Gaspar A, Marchetto R, Ribeiro SJL, Messaddeq Y. Bacterial cellulose-hydroxyapatite nanocomposites for bone regeneration. *International Journal of Biomaterials*. 2011;2011.

(30) Hutchens SA, Benson RS, Evans BR, O'Neill HM, Rawn CJ. Biomimetic synthesis of calcium-deficient hydroxyapatite in a natural hydrogel. *Biomaterials*. 2006;27(26):4661–70.

(31) Yano S, Maeda H, Nakajima M, Hagiwara T, Sawaguchi T. Preparation and mechanical properties of bacterial cellulose nanocomposites loaded with silica nanoparticles. *Cellulose*. 2008;15(1):111–20.

(32) Maeda H, Nakajima M, Hagiwara T, Sawaguchi T, Yano S. Bacterial cellulose/silica hybrid fabricated by mimicking biocomposites. *Journal of Materials Science.* 2006;41(17):5646–56.

(33) Sai H, Xing L, Xiang J, Cui L, Jiao J, Zhao C, et al. Flexible aerogels based on an interpenetrating network of bacterial cellulose and silica by a non-supercritical drying process. *Journal of Materials Chemistry A.* 2013;1(27):7963–70.

(34) Cai J, Liu S, Feng J, Kimura S, Wada M, Kuga S, et al. Cellulose—silica nanocomposite aerogels by in situ formation of silica in cellulose gel. *Angewandte Chemie International Edition.* 2012;51(9):2076–9.

(35) Barud H, Assunção R, Martines M, Dexpert-Ghys J, Marques R, Messaddeq Y, et al. Bacterial cellulose—silica organic—inorganic hybrids. *Journal of Sol-Gel Science and Technology.* 2008;46(3):363–7.

(36) Stoica-Guzun A, Stroescu M, Jinga S, Jipa I, Dobre T, Dobre L. Ultrasound influence upon calcium carbonate precipitation on bacterial cellulose membranes. *Ultrasonics Sonochemistry.* 2012;19(4):909–15.

(37) Shi S, Chen S, Zhang X, Shen W, Li X, Hu W, et al. Biomimetic mineralization synthesis of calcium-deficient carbonate-containing hydroxyapatite in a three-dimensional network of bacterial cellulose. *Journal of Chemical Technology & Biotechnology: International Research in Process, Environmental & Clean Technology.* 2009;84(2):285–90.

(38) Serafica G, Mormino R, Bungay H. Inclusion of solid particles in bacterial cellulose. *Applied Microbiology and Biotechnology.* 2002;58(6):756–60.

(39) Stoica-Guzun A, Stroescu M, Jinga SI, Jipa IM, Dobre T. Microwave assisted synthesis of bacterial cellulose-calcium carbonate composites. *Industrial Crops and Products.* 2013;50:414–22.

(40) Ray SS, Okamoto M. Polymer/layered silicate nanocomposites: A review from preparation to processing. *Progress in Polymer Science.* 2003;28(11):1539–641.

(41) Algar I, Garcia-Astrain C, Gonzalez A, Martin L, Gabilondo N, Retegi A, et al. Improved permeability properties for bacterial cellulose/montmorillonite hybrid bionanocomposite membranes by in-situ assembling. *Journal of Renewable Materials.* 2016;4(1):57–65.

(42) Perotti GF, Barud HS, Messaddeq Y, Ribeiro SJ, Constantino VR. Bacterial cellulose—laponite clay nanocomposites. *Polymer.* 2011;52(1):157–63.

(43) Ul-Islam M, Khan T, Khattak WA, Park JK. Bacterial cellulose-MMTs nanoreinforced composite films: Novel wound dressing material with antibacterial properties. *Cellulose.* 2013;20(2):589–96.

(44) Butcher M. PHMB: An effective antimicrobial in wound bioburden management. *British Journal of Nursing.* 2012;21(Sup12):S16–S21.

(45) Lipsky BA, Hoey C. Topical antimicrobial therapy for treating chronic wounds. *Clinical Infectious Diseases.* 2009;49(10):1541–9.

(46) Figueiredo AR, Figueiredo AG, Silva NH, Barros-Timmons A, Almeida A, Silvestre AJ, et al. Antimicrobial bacterial cellulose nanocomposites prepared by in situ polymerization of 2-aminoethyl methacrylate. *Carbohydrate Polymers.* 2015;123:443–53.

(47) Jiang T, James R, Kumbar SG, Laurencin CT. Chitosan as a biomaterial: Structure, properties, and applications in tissue engineering and drug delivery. In *Natural and synthetic biomedical polymers.* Elsevier; 2014. pp. 91–113.

(48) Basmaji P, de Olyveira GM, dos Santos ML, Guastaldi AC. Novel antimicrobial peptides bacterial cellulose obtained by symbioses culture between polyhexanide biguanide (PHMB) and green tea. *Journal of Biomaterials and Tissue Engineering.* 2014;4(1):59–64.

(49) Kukharenko O, Bardeau J-F, Zaets I, Ovcharenko L, Tarasyuk O, Porhyn S, et al. Promising low cost antimicrobial composite material based on bacterial cellulose and polyhexamethylene guanidine hydrochloride. *European Polymer Journal.* 2014;60:247–54.

(50) Serafica G, Mormino R, Oster GA, Lentz KE, Koehler KP. *Microbial cellulose wound dressing for treating chronic wounds.* Google Patents; 2010.

(51) Laçin NT. Development of biodegradable antibacterial cellulose based hydrogel membranes for wound healing. *International Journal of Biological Macromolecules.* 2014;67:22–7.

(52) Shao W, Liu H, Liu X, Sun H, Wang S, Zhang R. pH-responsive release behavior and antibacterial activity of bacterial cellulose-silver nanocomposites. *International Journal of Biological Macromolecules.* 2015;76:209–17.

(53) Shao W, Liu H, Liu X, Wang S, Zhang R. Anti-bacterial performances and biocompatibility of bacterial cellulose/graphene oxide composites. *Rsc Advances.* 2015;5(7):4795–803.

(54) Ul-Islam M, Khattak WA, Ullah MW, Khan S, Park JK. Synthesis of regenerated bacterial cellulose-zinc oxide nanocomposite films for biomedical applications. *Cellulose.* 2014;21(1):433–47.

(55) Kingkaew J, Kirdponpattara S, Sanchavanakit N, Pavasant P, Phisalaphong M. Effect of molecular weight of chitosan on antimicrobial properties and tissue compatibility of chitosan-impregnated bacterial cellulose films. *Biotechnology and Bioprocess Engineering.* 2014;19(3):534–44.

(56) Phisalaphong M, Jatupaiboon N. Biosynthesis and characterization of bacteria cellulose—chitosan film. *Carbohydrate Polymers.* 2008;74(3):482–8.

(57) van t Hof W, Veerman EC, Helmerhorst EJ, Amerongen AVN. Antimicrobial peptides: Properties and applicability. *Biological Chemistry.* 2001;382(4):597–619.

(58) Hilpert K, Hancock RE. Use of luminescent bacteria for rapid screening and characterization of short cationic antimicrobial peptides synthesized on cellulose using peptide array technology. *Nature Protocols.* 2007;2(7):1652.

(59) Hilpert K, Elliott M, Jenssen H, Kindrachuk J, Fjell CD, Körner J, et al. Screening and characterization of surface-tethered cationic peptides for antimicrobial activity. *Chemistry & Biology.* 2009;16(1):58–69.

(60) Nikaido H. Prevention of drug access to bacterial targets: Permeability barriers and active efflux. *Science.* 1994;264(5157):382–8.

(61) Wei B, Yang G, Hong F. Preparation and evaluation of a kind of bacterial cellulose dry films with antibacterial properties. *Carbohydrate Polymers.* 2011;84(1):533–8.

(62) Moritz S, Wiegand C, Wesarg F, Hessler N, Müller FA, Kralisch D, et al. Active wound dressings based on bacterial nanocellulose as drug delivery system for octenidine. *International Journal of Pharmaceutics.* 2014;471(1–2):45–55.

(63) Xaplanteri MA, Andreou A, Dinos GP, Kalpaxis DL. Effect of polyamines on the inhibition of peptidyltransferase by antibiotics: Revisiting the mechanism of chloramphenicol action. *Nucleic Acids Research.* 2003;31(17):5074–83.

(64) Moghimi SM, Farhangrazi ZS. Nanoparticles in medicine: Nanoparticle engineering for macrophage targeting and nanoparticles that avoid macrophage recognition. In *Nanoparticles and the immune system.* Elsevier; 2014. pp. 77–89.

(65) Rai M, Yadav A, Gade A. Silver nanoparticles as a new generation of antimicrobials. *Biotechnology Advances.* 2009;27(1):76–83.

(66) Guo L, Yuan W, Lu Z, Li CM. Polymer/nanosilver composite coatings for antibacterial applications. *Colloids and Surfaces A: Physicochemical and Engineering Aspects.* 2013;439:69–83.

(67) Geng J, Yang D, Zhu Y, Cao L, Jiang Z, Sun Y. One-pot biosynthesis of polymer—inorganic nanocomposites. *Journal of Nanoparticle Research.* 2011;13(6):2661–70.

(68) Zou H, Wu S, Shen J. Polymer/silica nanocomposites: Preparation, characterization, properties, and applications. *Chemical Reviews.* 2008;108(9):3893–957.

(69) Dellera E, Bonferoni MC, Sandri G, Rossi S, Ferrari F, Del Fante C, et al. Development of chitosan oleate ionic micelles loaded with silver sulfadiazine to be associated with platelet lysate for application in wound healing. *European Journal of Pharmaceutics and Biopharmaceutics.* 2014;88(3):643–50.

(70) Piatkowski A, Drummer N, Andriessen A, Ulrich D, Pallua N. Randomized controlled single center study comparing a polyhexanide containing bio-cellulose dressing with silver sulfadiazine cream in partial-thickness dermal burns. *Burns.* 2011;37(5):800–4.

(71) Luan J, Wu J, Zheng Y, Song W, Wang G, Guo J, et al. Impregnation of silver sulfadiazine into bacterial cellulose for antimicrobial and biocompatible wound dressing. *Biomedical Materials.* 2012;7(6):065006.

(72) Katepetch C, Rujiravanit R, Tamura H. Formation of nanocrystalline ZnO particles into bacterial cellulose pellicle by ultrasonic-assisted in situ synthesis. *Cellulose.* 2013;20(3):1275–92.

(73) Xiong G, Luo H, Zhang J, Jin J, Wan Y. Synthesis of ZnO by chemical bath deposition in the presence of bacterial cellulose. *Acta Metallurgica Sinica (English Letters).* 2014;27(4):656–62.

(74) Blake DM, Maness P-C, Huang Z, Wolfrum EJ, Huang J, Jacoby WA. Application of the photocatalytic chemistry of titanium dioxide to disinfection and the killing of cancer cells. *Separation and Purification Methods.* 1999;28(1):1–50.

(75) Limaye SY, Subramanian S, Evans BR, O'neill HM. *Photoactivated antimicrobial wound dressing and method relating thereto.* Google Patents; 2009.

(76) Khan S, Ul-Islam M, Khattak WA, Ullah MW, Park JK. Bacterial cellulose-titanium dioxide nanocomposites: Nanostructural characteristics, antibacterial mechanism, and biocompatibility. *Cellulose.* 2015;22(1):565–79.

(77) Hong S-I, Rhim J-W. Antimicrobial activity of organically modified nano-clays. *Journal of Nanoscience and Nanotechnology.* 2008;8(11):5818–24.

(78) Heinze T, Liebert T. Unconventional methods in cellulose functionalization. *Progress in Polymer Science.* 2001;26(9):1689–762.

(79) Carambassis A, Rutland MW. Interactions of cellulose surfaces: Effect of electrolyte. *Langmuir.* 1999;15(17):5584–90.

(80) Pertile RA, Andrade FK, Alves Jr C, Gama M. Surface modification of bacterial cellulose by nitrogen-containing plasma for improved interaction with cells. *Carbohydrate Polymers.* 2010;82(3):692–8.

(81) Zaborowska M, Bodin A, Bäckdahl H, Popp J, Goldstein A, Gatenholm P. Microporous bacterial cellulose as a potential scaffold for bone regeneration. *Acta Biomaterialia.* 2010;6(7):2540–7.

(82) Bora U, Kannan K, Nahar P. A simple method for functionalization of cellulose membrane for covalent immobilization of biomolecules. *Journal of Membrane Science.* 2005;250(1–2):215–22.

(83) de Souza CF, Lucyszyn N, Woehl MA, Riegel-Vidotti IC, Borsali R, Sierakowski MR. Property evaluations of dry-cast reconstituted bacterial cellulose/tamarind xyloglucan biocomposites. *Carbohydrate Polymers.* 2013;93(1):144–53.

(84) Chiaoprakobkij N, Sanchavanakit N, Subbalekha K, Pavasant P, Phisalaphong M. Characterization and biocompatibility of bacterial cellulose/alginate composite sponges with human keratinocytes and gingival fibroblasts. *Carbohydrate Polymers.* 2011;85(3):548–53.

(85) Da Róz A, Leite F, Pereiro L, Nascente P, Zucolotto V, Oliveira Jr O, et al. Adsorption of chitosan on spin-coated cellulose films. *Carbohydrate Polymers.* 2010;80(1):65–70.

(86) Dayal MS, Catchmark JM. Mechanical and structural property analysis of bacterial cellulose composites. *Carbohydrate Polymers.* 2016;144:447–53.

(87) Juncu G, Stoica-Guzun A, Stroescu M, Isopencu G, Jinga SI. Drug release kinetics from carboxymethylcellulose-bacterial cellulose composite films. *International Journal of Pharmaceutics.* 2016;510(2):485–92.

(88) Czaja W, Krystynowicz A, Bielecki S, Brown Jr RM. Microbial cellulose—the natural power to heal wounds. *Biomaterials.* 2006;27(2):145–51.

(89) Maneerung T, Tokura S, Rujiravanit R. Impregnation of silver nanoparticles into bacterial cellulose for antimicrobial wound dressing. *Carbohydrate Polymers.* 2008;72(1):43–51.

(90) Eardley WG, Watts SA, Clasper JC. Extremity trauma, dressings, and wound infection: Should every acute limb wound have a silver lining? *The International Journal of Lower Extremity Wounds.* 2012;11(3):201–12.

(91) Schluesener JK, Schluesener HJ. Nanosilver: Application and novel aspects of toxicology. *Archives of Toxicology.* 2013;87(4):569–76.

(92) Li Y, Jiang H, Zheng W, Gong N, Chen L, Jiang X, et al. Bacterial cellulose—hyaluronan nanocomposite biomaterials as wound dressings for severe skin injury repair. *Journal of Materials Chemistry B.* 2015;3(17):3498–507.

(93) Qiu Y, Qiu L, Cui J, Wei Q. Bacterial cellulose and bacterial cellulose-vaccarin membranes for wound healing. *Materials Science and Engineering: C.* 2016;59:303–9.

(94) Saibuatong O-a, Phisalaphong M. Novo aloe vera—bacterial cellulose composite film from biosynthesis. *Carbohydrate Polymers.* 2010;79(2):455–60.

(95) Oshima T, Taguchi S, Ohe K, Baba Y. Phosphorylated bacterial cellulose for adsorption of proteins. *Carbohydrate Polymers.* 2011;83(2):953–8.

(96) Picheth GF, Sierakowski MR, Woehl MA, Pirich CL, Schreiner WH, Pontarolo R, et al. Characterisation of ultra-thin films of oxidised bacterial cellulose for enhanced anchoring and build-up of polyelectrolyte multilayers. *Colloid and Polymer Science.* 2014;292(1):97–105.

(97) Krogman KC, Lowery JL, Zacharia NS, Rutledge GC, Hammond PT. Spraying asymmetry into functional membranes layer-by-layer. *Nature Materials.* 2009;8(6):512.

(98) Kalosakas G, Martini D. Drug release from slabs and the effects of surface roughness. *International Journal of Pharmaceutics.* 2015;496(2):291–8.

(99) Silva NH, Rodrigues AF, Almeida IF, Costa PC, Rosado C, Neto CP, et al. Bacterial cellulose membranes as transdermal delivery systems for diclofenac: In vitro dissolution and permeation studies. *Carbohydrate Polymers.* 2014;106:264–9.

(100) Shao W, Liu H, Wang S, Wu J, Huang M, Min H, et al. Controlled release and antibacterial activity of tetracycline hydrochloride-loaded bacterial cellulose composite membranes. *Carbohydrate Polymers.* 2016;145:114–20.

(101) Cacicedo ML, León IE, Gonzalez JS, Porto LM, Alvarez VA, Castro GR. Modified bacterial cellulose scaffolds for localized doxorubicin release in human colorectal HT-29 cells. *Colloids and Surfaces B: Biointerfaces.* 2016;140:421–9.

(102) Trovatti E, Freire CS, Pinto PC, Almeida IF, Costa P, Silvestre AJ, et al. Bacterial cellulose membranes applied in topical and transdermal delivery of lidocaine hydrochloride and ibuprofen: In vitro diffusion studies. *International Journal of Pharmaceutics.* 2012;435(1):83–7.

(103) Zhou Q, Malm E, Nilsson H, Larsson PT, Iversen T, Berglund LA, et al. Nanostructured biocomposites based on bacterial cellulosic nanofibers compartmentalized by a soft hydroxyethylcellulose matrix coating. *Soft Matter.* 2009;5(21):4124–30.

(104) Stoica-Guzun A, Stroescu M, Tache F, Zaharescu T, Grosu E. Effect of electron beam irradiation on bacterial cellulose membranes used as transdermal drug delivery systems. *Nuclear Instruments and Methods in Physics Research Section B: Beam Interactions with Materials and Atoms.* 2007;265(1):434–8.

(105) Wang J, Gao C, Zhang Y, Wan Y. Preparation and in vitro characterization of BC/PVA hydrogel composite for its potential use as artificial cornea biomaterial. *Materials Science and Engineering: C.* 2010;30(1):214–8.

(106) Goncalves S, Padrao J, Rodrigues IP, Silva JoP, Sencadas V, Lanceros-Mendez S, et al. Bacterial cellulose as a support for the growth of retinal pigment epithelium. *Biomacromolecules.* 2015;16(4):1341–51.

(107) Dutton JJ. Coralline hydroxyapatite as an ocular implant. *Ophthalmology.* 1991;98(3):370–7.

(108) Picheth GF, Pirich CL, Sierakowski MR, Woehl MA, Sakakibara CN, de Souza CF, et al. Bacterial cellulose in biomedical applications: A review. *International Journal of Biological Macromolecules*. 2017;104:97–106.

(109) Wan W-K, Millon L. *Poly (vinyl alcohol)-bacterial cellulose nanocomposite*. Google Patents; 2014.

(110) Cavicchioli M, Corso C, Coelho F, Mendes L, Saska S, Soares C, et al. Characterization and cytotoxic, genotoxic and mutagenic evaluations of bacterial cellulose membranes incorporated with ciprofloxacin: A potential material for use as therapeutic contact lens. *World Journal of Pharmacy and Pharmaceutical Sciences*. 2015;4(7).

(112) Xu C, Ma X, Chen S, Tao M, Yuan L, Jing Y. Bacterial cellulose membranes used as artificial substitutes for dural defection in rabbits. *International Journal of Molecular Sciences*. 2014;15(6):10855–67.

(113) Goldschmidt E, Cacicedo M, Kornfeld S, Valinoti M, Ielpi M, Ajler PM, et al. Construction and in vitro testing of a cellulose dura mater graft. *Neurological Research*. 2016;38(1):25–31.

(114) Fricain J, Granja P, Barbosa M, De Jéso B, Barthe N, Baquey C. Cellulose phosphates as biomaterials. In vivo biocompatibility studies. *Biomaterials*. 2002;23(4):971–80.

(115) Oshima T, Kondo K, Ohto K, Inoue K, Baba Y. Preparation of phosphorylated bacterial cellulose as an adsorbent for metal ions. *Reactive and Functional Polymers*. 2008;68(1):376–83.

(116) Favi P, Benson R, Neilsen N, Ehinger C, Dhar M. *Novel biodegradable microporous bacterial cellulose scaffolds engineered for bone and cartilage regeneration*. Federation of American Societies for Experimental Biology; 2013.

(117) Favi PM, Ospina SP, Kachole M, Gao M, Atehortua L, Webster TJ. Preparation and characterization of biodegradable nano hydroxyapatite—bacterial cellulose composites with well-defined honeycomb pore arrays for bone tissue engineering applications. *Cellulose*. 2016;23(2):1263–82.

(118) Hu Y, Zhu Y, Zhou X, Ruan C, Pan H, Catchmark JM. Bioabsorbable cellulose composites prepared by an improved mineral-binding process for bone defect repair. *Journal of Materials Chemistry B*. 2016;4(7):1235–46.

(119) Klemm D, Schumann D, Udhardt U, Marsch S. Bacterial synthesized cellulose—artificial blood vessels for microsurgery. *Progress in Polymer Science*. 2001;26(9):1561–603.

(120) Schumann DA, Wippermann J, Klemm DO, Kramer F, Koth D, Kosmehl H, et al. Artificial vascular implants from bacterial cellulose: Preliminary results of small arterial substitutes. *Cellulose*. 2009;16(5):877–85.

(121) Fink H, Faxälv L, Molnar GF, Drotz K, Risberg B, Lindahl TL, et al. Real-time measurements of coagulation on bacterial cellulose and conventional vascular graft materials. *Acta Biomaterialia*. 2010;6(3):1125–30.

(122) Tang J, Bao L, Li X, Chen L, Hong FF. Potential of PVA-doped bacterial nano-cellulose tubular composites for artificial blood vessels. *Journal of Materials Chemistry B*. 2015;3(43):8537–47.

(123) da Silva R, Sierakowski MR, Bassani HP, Zawadzki SF, Pirich CL, Ono L, et al. Hydrophilicity improvement of mercerized bacterial cellulose films by polyethylene glycol graft. *International Journal of Biological Macromolecules*. 2016;86:599–605.

(124) Cai Z, Kim J. Bacterial cellulose/poly (ethylene glycol) composite: Characterization and first evaluation of biocompatibility. *Cellulose*. 2010;17(1):83–91.

(125) Cho SS, Almeida N. *Dietary fiber and health*. CRC Press; 2012.

(126) Lin SB, Chen LC, Chen HH. Physical characteristics of surimi and bacterial cellulose composite gel. *Journal of Food Process Engineering*. 2011;34(4):1363–79.

(127) Chau C-F, Yang P, Yu C-M, Yen G-C. Investigation on the lipid-and cholesterol-lowering abilities of biocellulose. *Journal of Agricultural and Food Chemistry*. 2008;56(6):2291–5.

(128) Okiyama A, Motoki M, Yamanaka S. Bacterial cellulose IV. Application to processed foods. *Food Hydrocolloids.* 1993;6(6):503–11.

(129) Padrão J, Gonçalves S, Silva JP, Sencadas V, Lanceros-Méndez S, Pinheiro AC, et al. Bacterial cellulose-lactoferrin as an antimicrobial edible packaging. *Food Hydrocolloids.* 2016;58:126–40.

(130) Jipa IM, Stoica-Guzun A, Stroescu M. Controlled release of sorbic acid from bacterial cellulose based mono and multilayer antimicrobial films. *LWT-Food Science and Technology.* 2012;47(2):400–6.

(131) Kuswandi B, Oktaviana R, Abdullah A, Heng LY. A novel on-package sticker sensor based on methyl red for real-time monitoring of broiler chicken cut freshness. *Packaging Technology and Science.* 2014;27(1):69–81.

(132) Kuswandi B, Restyana A, Abdullah A, Heng LY, Ahmad M. A novel colorimetric food package label for fish spoilage based on polyaniline film. *Food Control.* 2012;25(1):184–9.

(133) Pourjavaher S, Almasi H, Meshkini S, Pirsa S, Parandi E. Development of a colorimetric pH indicator based on bacterial cellulose nanofibers and red cabbage (Brassica oleraceae) extract. *Carbohydrate Polymers.* 2017;156:193–201.

(134) Hu W, Chen S, Zhou B, Liu L, Ding B, Wang H. Highly stable and sensitive humidity sensors based on quartz crystal microbalance coated with bacterial cellulose membrane. *Sensors and Actuators B: Chemical.* 2011;159(1):301–6.

(135) Hu W, Chen S, Liu L, Ding B, Wang H. Formaldehyde sensors based on nanofibrous polyethyleneimine/bacterial cellulose membranes coated quartz crystal microbalance. *Sensors and Actuators B: Chemical.* 2011;157(2):554–9.

(136) Wang W, Zhang T-J, Zhang D-W, Li H-Y, Ma Y-R, Qi L-M, et al. Amperometric hydrogen peroxide biosensor based on the immobilization of heme proteins on gold nanoparticles—bacteria cellulose nanofibers nanocomposite. *Talanta.* 2011;84(1):71–7.

(137) Wang W, Li HY, Zhang DW, Jiang J, Cui YR, Qiu S, et al. Fabrication of bienzymatic glucose biosensor based on novel gold nanoparticles-bacteria cellulose nanofibers nanocomposite. *Electroanalysis.* 2010;22(21):2543–50.

(138) He J, Zhao H, Li X, Su D, Zhang F, Ji H, et al. Superelastic and superhydrophobic bacterial cellulose/silica aerogels with hierarchical cellular structure for oil absorption and recovery. *Journal of Hazardous Materials.* 2018;346:199–207.

(139) Dubey V, Pandey LK, Saxena C. Pervaporative separation of ethanol/water azeotrope using a novel chitosan-impregnated bacterial cellulose membrane and chitosan—poly (vinyl alcohol) blends. *Journal of Membrane Science.* 2005;251(1–2):131–6.

(140) Choi YJ, Ahn Y, Kang MS, Jun HK, Kim IS, Moon SH. Preparation and characterization of acrylic acid-treated bacterial cellulose cation-exchange membrane. *Journal of Chemical Technology & Biotechnology: International Research in Process, Environmental & Clean Technology.* 2004;79(1):79–84.

(141) Novak BM. Hybrid nanocomposite materials—between inorganic glasses and organic polymers. *Advanced Materials.* 1993;5(6):422–33.

(142) Qiu K, Netravali AN. A review of fabrication and applications of bacterial cellulose based nanocomposites. *Polymer Reviews.* 2014;54(4):598–626.

(143) Yano H, Sugiyama J, Nakagaito AN, Nogi M, Matsuura T, Hikita M, et al. Optically transparent composites reinforced with networks of bacterial nanofibers. *Advanced Materials.* 2005;17(2):153–5.

(144) Nogi M, Ifuku S, Abe K, Handa K, Nakagaito AN, Yano H. Fiber-content dependency of the optical transparency and thermal expansion of bacterial nanofiber reinforced composites. *Applied Physics Letters.* 2006;88(13):133124.

(145) Ifuku S, Nogi M, Abe K, Handa K, Nakatsubo F, Yano H. Surface modification of bacterial cellulose nanofibers for property enhancement of optically transparent composites: Dependence on acetyl-group DS. *Biomacromolecules.* 2007;8(6):1973–8.

(146) Fernandes SC, Oliveira L, Freire CS, Silvestre AJ, Neto CP, Gandini A, et al. Novel transparent nanocomposite films based on chitosan and bacterial cellulose. *Green Chemistry*. 2009;11(12):2023–9.

(147) Evans BR, O'Neill HM, Malyvanh VP, Lee I, Woodward J. Palladium-bacterial cellulose membranes for fuel cells. *Biosensors and Bioelectronics*. 2003;18(7):917–23.

(148) Yang J, Sun D, Li J, Yang X, Yu J, Hao Q, et al. In situ deposition of platinum nanoparticles on bacterial cellulose membranes and evaluation of PEM fuel cell performance. *Electrochimica Acta*. 2009;54(26):6300–5.

(149) Wang B, Li X, Luo B, Yang J, Wang X, Song Q, et al. Pyrolyzed bacterial cellulose: A versatile support for lithium ion battery anode materials. *Small*. 2013;9(14):2399–404.

(150) Huang J, Li D, Zhao M, Lv P, Lucia L, Wei Q. Highly stretchable and bio-based sensors for sensitive strain detection of angular displacements. *Cellulose*. 2019;26(5):3401–13.

(151) Chen LF, Huang ZH, Liang HW, Guan QF, Yu SH. Bacterial-cellulose-derived carbon nanofiber@ MnO2 and nitrogen-doped carbon nanofiber electrode materials: An asymmetric supercapacitor with high energy and power density. *Advanced Materials*. 2013;25(34):4746–52.

(152) Shah N, Ul-Islam M, Khattak WA, Park JK. Overview of bacterial cellulose composites: A multipurpose advanced material. *Carbohydrate Polymers*. 2013;98:1585–98.

(153) Khan S, Ul-Islam M, Ullah MW, Kim Y, Park JK. Synthesis and characterization of a novel bacterial cellulose–poly (3, 4-ethylenedioxythiophene)–poly (styrene sulfonate) composite for use in biomedical applications. *Cellulose*. 2015;22:2141–48.

(154) Ul-Islam M, Subhan F, Ul Islam S, Khan S, Shah N, Manan S, Ullah SW, Yang G. Development of three-dimensional bacterial cellulose/chitosan scaffolds: Analysis of cell-scaffold interaction for potential application in the diagnosis of ovarian cancer. *International Jouranl of Biological Macromolecules*. 2019;137:1050–59.

4 Bacterial Cellulose as Catalyst Support

Ashi Khalil, Sher Bahadar Khan,
and Tahseen Kamal

CONTENTS

4.1 CATALYSIS

The term catalysis refers to a major field of nanotechnology, describing the influence of a catalyst on the reaction outcome. Researchers seek an efficient and highly selective catalyst that can be recycled easily to enhance the reaction rate and is not consumed in the chemical reaction. The catalyst does not alter thermodynamic properties but only influences the kinetics of the reaction. The phenomenon of catalysis is studied under two broad headings: homogeneous catalysis and heterogeneous catalysis. Homogeneous catalysis refers to the class of catalytic systems where a catalyst is present in the same phase as the reacting species. In other words, one can say that this type of catalyst is soluble in the reaction solvent. A homogeneous catalyst undergoes action in the same step as the reactants do. The mechanism of this process follows intermediate complex formation theory. According to this theory, the catalyst first reacts with one of the reactants to form an intermediate compound, consuming less energy than that required for the actual reaction. The high-energy intermediate compound thus formed, being unstable, reacts with other reactants, resulting in product formation. Figure 4.1 illustrates a general example for homogeneous catalysis. The lead chamber process better illustrates homogeneous catalysis. Nitric oxide (NO) is used as a homogeneous catalyst for the conversion of sulphur dioxide into sulphur trioxide. NO first combines with O_2 to form NO_2, an intermediate compound, which in turn reacts with SO_2, forming SO_3. Both the reacting substances and catalyst are

DOI: 10.1201/9781003118756-4

FIGURE 4.1 Mechanism of homogeneous and heterogeneous catalysis.

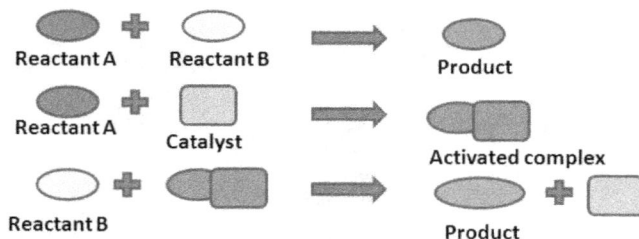

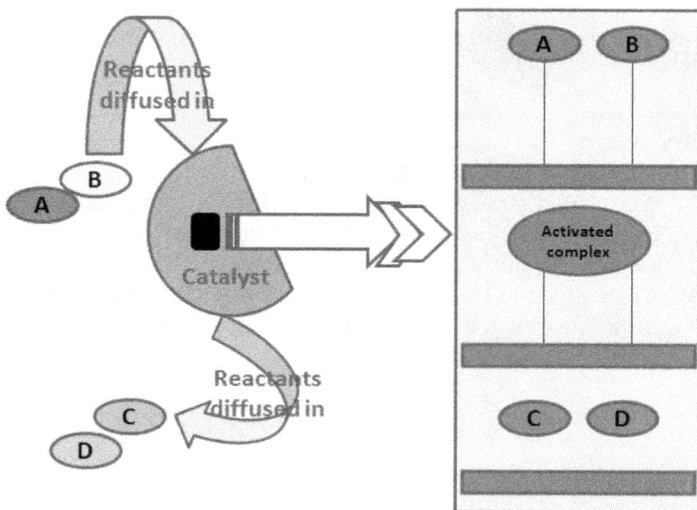

found in the common gaseous phase. In Deacon's process, cuprous chloride salt is used as a homogeneous catalyst for the synthesis of chlorine. Organometallic catalysts contribute greatly to homogeneous catalysis. Eremin et al. studied N-heterocyclic carbine–based palladium catalyst for its application in the Mizoroki–Heck reaction as a homogeneous catalyst (1). Owing to well dispersion of the catalyst in the homogeneous catalytic system, ensuring their even and excellent association with the reaction media, they show high catalytic efficiency. However, the same uniform dispersion hinders separation, recyclability, and reusability (2).

Another type of catalysis, called heterogeneous catalysis, refers to catalytic reaction systems in which the catalyst is present in a physical state other than that of the reactants, mostly solid in liquid and/or gas phase. Heterogeneous catalysis has the advantage of ease of recovery from the reaction system owing to its presence in solid state. The catalytic activity of a heterogeneous catalyst is highly dependent upon the availability of active sites on the surface; therefore, it is desirable to dig for a maximum number of active sites, hence surface area. This surface-dependent property has drawn scientists' attention to nanocatalysts with a particle size in the range of nanometers and that provide high surface area. Herein, the catalysis process follows an adsorption mechanism. The reactant molecules first surround the catalyst particles on the surface and then diffuse into the interior of the catalyst by intra-particle diffusion, and then they are adsorbed on the active sites by weak van der Waals attractive forces or partial chemical bonds. The adjacent reactant molecules interact with each other to form a highly unstable intermediate compound. The activated complex then decomposes to form product, which is instantly desorbed from the catalyst surface. Figure 4.1b graphically represents heterogeneous catalysis. Owing to its advantageous behavior, heterogeneous catalysis has found applications in various fields such as the food, chemical, pharmaceutical, petrochemical, and automobile industries. An estimated 90% of reported chemical processes utilize heterogeneous catalysts. The well-known Born–Haber process for synthesizing ammonia utilizes iron powder as a heterogeneous catalyst. The chemical conversion of ammonia into nitric acid by the Ostwald process follows the principle of heterogeneous catalysis, using platinized asbestos as a catalyst. Although heterogeneous catalysis shows good catalytic activity, the efficient removal of catalyst from the reaction system and its recyclability has created hurdles for the researchers (3).

4.2 CATALYST SUPPORT

With growing interest in nanocatalysts in the field of heterogeneous catalysis, researchers have addressed the problem of recovering catalyst from the reaction medium. To deal with these problems, different inorganic and organic supports are used. Inorganic catalytic supports may fail to meet the requirement of a good catalyst support in various ways. Some of them reported by various research groups will be discussed briefly in this chapter. Carbon materials like carbon black, pyrolytic carbon, glassy carbon, graphite, activated carbon, and polymer-derived carbon with particular characteristics have been used as catalyst supports. Despite having advantageous characteristics for use as a catalyst support, such as tunable shapes, controllable porosity, a hydrophobic nature, resistance to acids and bases, thermal and mechanical stability in inert atmospheres, little susceptibility to corrosion, and low cost, they also have some disadvantages. For instance, they show instability under higher temperature conditions, which is a basic requirement in oxidation and hydrogenation reactions where they may be subjected to gasification. In addition, intense microporosity because of activated carbon might be an issue and prompt mass transfer limitations, since it impedes the availability of reactant molecules in the active phase. Another associated problem is that of high preparation constancy, as varying carbon proportions are procured in various series of similar material. Carbon nanotubes (CNTs), hollow microcrystals of graphite with diameters in the range of nanoscale and length extending to centimeters, are reported as a catalyst support owing to their high surface area, excellent chemical and thermal stability, and exceptional electrical conductivity. The foremost drawbacks of CNTs for their use as support are their relatively high cost and the requirement of an inert atmosphere at higher temperatures.

Moreover, they require pretreatment such as carboxylation, hydroxylation, or oxidation in order to be utilized as a catalyst support. Alumina, Al_2O_3, with a high surface area, well-synchronized pores of large diameter, good thermal stability, and high resistance to attenuation appeals to many researchers for its use as a catalyst support; its main drawback is its high cost. Also, the hydrolysis rate of aluminum precursors is difficult to control. Although the surface acidic properties of alumina significantly enhance catalytic activity, it is prone to coke deposition when used as a catalyst support in the methanation of CO_2, as catalytic activity and carbon deposition are greatly affected by acidity/basicity of the supporting material. Another inorganic support, silica, has garnered research attention owing to its relatively inert and mildly acidic nature, chemical and thermal stability, high mechanical strength, high efficiency and selectivity, and highly porous structure. Nevertheless, silica has some drawbacks. Utilization of silica as a catalyst surface has associated problems such as low compatibility and micro-agglomerates/aggregate formation from pristine silica. Zeolites, being microporous and inert three-dimensional crystalline solids of aluminum silicates, are used as a catalyst support. They have the advantages of high electron conductivity, no waste problems, few or almost no chance of corrosion, high thermal stability, and high effectiveness in nature. However, they may irreversibly adsorb secondary pollutants, and it is difficult to exploit their microporosity and shape selectivity. Clays, ubiquitous and less costly, have found application as catalysts and catalyst supports owing to their acidic characteristics. However, their application as catalyst support can be utilized under mild conditions only, as their chemical nature might change in wild reaction conditions. Furthermore, because of low hydrothermal stability, they are incapable of maintaining their large surface area and high porosity, which limits their application as supporting material for catalyst. Instead, cost-effective modifications or pretreatment are required before their practical use. Owing to ultrathin thickness, larger surface area, and strong interactions with the supported metal nanocatalysts, graphene oxide is marked as an excellent support for immobilizing catalysts. However, graphene oxide has the disadvantage of feathery light weight. Also, its oxidative decomposition at higher temperature limits its applications as a catalyst support.

In light of the previous discussion, inorganic materials may fail in their use as an efficient catalyst support. Instead, biopolymer-based catalyst supports have garnered enormous attention for their easy availability, lower cost, and nontoxic and benign behavior toward the environment. Different biopolymers have been reported on such as starch, chitosan, cellulose, carboxymethyl cellulose, gelatin, aloe vera gel, alginate, and agar. In this chapter, we will focus on the use of bacterial cellulose (BC), a biocompatible and biodegradable polysaccharide with high mechanical strength and a large number of anchor sites for the attachment of a catalyst.

4.3 BACTERIAL CELLULOSE AS CATALYST SUPPORT

Bacterial cellulose is recognized as a biosynthetic substitute for naturally occurring cellulose, a major constituent of cell walls. Cellulose polymer obtained from plant extracts contains hemicelluloses and lignin. Several treatments are required to obtain cellulose in pure form. Instead, biosynthesized BC, which is obtained in highly pure form and can be used without any further processing, is preferred over naturally occurring cellulose. Biosynthesis of BC was first reported by Brown in 1886. He investigated cellulose biosynthesis from glucose as a carbon source, using Gram-negative bacteria, *Acetobacter xylinum*, under aerobic conditions. Later on, BC was biosynthesized by culturing a strain of Gram-positive bacteria of the genera *Gluconacetobacter, Sarcina, Rhizobium, Alcaligenes, Agrobacterium, Azotobacter, Aerobacter*, and *Salmonella*. Currently, *Gluconacetobacter*, acetic acid bacteria, is considered the highest yielding strain for BC biosynthesis by utilizing a wide variety of carbon and nitrogen sources. BC is the linear polymer of glucose subunits linked by β-(1→4) glycosidic linkages (Figure 4.3) with inter- and intramolecular hydrogen bonds. Both BC and plant cellulose have the same molecular formula, but BC is found in high purity, free of hemicelluloses and lignin. Additionally, BC shows a high degree of mechanical stability, crystallinity, polymerization, and water retention capacity. Furthermore, BC fibers can be scaled down to the nanoscale range, owing to its extremely thin texture, about 100 times finer than that of vegetable cellulose. Moreover, BC is known as a biocompatible, biodegradable, and nonallergenic polymer. BC has been extensively studied in a wide variety of fields such as those of paper, textiles, cosmetics, food, and medicine. For example, Fillat et al. investigated the use of BC combined with wood cellulose, which showed enhanced barrier properties with increased smoothness and surface gloss and exceptionally higher resistance to water and air. Thus, BC can be the best option for filter paper, food-packaging materials, and other paper products (4). Its high purity, *in situ* color and flavor change, and capability of forming different textures have made BC significant as a food ingredient. Fernandes et al. reported the saturation of BC membranes with Baygard EFN (perfluorocarbon) and Persoftal MS (polydimethylsiloxane), known hydrophobic polymers utilized in the textile industry for finishing processes (5). The vapor permeability values, drop absorption over time, water static contact angles, and improved mechanical strength of the hydrophobized, malleable, robust, and breathable nanocomposites based on the BC thus obtained, featured all the favorable properties applicable in textile industries (5). Hasan and his colleagues reported the formulation of facial scrub with BC as the main ingredient. The viscosity studies of the formulated facial scrub in comparison with the commercial scrub proved it safe for use on skin (6). Maneerung et al. impregnated BC with silver nanoparticles and studied its antimicrobial activity for use in dressing wounds. Although BC does not possess any antibacterial property, it provides a moist environment suitable for quick wound healing (7).

Apart from its industrial and medical applications, BC has also gained attention as a catalyst support for other purposes, which will be discussed later in this chapter. Morphological studies of BC show the presence of a nanofibrous structure ranging in diameter from 50 to 100 nm as shown in Figure 4.1a. The presence of these nanofibers gives BC increased surface area and a nanoporous structure, hence making it a good option for use as a catalyst support. As catalysis is a surface-dependent phenomenon, nanofibers provide extensive surface area, which in turn makes possible the attachment of a greater number of hydroxyl groups for binding the catalyst and hence increases the number of active sites for a reaction to take place. The presence of inter- and intra-molecular hydrogen bonding in its chemical structure are responsible for the stability of BC-based hydrogels in aqueous solutions. Thus, their water retention capacity and insolubility in water make them applicable as a hydrogel-based catalyst support in water-based reactions. Aqueous solution can pass through the interior of the catalytic system and hence ensures greater interaction of the reactant molecules with the active sites. Additionally, BC does not need any further chemical or physical processing before application but is synthesized in highly pure form under the action of aerobic microbes. BC comprises flexible, organic cellulose chains with a number of functional groups, where nanoparticles have to be supported. Given the presence of many hydroxyl groups on BC surfaces, it can be effortlessly modified chemically and has an ultrafine and orderly assembled structure, high water retention potential, and an eco-friendly and nontoxic nature. Due to intra- and intermolecular interactions of the cellulose chains, they are at high risk of entanglement, hence the anchor sites for the binding of nanoparticles are declined. In order to obtain individually dispersed cellulose chains with more binding sites, different physical and chemical approaches have been made such as 2, 2, 6, 6-tetramethylpylperidine-L-oxyl (TEMPO) oxidation, monomer grafting, and physical modification. Chen et al. studied the conversion of entangled BC nanofibers into individually dispersed fibrils with high surface area and increased binding sites by TEMPO oxidation. TEMPO oxidation involves the oxidation of C6 primary hydroxyl groups into aldehydes and carboxylic groups mediated by a 2, 2, 6, 6-tetramethylpylperidine-L-oxyl radical. The mutual repulsion between carboxylic groups results in the formation of individually dispersed uniform-sized nanofibrils. Second, carboxylate ions provide high-density anchor sites for nanoparticles. FT-IR spectra confirm the presence of carboxylate ions, indicated by the appearance of new peak at 1,727 cm^{-1}, associated with stretching vibration of C=O group As shown in Figure 4.2b (8). Elayaraja et al. reported BC synthesis by *Gluconacetobacter xylinus, followed by activation of the carboxyl groups–mediated TEMPO oxidation, ensuring better adherence of silver nanoparticles to the surface.* The high degree of hydrophilicity of BC rendered to it

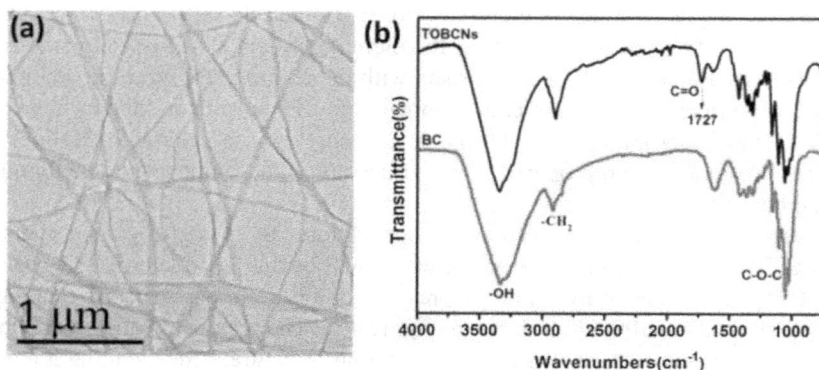

FIGURE 4.2   (a) Typical TEM image of BC, (b) FT-IR spectra of BC and TOBCNs (Chen et al. 2017) from reference (8).

Source: Chen, Y.; Chen, S.; Wang, B.; Yao, J.; Wang, H. TEMPO-Oxidized Bacterial Cellulose Nanofibers-Supported Gold Nanoparticles with Superior Catalytic Properties. *Carbohydr. Polym.* **2017**, *160*, 34–42. https://doi.org/10.1016/j.carbpol.2016.12.020.

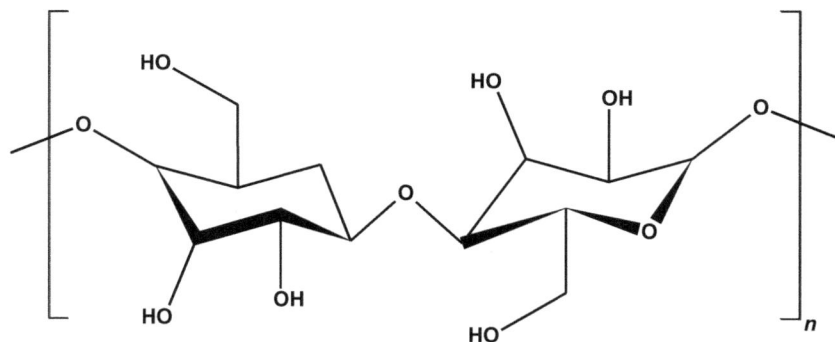

FIGURE 4.3 Chemical structure of BC.

by a large number of −OH groups has led to effect its application as a catalyst support
(9). For increased nanoparticle adhesion to the BC surface, hydrophilicity is reduced
by taking into account the advantages of graft copolymerization. Graft copolymerization can be achieved in three different ways: (a) grafting with a single monomer,
(b) grafting with a mixture of monomers, and (c) grafting with a polymer directly.
Li and colleagues investigated the surface modification of BC with polylactide-graft-
γ-methacryloxypropyltrimethoxysilane (MPS-g-PLA) to enhance the hydrophobic
nature of BC for increased compatibility (10). Fijałkowski et al. studied the increased
absorption capacity of bacterial cellulose modified by exposure to a rotating magnetic
field. RMF-modified BC showed less interassociated microfibrils and higher thermal
stability (11).

4.3.1 PREPARATION STRATEGIES OF BC AND METAL NANOPARTICLES CATALYST

The low-cost, environmentally benign, highly pure (free of hemicellulose, pectin,
and lignin as in the case of plant cellulose), and crystalline bacterial cellulose presents a highly attractive example as a support for metal nanoparticle (MNP) catalysts.
The presence of a large number of OH groups in its chemical structure provides a
skeleton for the binding of MNPs. The highly hydrophilic nature of BC (rendered to
it by the presence of hydroxyl functional groups), the presence of non-woven, three-
dimensional nanofibrillar structures, and its porous morphology permit the efficient
diffusion of incoming molecules throughout its inner surface. Owing to its peculiar
features, BC has thus been reported as a matrix for the immobilization of a variety of
MNP catalysts, such as gold, palladium, silver, magnetic oxide, titanium/vanadium
oxide, ZnO, and MgO NPs. Wu et al. prepared AuNPs/BC dip-catalyst and evaluated
its activity for the reductive decolorization of 4-nitrophenol (12). Teixeira et al. biosynthesized BC membranes and incorporated cerium nitrate and silver nanoparticles
on to them (13). Chanthiwong et al. investigated BC as a support for the controlled
coprecipitation of iron oxide (14). Two different strategies are employed for the preparation of BC/metal nanocomposites: *ex situ* and *in situ* synthesis methods. Both
of these methods are schematically shown in Figure 4.4. The *ex situ* method, also

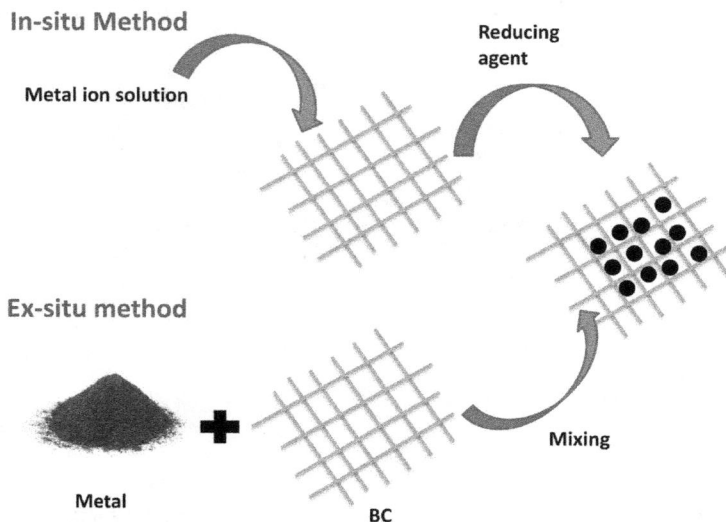

FIGURE 4.4   Methods of preparation of BC-metal nanoparticle composites.

known as the direct mixing technique, involves pre-preparation of nanoparticles and BC support, separately, followed by mixing wherein they interact with each other by different approaches. The porous structure and hydrophilic property of BC allows the predispersed metal nanoparticles to penetrate and physically absorb on the surface of nanofibrillar structures inside the BC matrix. Additionally, the metal nanoparticles interact with the OH groups present on the BC matrix by the virtue of hydrogen bonding. After the synthesis process is completed, any NPs left unbounded on the surface of BC/metal nanocomposites are washed up. *The ex situ* technique is known as a simple and valuable approach to synthesizing BC/metal nanocomposites. It has an advantage of conserving all the characteristic features of the BC matrix without changing its original structure. Moreover, pore size and pore volume are maintained in this approach. Owing to its superior factors, numerous researchers employ the *ex situ* approach for introducing different catalytic systems. For instance, Gupta et al. synthesized zero valent silver nanoparticles (AgNPs) from silver nitrate solution by the green approach using curcumin, a natural polyphenol as a reducing agent. The *ex situ* synthesized AgNPs were then incorporated on to a BC hydrogel membrane, already synthesized by Hestrin and Schramm (HS) culture medium under static conditions (15). Shahmohammadi Jebel and Almasi studied *ex situ* synthesis of ZnO/BC hybrid films via ultrasonic treatment. The size of ZnO NPs was observed to decrease upon ultrasonic irradiation as revealed by the SEM results (16). Khan et al. incorporated *ex situ* synthesized gold nanoparticles (AuNPs) and poly(3,4-ethylenedioxythiophene) polystyrene sulfonate (PEDOT:PSS) on to a BC matrix for the fabrication of biocompatible and highly conducting BC-AuNPs-PEDOT:PSS and composite. The uniform distribution over the BC matrix was confirmed by SEM and TEM images

(17). Apart from its beneficial aspects, this approach has certain limitations. First, only particles within the size range of sub-microns are capable of penetrating into the BC matrix. Second, it is difficult to achieve a homogeneous surface distribution of particles; instead, it is highly subjected to aggregate formation. Another approach, namely *in situ* synthesis, involves the pre-loading of metal ions of the desired MNPs to the BC media prior to matrix fabrication. The loaded metal ions are then converted to their respective MNPs by exposure to the corresponding conditions. The *in situ* approach has the advantages of easy processing, uniform distribution of particles, and proposed control overgrowth of NPs size. Yang et al. utilized the *in situ* approach for the preparation of CdS/BC nanohybrids by hydrothermal method. Prior to their addition to the reaction mixture containing $CdCl_2$ and thiourea, BC fibers were first saturated with ethanol (18). Ma et al. utilized carbonized BC for one-pot synthesis of CuNPs by dip-catalyst method. Carbonized bacterial cellulose (CBC) provided excellent stability and resistance to oxidation by CuNPs (19).

4.3.2 BACTERIAL CELLULOSE SUPPORT FOR PHOTOCATALYSTS

Generally, the absorption of photons in photocatalysis is interpreted as a property of volume; therefore, the contribution of the absorption of the surface layers may become important in NPs. As the diameter of the semiconductor particles decreases, the number of surface atoms considerably increases. Consequently, the photocatalytic activity increases with the decreasing size of the particles due to the larger surface area available for reaction. Problems associated with photocatalysts, such as their high tendency toward aggregation leading to declination of surface area and that they are difficult to recycle and reuse, has restricted their large-scale application. Therefore, researchers are seeking a potential catalyst support that can overcome the related hurdles. Among different known catalyst supports, BC has attained considerable attention owing to its high crystallinity, excellent tensile strength, high Young's modulus (approximately 78±17 GPa), considerable water retention capacity, and biocompatible nature. Due to high adsorption features and hydrophilic nature rendered to it by the presence of a large number of hydroxyl groups, a photocatalyst is well dispersed and firmly anchored on to its surface, thereby ensuring its ease of separation and recycling for another assembly of reaction. In addition to convenient separation and recycling, its highly porous nature provides a high surface area, the primary requirement of catalysis. BC-composite membranes are also studied for their long-term stability at varying flow rates. Liu et al. utilized BC as a matrix for the synthesis of graphene oxide/titanium dioxide-based hybrid nanomaterials. The as-synthesized $GO/TiO_2/BC$ was studied for its photocatalytic activity against methyl orange. High photocatalytic activity was observed in the near-UV region. BC was found to preserve the photocatalytic activity of the GO/TiO_2 nanocomposite (20). Phutanon and his colleagues reported the use of BC sheets as a support for the preparation of CuO nanoparticles, taking advantage of the free hydrolysis technique. The uniform distribution of CuO on the BC paper was confirmed by EDX mapping analysis. SEM analysis revealed the filling of CuO into the network spaces distributed along the structure of BC. The CuO/BC nanocomposite showed thermal stability up to 200 °C as indicated by TGA results. CuO/BC nanocomposite paper was observed to exhibit both photocatalytic and antibacterial properties (21). Wahid et al. reported the fabrication of BC with ZnO nanoparticles, of high uniformity and smooth texture. The facile method was implemented for the incorporation of ZnO nanoparticles on to the BC films. ZnO/BC nanocomposite films exhibited high thermal resistance with high photocatalytic potential for the degradation of methyl orange (up to 91%) under UV light. Antibacterial activity was also observed by ZnO/BC (22). Jiang et al. investigated the use of BC as a scaffold for controlling the shape and morphology of CdS photocatalyst. The comparison of BC-embedded CdS nanoparticles (labeled as b-CdS) and those synthesized without BC scaffolds revealed greater H_2 production by b-CdS. Moreover, photocatalytic activity of b-CdS was further increased by the addition of MoS_2, thus forming a hetero-junction-based photocatalyst with closely acquainted interfacial contacts (23). Similarly, Zhou et al. studied the synthesis of 2D porous CdS on to the surface of bacterial cellulose by an epitaxial growth route. BC was used as a biotemplate and scaffold of BC@CdS nanosheets with different window sizes, controlled by varying coating time. The photocatalytic H_2 production by BC@CdS was increased by 3.5 times as compared to pure CdS. Furthermore, photodegradation of methyl orange was also influenced by varying connecting window sizes, showing that BC has an important role in photocatalysis (24). Janpetch et al. reported the visible light–driven disinfection of water by titanium dioxide (TiO_2) doped with nitrogen (N) and fluorine (F). The surface area and hence the photocatalytic activity of N–F-co-doped TiO_2 was enhanced by embedding it into BC nanofibers. Besides its role in reducing particle size, BC/N–F-co-doped TiO_2 also reduced recovery requirements (25). Liu et al. reported vinyl-functionalized bacterial cellulose as a matrix for facile synthesis of TiO_2 photocatalyst with different sizes by using a thiol-ene click reaction. Further, the photocatalytic activity of the BC/TiO_2 was enhanced by treating with urea to achieve nitrogen doping. Direct correlation was observed between the size of TiO_2 nanoparticles and OH$^•$ production and hence photocatalytic activity. Thus, BC responsible for the controllable size of TiO_2 nanoparticles is of noticeable importance in photocatalysis (26). Almasi et al. studied the synthesis of copper oxide nanoparticles supported on BC by using three different approaches; *in situ* synthesis by precipitation and sonochemical methods and *ex situ* synthesis. BC/CuO nanohybrids thus synthesized were evaluated for their application as a photocatalyst for removal of heavy metal, microbial, and dye pollutants from water bodies (27). Bhatt and Patel investigated the immobilization of Ag^0/AgBr on BC and studied it for its activity for photocatalytic oxidation of reactive black 5. The reusability of Ag^0/AgBr/BC for up to five cycles reveals it as a potential recyclable photocatalyst (28). Jiang and colleagues reported synthesis of Ag_2O/zwitterionic compound, immobilized on bacterial cellulose aerogel and prepared by a freeze-drying technique. BCAZ aerogels exhibited super hydrophilic and superoleophilic properties. The BCAZ aerogels showed photocatalytic activity for degradation methylene blue dye (29).

4.3.3 Application in Hydrogen Evolution Reaction

To meet with the increasing demand for energy, different fossil fuels and petroleum fuel are utilized as energy sources. However, excessive use of biomass, fossil fuels, natural gas, coal and so on, being nonrenewable sources, has led to the production of carbon dioxide, which in turn has an adverse effect on climate, air, soil, and water bodies. Therefore, scientists are seeking an eco-friendly and clean energy source. Hydrogen, known as a clean, environmentally friendly, and sustainable energy source, is considered a propitious alternative to conventional energy sources. The electrochemically induced water-splitting reaction is evaluated as a promising approach toward H_2 production. The water-splitting reaction consists of two half reactions: oxygen evolution reaction and hydrogen evolution reaction. Hydrogen is produced in the hydrogen evolution reaction (HER). In order to overcome the kinetic barrier for the HER, different noble metal–based and transition metal–based semiconductors are used as catalysts. The lower specific surface area and nonrecyclability of the catalyst from the reaction medium thus make the approach costly and environmentally unfriendly, which has limited its application on larger scale. One of the strategies, employed for low loading and high efficiency of the catalyst, is the use of a catalyst support capable of providing a larger surface area, with high electrochemical stability and good conductivity. BC, a carbon nanotube–like structure and a cost-effective biopolymer, produced by microbial fermentation process at industrial scale, is a good option for use as a catalyst support. In addition to its highly crystalline nature, high tensile strength, hydrophilicity, and good water retention capacity, as discussed earlier in this chapter, BC shows excellent electrochemical stability, as per the requirement of a promising catalyst support for electrochemical processes. The poor conductivity of BC has been resolved by doping with a heteroatom such as nitrogen and fluorine. For instance, Zhang et al. utilized nitrogen-doped carbon nanofibers derived from BC, doped with polyaniline as a nitrogen source, as a support for platinum nanoparticles. The HER activity of the electrocatalysts, labeled as xPt@LN-BC and xPt@HN-BC (referring to low and high nitrogen proportions, respectively), were investigated for HER activity. The low Pt loading and well durability mark the catalytic system as promising for hydrogen evolution reactions (30). Lai et al. studied the production of nitrogen-based carbon nanofibers derived from aniline-doped BC by the combined effect of oxidative polymerization, followed by a high temperature carbonization (HTC) process. pBC-N thus prepared were used as a support for hydrothermal growth of MoS_2. Excellent stability, small overpotential value of 108 mV, and high current density of 8.7 mA cm^{-2} at η = 200 mV, highly exposed active sites with uniform distribution of MoS_2, and collaboration between MoS_2 and doped nitrogen atom make pBC-N/MoS_2 an efficient electrocatalyst for energy production and storage plants (31). Similarly, Wu and colleagues reported nitrogen-doped carbon nanofibers derived from BC as a support for MoC_2. BC-derived $Mo_2C@N$-CNFs showed remarkable electrocatalytic activity for HER. The synergistic effect of N-CNFs and Mo_2C for HER activity was revealed by theoretical calculations (32). Mi et al. reported an ultra-low amount about 0.87 wt% loading of Pt nanoparticles on to carbon nanofibers derived from 3D-BC, taking advantage of the atomic layer deposition method. The HER results revealed Pt/BCF as a highly stable and efficient catalytic system with average particle size of 2 nm (33). Zhang et al. reported the synthesis of zeolitic imidazolate framework–derived cobalt oxide supported on BC, followed by electrodeposition with Pt/Pd nanoparticles, and studied its applications as an HER electrocatalyst. The good durability displayed by PtPd@ZIF-67/BC marked BC a better option as a support for the HER electrocatalyst (34). Lai et al. presented few-layered, uniform

MoSe$_2$ nanosheets growth assisted by BC-derived cellulose nanofibers, which are found to increase active sites density of MoSe$_2$ and also accelerate electron transfer by providing a 3D network for the penetration of electrolyte into the inner space, thus resulting in a pronounced increase in the HER activity (35). Chen et al. investigated the utilization of BC-derived nitrogen-doped carbon nanofiber as an anchor for MoC and MoP nanoparticles–based heterostructures and studied its application as an electrocatalyst for HER reactions. The poor conductivity of hetero-structured MoC/MoP was overcome by BCNC NFs (36).

Besides electrochemically induced HER, another strategy, the visible light–driven photocatalytic approach, has also been employed for the successful H$_2$ production from water splitting. The said strategy has high potential for conversion of solar energy into chemical energy. Interest in visible light–driven photocatalysts for H$_2$ production was aroused after Fujishima and Honda carried out the first photocatalytic-induced water-splitting reaction on TiO$_2$ electrodes in 1972. Since then, many semiconductor-based photocatalysts have been developed, among which integrated catalytic systems have garnered much attention, owing to the synergistic effect shown by multiple desirable semiconductors. As a general rule, in a multi-semiconductor-based catalytic system, a photoexcited electron is allowed to flow from a semiconductor with a higher conduction band minimum to one with a lower conduction band minimum. Strongly excited electrons and holes are retained by the recombination of photo-excited electrons from a semiconductor with a lower conduction band minimum with holes from the semiconductor with a higher conduction band minimum. The photocatalytic activity can be heightened by enhancing the surface-to-volume ratio. For this purpose, cocatalysts such as noble metals or graphene are reported, but they have the drawback of high cost and rare availability. The low-cost and eco-friendly BC can be used as a scaffold for controlling surface area of the photocatalyst. For example, Wang et al. studied the enhancement of photocatalytic H$_2$ production activity of ZnxCd$_{1-x}$S nanoparticles by organizing them into bacterial cellulose with highly ordered architecture (37). Jiang et al. investigated the preparation of berry-shaped cadmium sulphide (b-CdS) with the help of BC used as a scaffold. The comparison of b-CdS and CdS synthesized without BC scaffold showed a higher photocatalytic H$_2$ production by b-CdS (23). Dal'Acqua et al. utilized BC as a support for the synthesis of self-assembled thin polyelectrolyte films of polyacrylic acid and polyallylamine hydrochloride by using a layer-by-layer technique with TiO$_2$ nanoparticles and varying gold concentration. The obtained BC films were found to be an efficient photocatalyst for H$_2$ production (38).

4.3.4 Applications in Dye Reduction and Useful Chemical Synthesis

Dye reduction involves converting the toxic functional group in the dye structure into a nontoxic reduced form in the presence of a suitable reducing agent, mostly sodium borohydride. Although the reduction of dyes with NaBH$_4$ alone is a thermodynamically feasible process, it possesses a high kinetic barrier in the path of the reduction process. Therefore, metal nanoparticles are used as catalysts to make the reduction process kinetically feasible. The use of different noble metals and transition metals have been reported so far. Besides dye reduction, catalytic reduction can also be applied to convert various nitroaromatic compounds into their corresponding amine derivatives. Even though MNPs exhibit high catalytic activity, the recovery of the catalyst from the reaction medium and its recyclability with conserved efficiency, to make the process cost effective and tidier, has become the question of interest. Various synthetic and natural polymers have been reported as a support. For example, BC, a biopolymer synthesized via microbial-mediated fermentation and with interlaced cellulose nanofibers consisting of a large number of hydroxyl groups for the attachment of MNPs, has been reported on as a catalyst support by many research groups. For instance, Chen et al. utilized Au nanoparticles immobilized on TEMPO-oxidized BC nanofibers for the catalytic reduction of 4-nitrophenol into 4-aminophenol. AuNPs/TOBCNs catalytic system showed excellent stability and superior catalytic activity at varying concentrations of reducing agent and temperature conditions (8). Kamal et al. reported a coating of carboxymethyl cellulose–supported cobalt nanoparticles on to the BC surface and investigated their catalytic properties for the reduction of methylene blue and 2,6-dinitrophenol. CMC-Co-BCN exhibited increased surface area, with an advantage of ease of recovery (39). In one of their other studies, they reported a retrievable CMC-Ni-BC dip-catalyst using BC as a supporting material for Ni nanoparticles stabilized on carboxymethyl cellulose, an anionic polysaccharide, and investigated their catalytic performance for 2-nitrophenol and methylene blue dye reduction (40). Song et al. investigated the 4-nitrophenol reduction with Cu and Ni nanoparticles using BC aerogels as a supporting and dispersion matrix. High stability and recyclability of Cu/BC and Ni/BC aerogels, thus proves BC as a potential support for MNPs (41). Dou et al. reported use of BC as pore-former for the synthesis of hierarchical porous CuO—CeO_2—ZrO_2 catalyst. the catalytic efficiency of the obtained catalyst was evaluated against volatile organic compounds, ethyl acetate and toluene (42). Wang et al. reported utilization of polyethyleneimine coated BC as an adsorbent and catalytic support for Cu^{2+} and Pb^{2+}, and Cu nanoparticles. They further studied the catalytic reduction of methylene blue with BC@PEI/Cu-NPs (43).

Along with applications of BC in reduction reactions, it has also been reported as a supporting material in useful chemical synthesis processes. Zhou et al. reported use of BC supported palladium nanocomposite as a recyclable catalyst in Heck coupling reactions (44). Jeremic et al. utilized Pd/BC and Cu/BC nanocomposite catalysts for Suzuki-Miyaura and Chan-Lam coupling reactions, respectively (45). Patel et al. introduced rotating catalyst contact reactor (RCCR) designed with palladized BC incorporated on acrylic discs for dechlorination of pentachlorophenol in the presence of hydrogen gas (46).

4.3.5 BC AS SUPPORT FOR CATALYSTS USED IN ELECTROCHEMISTRY

The increasing energy consumption and environment deterioration worldwide have recently aroused the enormous interests of research in the development of sustainable energy conversion and storage systems with green and efficient electrochemical techniques. The development of novel multi-functional nanocomposites has gained tremendous research interest during the last decade. In this direction, low cost resources, with renewable and biodegradable characteristics, are aimed for light weight, flexible and eco-friendly biomaterials intended to wearable electronics, biosensors or energy storage device. Among the different biomaterials, bacterial cellulose is one kind of nanocellulose structure that shows outstanding properties together with a complete biodegradable and biocompatible nature. The excellent physical-mechanical properties of BC matrix combined with highly conducting electroactive polymers are expected to provide nanocomposites with the electrical and mechanical characteristics that cannot be reached by the single materials. Owing to their inherent structural particularity, biocompatibility, and applicability, BC and BC-derived materials can be used for the preparation of different components (electrode materials and electrolytes) of energy-related devices. The unique nanostructures that originate from the assembly of low-dimensional nanoscale fibers can retain the advantages of the building blocks, which enlarges the total active volume of the electrodes for various energy storage systems, such as supercapacitors, lithium ion batteries (LIBs), sodium ion batteries (SIBs), potassium-ion batteries (PIBs), lithium—sulfur batteries (LSBs), and fuel cells. For these electrochemical energy storage systems, nano-structuring of electro-active materials (e.g., electrodes, separators, current collectors) introduces significant variations in the electrochemical properties of the devices in which they are a core component. Zhou et al. incorporated CuO/Cu on BC membrane by *in situ* chemical method and evaluated its applications as an electrode in CO_2 electrode reduction reactions. Due to ascendancy of porous, web like 3D structure of BC, providing enormous active sites for accessibility of electrolytes and corroborating steady movement of reactants and products, CuO/Cu@BC displayed magnificent charge transport property and higher current density for CO_2 reduction (47). Evans et al. prepared palladium based BC membranes and evaluated it as a promising catalyst in hydrogen generation reactions. The electrochemical potential of Pd-BC in catalyzing H_2 evolution when incubated in sodium thionite, suggested its use in anodic oxidation of hydrogen, and thus can be functionalized as energy storage device (48). Mei et al. prepared vanadium pentoxide (V_2O_5) based carbonized BC nanocomposite and investigated its specific capacitance by using galvanostatic charge-discharge (GCD) and cyclic voltammetry (CV). The enhanced electrochemical properties were thought to be associated with the CBC boating on the V2O5 nanobelts. Thus, revealing application of V2O5-CBC as potential electrodes for aqueous supercapacitors (49). Li et al. synthesized copper and phosphorus doped BC electrode, an alternative for expensive Pt catalysts and utilized in microbial fuel cell cathode catalyst. The improved catalytic potential was correlated to the Cu and P doping on BC surface, which resulted in an increased number of active sites (50). Huang et al. proposed utilization of N-PC@CBC synthesized by carbonization strategy as a template for growth of ZIF-8 nanocrystals and evaluated its catalytic efficiency in oxygen reduction reactions and sodium-ion storage batteries (51). Song et al. deposited polypyrole (PPy), holding efficient electrochemical potential with the limitations of poor cyclabilty and densely packed structure, on BC nanofibers by *in situ* polymerization, followed by intercalation with MXene nanoflakes exhibiting high conductivity. The enhanced electrochemical performance, high specific capacitance, and excellent cycling stability displayed by PPy@BC/MXe nanocomposite films reveal them as potential candidates for use in free-standing supercapacitor electrodes (52). Rebelo et al. functionalized BC nanofibers with polyvinylaniline/polyaniline, an electrically conducting bilayer. The BC/PVAN/PANI displayed applications as electrochemical biosensors for the differentiation and viability assays for nerve stem cells (53). Yao et al. designed fiber-based supercapacitors by flat packing PPy@CNT/BC electrodes. In addition to avoiding agglomeration of CNTs and enhancing the wettability of supercapacitor, BC serves as a nano-reservoir to promote efficient electrolyte ion diffusion and enhance the electrochemical output (54). Huang et al. assembled CBC aerogels bearing high electrical conductivity and mechanical stability with amorphous Fe_3O_4 nanostructures via *in situ* thermal decomposition technique. The uniform distribution of Fe_3O_4 nanoshells and the presence of well-ordered pores in CBC benefits Fe_2O_3@CBC with high rate capability and stable cycling performance, for use as an anode in lithium ion storage batteries (55).

REFERENCES

(1) Eremin, D. B.; Denisova, E. A.; Yu. Kostyukovich, A.; Martens, J.; Berden, G.; Oomens, J.; Khrustalev, V. N.; Chernyshev, V. M.; Ananikov, V. P. Ionic Pd/NHC Catalytic System Enables Recoverable Homogeneous Catalysis: Mechanistic Study and Application in the Mizoroki—Heck Reaction. *Chem.—Eur. J.* **2019**, *25* (72), 16564–16572. https://doi.org/10.1002/chem.201903221.

(2) Bender, T. A.; Dabrowski, J. A.; Gagné, M. R. Homogeneous Catalysis for the Production of Low-Volume, High-Value Chemicals from Biomass. *Nat. Rev. Chem.* **2018**, *2* (5), 35–46. https://doi.org/10.1038/s41570-018-0005-y.

(3) Wacławek, S.; Padil, V. V. T.; černík, M. Major Advances and Challenges in Heterogeneous Catalysis for Environmental Applications: A Review. *Ecol. Chem. Eng. S* **2018**, *25* (1), 9–34. https://doi.org/10.1515/eces-2018-0001.

(4) Fillat, A.; Martínez, J.; Valls, C.; Cusola, O.; Roncero, M. B.; Vidal, T.; Valenzuela, S. V.; Diaz, P.; Pastor, F. I. J. Bacterial Cellulose for Increasing Barrier Properties of Paper Products. *Cellulose* **2018**, *25* (10), 6093–6105. https://doi.org/10.1007/s10570-018-1967-0.

(5) Fernandes, M.; Gama, M.; Dourado, F.; Souto, A. P. Development of Novel Bacterial Cellulose Composites for the Textile and Shoe Industry. *Microb. Biotechnol.* **2019**, *12* (4), 650–661. https://doi.org/10.1111/1751-7915.13387.

(6) Hasan, N.; Biak, D. R. A.; Kamarudin, S. Application of Bacterial Cellulose (BC) in Natural Facial Scrub. *Int. J. Adv. Sci. Eng. Inf. Technol.* **2012**, *2* (4), 272–275.

(7) Maneerung, T.; Tokura, S.; Rujiravanit, R. Impregnation of Silver Nanoparticles into Bacterial Cellulose for Antimicrobial Wound Dressing. *Carbohydr. Polym.* **2008**, *72* (1), 43–51. https://doi.org/10.1016/j.carbpol.2007.07.025.

(8) Chen, Y.; Chen, S.; Wang, B.; Yao, J.; Wang, H. TEMPO-Oxidized Bacterial Cellulose Nanofibers-Supported Gold Nanoparticles with Superior Catalytic Properties. *Carbohydr. Polym.* **2017**, *160*, 34–42. https://doi.org/10.1016/j.carbpol.2016.12.020.

(9) Elayaraja, S.; Liu, G.; Zagorsek, K.; Mabrok, M.; Ji, M.; Ye, Z.; Zhu, S.; Rodkhum, C. TEMPO-Oxidized Biodegradable Bacterial Cellulose (BBC) Membrane Coated with Biologically-Synthesized Silver Nanoparticles (AgNPs) as a Potential Antimicrobial Agent in Aquaculture (In Vitro). *Aquaculture* **2021**, *530*, 735746. https://doi.org/10.1016/j.aquaculture.2020.735746.

(10) Li, Z.; Zhou, X.; Pei, C. Synthesis and Characterization of MPS-g-PLA Copolymer and Its Application in Surface Modification of Bacterial Cellulose. *Int. J. Polym. Anal. Charact.* **2010**, *15* (4), 199–209. https://doi.org/10.1080/10236661003681222.

(11) Fijałkowski, K.; Żywicka, A.; Drozd, R.; Junka, A. F.; Peitler, D.; Kordas, M.; Konopacki, M.; Szymczyk, P.; Rakoczy, R. Increased Water Content in Bacterial Cellulose Synthesized under Rotating Magnetic Fields. *Electromagn. Biol. Med.* **2017**, *36* (2), 192–201. https://doi.org/10.1080/15368378.2016.1243554.

(12) Wu, X.; Xiang, Z.; Song, T.; Qi, H. Wet-Strength Agent Improves Recyclability of Dip-Catalyst Fabricated from Gold Nanoparticle-Embedded Bacterial Cellulose and Plant Fibers. *Cellulose* **2019**, *26* (5), 3375–3386. https://doi.org/10.1007/s10570-019-02297-0.

(13) Teixeira, S. R. Z.; Reis, E. M. dos; Apati, G. P.; Meier, M. M.; Nogueira, A. L.; Garcia, M. C. F.; Schneider, A. L. dos S.; Pezzin, A. P. T.; Porto, L. M.; Teixeira, S. R. Z.; Reis, E. M. dos; Apati, G. P.; Meier, M. M.; Nogueira, A. L.; Garcia, M. C. F.; Schneider, A. L. dos S.; Pezzin, A. P. T.; Porto, L. M. Biosynthesis and Functionalization of Bacterial Cellulose Membranes with Cerium Nitrate and Silver Nanoparticles. *Mater. Res.* **2019**, *22*. https://doi.org/10.1590/1980-5373-mr-2019-0054.

(14) Chanthiwong, M.; Mongkolthanaruk, W.; Eichhorn, S. J.; Pinitsoontorn, S. Controlling the Processing of Co-Precipitated Magnetic Bacterial Cellulose/Iron Oxide Nanocomposites. *Mater. Des.* **2020**, *196*, 109148. https://doi.org/10.1016/j.matdes.2020.109148.

(15) Gupta, A.; Briffa, S. M.; Swingler, S.; Gibson, H.; Kannappan, V.; Adamus, G.; Kowalczuk, M.; Martin, C.; Radecka, I. Synthesis of Silver Nanoparticles Using Curcumin-Cyclodextrins Loaded into Bacterial Cellulose-Based Hydrogels for Wound Dressing Applications. *Biomacromolecules* **2020**, *21* (5), 1802–1811. https://doi.org/10.1021/acs.biomac.9b01724.

(16) Shahmohammadi Jebel, F.; Almasi, H. Morphological, Physical, Antimicrobial and Release Properties of ZnO Nanoparticles-Loaded Bacterial Cellulose Films. *Carbohydr. Polym.* **2016**, *149*, 8–19. https://doi.org/10.1016/j.carbpol.2016.04.089.

(17) Khan, S.; Ul-Islam, M.; Ullah, M. W.; Israr, M.; Jang, J. H.; Park, J. K. Nano-Gold Assisted Highly Conducting and Biocompatible Bacterial Cellulose-PEDOT: PSS Films for Biology-Device Interface Applications. *Int. J. Biol. Macromol.* **2018**, *107*, 865–873. https://doi.org/10.1016/j.ijbiomac.2017.09.064.

(18) Yang, J.; Yu, J.; Fan, J.; Sun, D.; Tang, W.; Yang, X. Biotemplated Preparation of CdS Nanoparticles/Bacterial Cellulose Hybrid Nanofibers for Photocatalysis Application. *J. Hazard. Mater.* **2011**, *189* (1), 377–383. https://doi.org/10.1016/j.jhazmat.2011.02.048.

(19) Ma, B.; Sun, B.; Huang, Y.; Chen, C.; Sun, D. Facile Synthesis of Cu Nanoparticles Encapsulated into Carbonized Bacterial Cellulose with Excellent Oxidation Resistance and Stability. *Colloids Surf. Physicochem. Eng. Asp.* **2020**, *590*, 124462. https://doi.org/10.1016/j.colsurfa.2020.124462.

(20) Liu, L.-P.; Yang, X.-N.; Ye, L.; Xue, D.-D.; Liu, M.; Jia, S.-R.; Hou, Y.; Chu, L.-Q.; Zhong, C. Preparation and Characterization of a Photocatalytic Antibacterial Material: Graphene Oxide/TiO2/Bacterial Cellulose Nanocomposite. *Carbohydr. Polym.* **2017**, *174*, 1078–1086. https://doi.org/10.1016/j.carbpol.2017.07.042.

(21) Phutanon, N.; Motina, K.; Chang, Y.-H.; Ummartyotin, S. Development of CuO Particles onto Bacterial Cellulose Sheets by Forced Hydrolysis: A Synergistic Approach for Generating Sheets with Photocatalytic and Antibiofouling Properties. *Int. J. Biol. Macromol.* **2019**, *136*, 1142–1152. https://doi.org/10.1016/j.ijbiomac.2019.06.168.

(22) Wahid, F.; Duan, Y.-X.; Hu, X.-H.; Chu, L.-Q.; Jia, S.-R.; Cui, J.-D.; Zhong, C. A Facile Construction of Bacterial Cellulose/ZnO Nanocomposite Films and Their Photocatalytic and Antibacterial Properties. *Int. J. Biol. Macromol.* **2019**, *132*, 692–700. https://doi.org/10.1016/j.ijbiomac.2019.03.240.

(23) Jiang, S.; Hu, Q.; Xu, M.; Hu, S.; Shi, X.-C.; Ding, R.; Tremblay, P.-L.; Zhang, T. Crystalline CdS/MoS2 Shape-Controlled by a Bacterial Cellulose Scaffold for Enhanced Photocatalytic Hydrogen Evolution. *Carbohydr. Polym.* **2020**, *250*, 116909. https://doi.org/10.1016/j.carbpol.2020.116909.

(24) Zhou, M.; Chen, J.; Hou, C.; Liu, Y.; Xu, S.; Yao, C.; Li, Z. Organic-Free Synthesis of Porous CdS Sheets with Controlled Windows Size on Bacterial Cellulose for Photocatalytic Degradation and H$_2$ Production. *Appl. Surf. Sci.* **2019**, *470*, 908–916. https://doi.org/10.1016/j.apsusc.2018.11.207.

(25) Janpetch, N.; Vanichvattanadecha, C.; Rujiravanit, R. Photocatalytic Disinfection of Water by Bacterial Cellulose/N—F Co-Doped TiO2 under Fluorescent Light. *Cellulose* **2015**, *22* (5), 3321–3335. https://doi.org/10.1007/s10570-015-0721-0.

(26) Liu, F.; Sun, Y.; Gu, J.; Gao, Q.; Sun, D.; Zhang, X.; Pan, B.; Qian, J. Highly Efficient Photodegradation of Various Organic Pollutants in Water: Rational Structural Design of Photocatalyst via Thiol-Ene Click Reaction. *Chem. Eng. J.* **2020**, *381*, 122631. https://doi.org/10.1016/j.cej.2019.122631.

(27) Almasi, H.; Mehryar, L.; Ghadertaj, A. Photocatalytic Activity and Water Purification Performance of in Situ and Ex Situ Synthesized Bacterial Cellulose-CuO Nanohybrids. *Water Environ. Res.* **2020**, wer.1331. https://doi.org/10.1002/wer.1331.

(28) Bhatt, D. K.; Patel, U. D. Mechanism Underlying Visible-Light Photocatalytic Activity of Ag/AgBr: Experimental and Theoretical Approaches. *J. Phys. Chem. Solids* **2019**, *135*, 109118. https://doi.org/10.1016/j.jpcs.2019.109118.

(29) Jiang, J.; Zhu, J.; Zhang, Q.; Zhan, X.; Chen, F. A Shape Recovery Zwitterionic Bacterial Cellulose Aerogel with Superior Performances for Water Remediation. *Langmuir* **2019**, *35* (37), 11959–11967. https://doi.org/10.1021/acs.langmuir.8b04180.

(30) Zhang, Y.; Tan, J.; Wen, F.; Zhou, Z.; Zhu, M.; Yin, S.; Wang, H. Platinum Nanoparticles Deposited Nitrogen-Doped Carbon Nanofiber Derived from Bacterial Cellulose for Hydrogen Evolution Reaction. *Int. J. Hydrog. Energy* **2018**, *43* (12), 6167–6176. https://doi.org/10.1016/j.ijhydene.2018.02.054.

(31) Lai, F.; Miao, Y.-E.; Huang, Y.; Zhang, Y.; Liu, T. Nitrogen-Doped Carbon Nanofiber/Molybdenum Disulfide Nanocomposites Derived from Bacterial Cellulose for High-Efficiency Electrocatalytic Hydrogen Evolution Reaction. *ACS Appl. Mater. Interfaces* **2016**, *8* (6), 3558–3566. https://doi.org/10.1021/acsami.5b06274.

(32) Wu, Z.-Y.; Hu, B.-C.; Wu, P.; Liang, H.-W.; Yu, Z.-L.; Lin, Y.; Zheng, Y.-R.; Li, Z.; Yu, S.-H. Mo2C Nanoparticles Embedded within Bacterial Cellulose-Derived 3D N-Doped Carbon Nanofiber Networks for Efficient Hydrogen Evolution. *NPG Asia Mater.* **2016**, *8* (7), e288—e288. https://doi.org/10.1038/am.2016.87.

(33) Mi, Y.; Wen, L.; Wang, Z.; Cao, D.; Zhao, H.; Zhou, Y.; Grote, F.; Lei, Y. Ultra-Low Mass Loading of Platinum Nanoparticles on Bacterial Cellulose Derived Carbon Nanofibers for Efficient Hydrogen Evolution. *Catal. Today* **2016**, *262*, 141–145. https://doi.org/10.1016/j.cattod.2015.08.019.

(34) Zhang, Y.; Zhu, M.; Xu, S.; Zhou, H.; Qi, H.; Wang, H. Zeolitic Imidazolate Framework Derived Cobalt Oxide Anchored Bacterial Cellulose: A Good Template with Improved H2O Adsorption Ability and Its Enhanced Hydrogen Evolution Performance. *Electrochimica Acta* **2020**, *353*, 136499. https://doi.org/10.1016/j.electacta.2020.136499.

(35) Lai, F.; Yong, D.; Ning, X.; Pan, B.; Miao, Y.-E.; Liu, T. Bionanofiber Assisted Decoration of Few-Layered MoSe2 Nanosheets on 3D Conductive Networks for Efficient Hydrogen Evolution. *Small* **2017**, *13* (7), 1602866. https://doi.org/10.1002/smll.201602866.

(36) Chen, N.; Mo, Q.; He, L.; Huang, X.; Yang, L.; Zeng, J.; Gao, Q. Heterostructured MoC-MoP/N-Doped Carbon Nanofibers as Efficient Electrocatalysts for Hydrogen Evolution Reaction. *Electrochimica Acta* **2019**, *299*, 708–716. https://doi.org/10.1016/j.electacta.2019.01.054.

(37) Wang, P.; Geng, Z.; Gao, J.; Xuan, R.; Liu, P.; Wang, Y.; Huang, K.; Wan, Y.; Xu, Y. Zn$_x$Cd$_{1-x}$S/Bacterial Cellulose Bionanocomposite Foams with Hierarchical Architecture and Enhanced Visible-Light Photocatalytic Hydrogen Production Activity. *J. Mater. Chem. A* **2015**, *3* (4), 1709–1716. https://doi.org/10.1039/C4TA05722H.

(38) Dal'Acqua, N.; Faria, A. C. R.; Vebber, M. C.; da Silva Barud, H.; Giovanela, M.; Machado, G.; da Silva Crespo, J. Hydrogen Photocatalytic Production from the Self-Assembled Films of PAH/PAA/TiO2 Supported on Bacterial Cellulose Membranes. *Int. J. Hydrog. Energy* **2018**, *43* (33), 15794–15806. https://doi.org/10.1016/j.ijhydene.2018.06.176.

(39) Kamal, T.; Ahmad, I.; Khan, S. B.; Asiri, A. M. Bacterial Cellulose as Support for Biopolymer Stabilized Catalytic Cobalt Nanoparticles. *Int. J. Biol. Macromol.* **2019**, *135*, 1162–1170. https://doi.org/10.1016/j.ijbiomac.2019.05.057.

(40) Kamal, T.; Ahmad, I.; Khan, S. B.; Asiri, A. M. Anionic Polysaccharide Stabilized Nickel Nanoparticles-Coated Bacterial Cellulose as a Highly Efficient Dip-Catalyst for Pollutants Reduction. *React. Funct. Polym.* **2019**, *145*, 104395. https://doi.org/10.1016/j.reactfunctpolym.2019.104395.

(41) Song, L.; Shu, L.; Wang, Y.; Zhang, X.-F.; Wang, Z.; Feng, Y.; Yao, J. Metal Nanoparticle-Embedded Bacterial Cellulose Aerogels via Swelling-Induced Adsorption for Nitrophenol Reduction. *Int. J. Biol. Macromol.* **2020**, *143*, 922–927. https://doi.org/10.1016/j.ijbiomac.2019.09.152.

(42) Dou, B.; Zhao, R.; Yan, N.; Zhao, C.; Hao, Q.; Hui, K. S.; Hui, K. N. A Facilitated Synthesis of Hierarchically Porous Cu—Ce—Zr Catalyst Using Bacterial Cellulose for VOCs Oxidation. *Mater. Chem. Phys.* **2019**, *237*, 121852. https://doi.org/10.1016/j.matchemphys.2019.121852.

(43) Wang, J.; Lu, X.; Ng, P. F.; Lee, K. I.; Fei, B.; Xin, J. H.; Wu, J. Polyethylenimine Coated Bacterial Cellulose Nanofiber Membrane and Application as Adsorbent and Catalyst. *J. Colloid Interface Sci.* **2015**, *440*, 32–38. https://doi.org/10.1016/j.jcis.2014.10.035.

(44) Zhou, P.; Wang, H.; Yang, J.; Tang, J.; Sun, D.; Tang, W. Bacteria Cellulose Nanofibers Supported Palladium(0) Nanocomposite and Its Catalysis Evaluation in Heck Reaction. *Ind. Eng. Chem. Res.* **2012**, *51* (16), 5743–5748. https://doi.org/10.1021/ie300395q.

(45) Jeremic, S.; Djokic, L.; Ajdačić, V.; Božinović, N.; Pavlovic, V.; Manojlović, D. D.; Babu, R.; Senthamaraikannan, R.; Rojas, O.; Opsenica, I.; Nikodinovic-Runic, J. Production of Bacterial Nanocellulose (BNC) and Its Application as a Solid Support in Transition Metal Catalysed Cross-Coupling Reactions. *Int. J. Biol. Macromol.* **2019**, *129*, 351–360. https://doi.org/10.1016/j.ijbiomac.2019.01.154.

(46) Patel, U. D.; Suresh, S. Complete Dechlorination of Pentachlorophenol Using Palladized Bacterial Cellulose in a Rotating Catalyst Contact Reactor. *J. Colloid Interface Sci.* **2008**, *319* (2), 462–469. https://doi.org/10.1016/j.jcis.2007.12.019.

(47) Zhou, Y.; Guo, X.; Li, X.; Fu, J.; Liu, J.; Hong, F.; Qiao, J. In-Situ Growth of CuO/Cu Nanocomposite Electrode for Efficient CO2 Electroreduction to CO with Bacterial Cellulose as Support. *J. CO2 Util.* **2020**, *37*, 188–194. https://doi.org/10.1016/j.jcou.2019.12.009.

(48) Evans, B. R.; O'Neill, H. M.; Malyvanh, V. P.; Lee, I.; Woodward, J. Palladium-Bacterial Cellulose Membranes for Fuel Cells. *Biosens. Bioelectron.* **2003**, *18* (7), 917–923. https://doi.org/10.1016/S0956-5663(02)00212-9.

(49) Mei, J.; Ma, Y.; Pei, C. V2O5 Nanobelt-Carbonized Bacterial Cellulose Composite with Enhanced Electrochemical Performance for Aqueous Supercapacitors. *J. Solid State Electrochem.* **2017**, *21* (2), 573–580. https://doi.org/10.1007/s10008-016-3392-3.

(50) Li, H.; Ma, H.; Liu, T.; Ni, J.; Wang, Q. An Excellent Alternative Composite Modifier for Cathode Catalysts Prepared from Bacterial Cellulose Doped with Cu and P and Its Utilization in Microbial Fuel Cell. *Bioresour. Technol.* **2019**, *289*, 121661. https://doi.org/10.1016/j.biortech.2019.121661.

(51) Huang, Y.; Tang, K.; Yuan, F.; Zhang, W.; Li, B.; Seidi, F.; Xiao, H.; Sun, D. N-Doped Porous Carbon Nanofibers Fabricated by Bacterial Cellulose-Directed Templating Growth of MOF Crystals for Efficient Oxygen Reduction Reaction and Sodium-Ion Storage. *Carbon* **2020**, *168*, 12–21. https://doi.org/10.1016/j.carbon.2020.06.052.

(52) Song, Q.; Zhan, Z.; Chen, B.; Zhou, Z.; Lu, C. Biotemplate Synthesis of Polypyrrole@ bacterial Cellulose/MXene Nanocomposites with Synergistically Enhanced Electrochemical Performance. *Cellulose* **2020**, *27* (13), 7475–7488. https://doi.org/10.1007/s10570-020-03310-7.

(53) Rebelo, A. R.; Liu, C.; Schäfer, K.-H.; Saumer, M.; Yang, G.; Liu, Y. Poly(4-Vinylaniline)/ Polyaniline Bilayer-Functionalized Bacterial Cellulose for Flexible Electrochemical Biosensors. *Langmuir* **2019**, *35* (32), 10354–10366. https://doi.org/10.1021/acs. langmuir.9b01425.

(54) Yao, J.; Ji, P.; Sheng, N.; Guan, F.; Zhang, M.; Wang, B.; Chen, S.; Wang, H. Hierarchical Core-Sheath Polypyrrole@carbon Nanotube/Bacterial Cellulose Macrofibers with High Electrochemical Performance for All-Solid-State Supercapacitors. *Electrochimica Acta* **2018**, *283*, 1578–1588. https://doi.org/10.1016/j.electacta.2018.07.086.

(55) Huang, Y.; Lin, Z.; Zheng, M.; Wang, T.; Yang, J.; Yuan, F.; Lu, X.; Liu, L.; Sun, D. Amorphous Fe2O3 Nanoshells Coated on Carbonized Bacterial Cellulose Nanofibers as a Flexible Anode for High-Performance Lithium Ion Batteries. *J. Power Sources* **2016**, *307*, 649–656. https://doi.org/10.1016/j.jpowsour.2016.01.026.

5 Synthesis of Bacterial Cellulose Sheets from Alternative Natural and Waste Resources

Salman Ul Islam, Laeeq Ahmad, Atiya Fatima, Shaukat Khan, Muhammad Wajid Ullah, Sehrish Manan, and Mazhar Ul-Islam

CONTENTS

5.1 INTRODUCTION

Cellulose is the most abundant polymer found on the earth's surface. Plants represent the biggest sources of cellulose utilized in a number of commercial products, specifically in the paper, pulp, and textile industries (1). Although it has numerous applications in multiple fields, plant cellulose also has some important limitations. The most crucial limitation is its impure nature as certain other compounds, including lignin, pectin, and hemicellulose, are linked into the framework of plant cellulose (2). If left

DOI: 10.1201/9781003118756-5

untreated, these compounds negatively affect cellulose applications. Therefore, plant cellulose must undergo thorough, intense pretreatment processes before it can be employed in specific applications (3).

Although plants represent the biggest sources of cellulose, they are not the only ones (4). Indeed, several microbes, including bacteria, fungi, and few algae, have the potential to synthesize cellulose from various synthetic and nonsynthetic media (3). The produced cellulose is termed microbial or, most commonly, bacterial cellulose. Several species of the genus *Acetobacter* are known to produce BC (5, 6). Unlike plant cellulose, BC is much purer and possesses superior physical mechanical and chemical properties. The fibril structure, size, porosity, and crystallinity of BC are different from plant cellulose despite having the same chemical structure (7).

Owing to pure nature and production ease, BC has been extensively used in different applications, especially medically for healing burned skin through skin grafts and for dressing wounds, as a filter thanks to its fine pores and small fibers, in cosmetics like facial masks, as food delicacies, in the textile industry, and for purses and designer bags (5, 7–9). Various BC-based food items and cosmetics are also available in the market (10). An important aspect related to BC is the synthesis of its composite. This composite helps in developing various advanced functional materials. Several limitations associated with pure BC are addressed in its composites synthesized through multiple approaches. BC composite synthesis utilizes nanomaterials, polymers, clay particles, natural products, and so on for bactericidal, electromagnetic, medical, and pharmaceutical applications (60–62). New composite strategies are being developed, novel applications are being explored, and several interesting findings have resulted from the approvals and applications of synthetics (6, 11, 12).

5.2 LIMITATIONS OF BC

Along with extensive practical applications, BC does come up against certain limitations (1, 13, 14) that comprise its properties and applications as well as its production costs. With the lack of certain features, including biocompatibility, electromagnetic properties, bactericidal features, and low thermo-mechanical properties, BC use is restricted for various applications. For instance, pure BC cannot be applied to the development of devices requiring magnetic features. Its non-bactericidal nature also restricts its applications in wound healing in infected environments (11). To cope with these limitations, the only possible approach is its combination with various materials in the form of composites. BC composites with numerous materials, including nanomaterials, polymers, natural products, and clay particles, have been developed successfully and have reduced the aforementioned limitations and enhanced its applications in various fields (15, 16).

Another main limitation of BC is its production costs and the resources involved to scale up the production; the high cost of synthetic media is causing serious hurdles in its commercialization (16). This limitation demands the use of waste/renewable carbon sources for producing BC. This will minimize the cost associated with expensive media and will enhance its upscaling potential (11, 17). The conversion of industrial-scale waste to BC through microbes will be a responsible production and consumption in line with the sustainable development goals of the United Nations

(UN) (18). In this chapter, we focus on the possibilities and strategies developed so far for cost-effective BC production. The next section mainly focuses on two main sources of BC—waste feed and low-cost natural resources.

5.3 BC FROM WASTE RESOURCES

As mentioned earlier, alternative waste and natural resources are the possible solution for reducing BC costs and boost its applications in different fields (19, 20). Waste resources, especially industrial ones, not only create huge economic losses but also significantly contribute to environmental issues (5, 21). Their management, treatment, and disposal are also governed by several economic and environmental issues. Therefore, a best management strategy could be the recycling/conversion of waste to BC. This would indeed reduce BC production costs as well as diminish the associated environmental threats. Among several types of waste, those produced by sugar/textile/paper mills and the food/breweries/agriculture sector are ideal for BC manufacture (18). These wastes are full of sugar resources, which can easily be utilized as carbon sources in growth culture media for BC production. Several studies have reported BC production from waste sources. Herein, we highlight the BC production capabilities of several waste resources based on published reports (20, 22).

5.3.1 BC from Industrial Waste

Using biotechnological principles, industrial waste can be economically converted to environmentally friendly products. The industrial wastes might comprise monosaccharides, including glucose and fructose, or polysaccharides such as starch and cellulose. Conversion of simple sugars to BC does not require complex processing and hydrolysis. The removal of contaminants through sterilization and solid separation is the main requirement for culture media preparation (11).

A significant portion of the waste produced by modern industries mainly contains cellulose that is not readily available for BC production. Conversion of such waste to BC requires initial breakdown of this cellulose to the D-glucose (a fragment of sugar) by employing an environmentally benign technique. Enzymatic hydrolysis is very suitable for processing cellulose and uses an enzyme called "cellulase" that splits the structure of cellulose into many glucose fragments and is a mechanism still in development (23).

One study reported waste of fiber sludge obtained from a pulp factory (European mills) as media for BC production (24). The sludge fiber hydrolysis was carried out using the Cellic CTec2 (Novozymes, Denmark) hydrolyzing enzyme under shaking conditions at 50 °C (150 rpm) for 2 days. The synthetic media developed from the hydrolyzed sludge, a major carbon source, was inoculated with *Acetobacter xylinus* (ATCC 23770). It was noteworthy that this waste media indicated very high BC production after 2 weeks of cultivation. Additionally, sludge media cultures diluted two and three times also produced sufficient quantities of BC.

Another interesting study conducted by a group of researchers illustrated the tremendous potential of waste from the beer fermentation industry for BC production (25). After the beer fermentation process is over, a significant amount of solid waste remains that mainly holds non-fermented sugar. Depending on feed sources, the left-over sugar can be polysaccharides/glucose. Due to the presence of fungus needed for ethanol production, the beer industry waste needs sterilization. The amount of BC produced from beer waste further increased under shaking condition (25). It is also important to mention that such waste can be used for other valuable products, including ethanol and butanol (26).

5.3.2 BC FROM MUNICIPAL WASTE

Wastewater from many industries is often loaded with harmful nutrients; if not extracted, they end up contaminating the environment. Exploitation of effluent from these industries can serve as good media for manufacturing bio products that could be valuable from both economic and environmental perspectives. Municipal wastes have been effectively utilized for BC production. BC production from candied jujube wastewater processing industries is also reported. The study utilized direct raw waste, as well as acid-diluted feed, as a source of carbon for BC-production using *G. xylinum CGMCC 295* as a source strain. It was observed that glucose content increased with acid pretreatment compared to water from raw candied jujube waste. The hydrolyzed products produce BC as much as 1.5 times higher (2.25 g/L) than the raw wastewater with a fiber size structure of 3–14 nm, with an average diameter of 5.9 nm. However, the effect on crystallinity was rather negative using acid pretreated substrate compared to wastewater from candied jujube (27).

Wastewater from an acetone/butanol/ethanol (ABE) fermentation factory has also been utilized for BC production in some reports. A study undertaken by Huang et al. (2015) aimed to reduce the BC production cost using ABE waste as a potential carbon source for BC production (28). The waste media was inoculated with *G. xylinum* CH001 (BC-producing bacterial strain). 1.34 g/L of BC was produced after one week of cultivation. The nutrients included sugars (glucose/xylose) and organics (butyric/acetic acid), as well as alcohols (ethyl/butyl). The nature of ABE is such that they do not affect BC structure compared with conventional Hestrin and Schramm (HS) medium as observed from the X-ray diffraction and FT-IR. ABE substrate media is a low-cost source for BC manufacture and can be a valuable future substitute for conventional synthetic media (28).

5.3.3 BC FROM WASTE FOOD

Wasted food items and effluents coming from food industry waste could prove an ideal source of cost-effective BC production. Indeed, all types of food industry waste will contain considerable amounts of sugar and other nutrients. Slight initial treatment to remove contamination will be enough to prove waste as ideal culture media for BC production. Various industries produce waste suitable for BC productions such as palm oil mill (effluent containing sucrose), pineapple (waste containing sucrose), cane (sugar cane molasses), agro (sweet potato pulp), and fruit-processing industries (rotten apples are juice industry waste).

To lower the BC production cost, various strategies are employed, including using dairy by-products and alcohol factory waste that need no chemical pretreatment.

Revin et al. (29) reported substantial BC production (6.19 g/L) from wheat thin still-age obtained through dynamic cultivation in 3 days that is three times more from that received using the standard HS medium (2.14 g/L). Using whey as a microbe nutrient gave 5.45 g/L of BC where the media type and initial pH were the main driving force during incubation. During bacteria cultivation on whey and thin still-age, divergence in the crystallinity/micromorphology in produced BC was observed (29). The experiment revealed the potential of wheat stillage and alcohol factory waste, which are available as a by-product, for significantly reducing the cost of BC production.

Waste from bakeries, expired flour/dough/bread that is generally thrown out, can be hydrolyzed into glucose using hydrochloric or sulfuric acid solutions in high pressure vessels and subsequently used as media for BC. Waste from confectionery industries produce better-quality BC when fermentation with *Komagataeibacter sucrofermentans* produce as much as 13 g/L of BC with a water retention capacity of 102–138 gm H_2O/gm dry BC, stress break 72.3–139.5 MPa with a Young's modulus of 0.97–1.61 GPa. The properties of BC from confectionary waste are similar to BC produced using synthetic media (30).

5.3.4 BC FROM AGRICULTURAL WASTE

Agricultural wastes are of vital economic significance worldwide. They are a source of renewable energy, and among all agro wastes a small proportion (less than 10%) is utilized as alternative industries' raw feed (10). Agricultural waste usually consists of significant quantities of cellulose, starch, and other polysaccharides, as well as some quantities of simple sugars. These wastes can be effectively utilized for generating various bioproducts, including BC.

Pretreatment of agricultural waste to a simpler sugar is an important step, which affects the overall cost of BC production. Various techniques have been introduced to establish simple strategies of pretreatment (hydrolysis) and conversion to glucose and other monosaccharides. Agricultural waste such as defatted grape seeds, coconut husks, and palm fibers can serve as a sugar source by supercritical water hydrolysis, in which the process temperature is raised to near the critical point of water (100 °C to 374 °C) for keeping water as liquid that would break down cellulose/lignocellulose into smaller constituents. This can render hydrolysis waste acid free into sugars that serve as media for microbes.

Corn stalk hydrolysate consisting of various sugars was used as media for BC production that resulted in 2.86 g/L of BC under optimized conditions (17). Goelzer et al. evaluated the BC production from wheat straw and after conditioning through chemical and thermal pretreatment, in which the wheat straw was hydrolyzed through xylanase/cellulase/b-glucosidase enzymes. The hydrolyzed media produced 10.6 g/L of BC. Rice barks obtained from agricultural residue was enzymatically hydrolyzed and utilized for BC production, and the results indicated 2.42 g/L (static) and 1.57 g/L (shaking) of BC production (31). The hydrolysates obtained from the alkali treatment of wheat straw were used as a culture media and produced 15.4 mg/ml of BC (32). These examples and many more like that augment that waste agricultural sources can be utilized as effective alternatives for cost-effective BC production.

5.4 BC FROM CHEAP RESOURCES

As mentioned earlier, the major limitations currently associated with BC are its production cost. The conventional media used for BC production is very expensive culture media, which robustly elevates the production costs and finally limits the use of BC to very high value-added applications. The approach to overcome this restriction is through the utilization of (1) cheap carbon and nutrient sources and (2) waste sources bearing carbon residue. A number of researchers have produced BC using various low-cost natural carbon/nitrogen sources (33, 34). Figure 5.1 illustrates the possibilities of BC production from several locally available cheap resources. Furthermore, several efforts made to produce BC from cheap resources have been summarized in Table 5.1. Here, we enumerate the production of BC from different low-cost natural carbon and nitrogen sources like coconut water, sugar cane molasses, fruit juices, wheat straw, corn steep liquor, and coffee cherry husk.

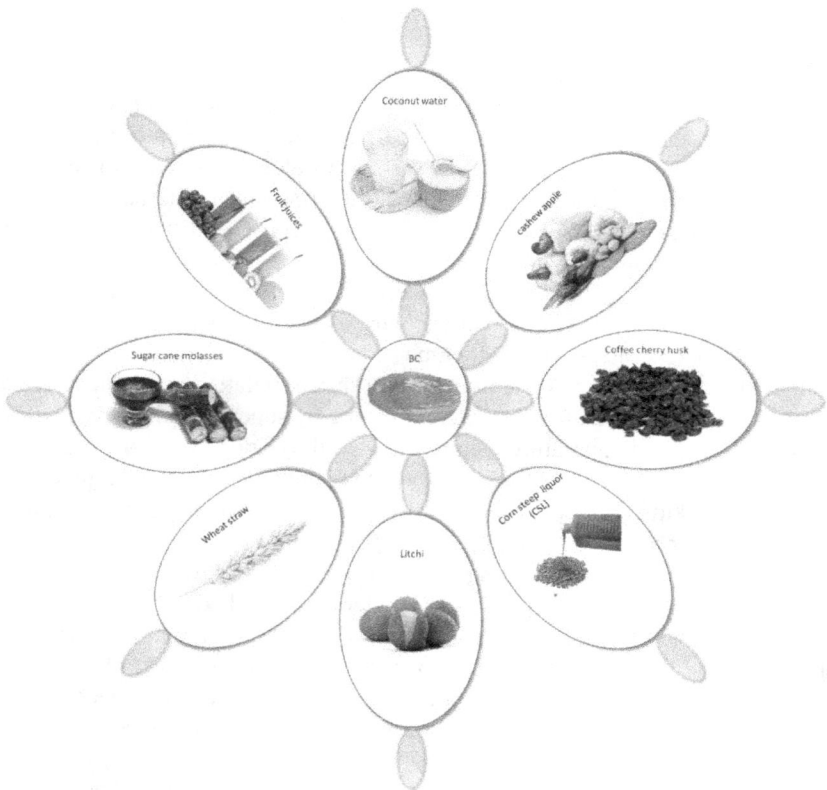

FIGURE 5.1 BC production from various cheap sources. Figure reproduced with permission from Springer (Ul-Islam et al., 2020).

5.4.1 BC FROM COCONUT WATER

Coconut is available in every season, and in Southeast Asian countries, many agro-industries discard coconut juice as an economical by-product. Abundant coconuts are produced in Indonesia and India and need to be used efficiently. It has been shown that the coconut residues could be utilized as an optimized substrate for producing a high-quality BC because these residues still contain carbon and nitrogen sources (35). An investigation was carried out to observe the production of BC of *Acetobacter xylinum*, cultured in coconut water and pineapple juice-based media, and to analyze its fermentation kinetics. BC production was observed in both shaking/static fermented culture. *A. xylinum* was observed to exhibit exponential phase at 48 h, which expressed that BC production was associated with its cell growth on both substrates. The recorded fermentation kinetics of *A. xylinum* using coconut water were as follows: Rp 0.117 g/l/h, Rx 0.309 g/l/h, Rs 0.079 g/l/h, Yx/s 1.408 biomass/g glucose, Yp/s 3.612 g cellulose/g biomass, Yp/x 2.235 g cellulose/g biomass, μmax 0.0132/h, and σ 0.028/h. On the other hand, fermentation kinetics of *A. xylinum* using pineapple juice were as follows: Rp 0.051 g/l/h, Rx 0.133 g/l/h, Rs 0.215 g/l/h, Yx/s 0.240 g biomass/g glucose, Yp/s 0.599 g cellulose/g glucose, Yp/x 2.452 g cellulose/g biomass, μmax 0.0082/h, and σ 0.0173/h. These fermentation kinetics parameters indicated an elevated rate of BC production and appreciable conversions of coconut water glucose to biomass and BC by *A. xylinum* and that coconut water is more efficient in producing BC than is pineapple juice (36). Another study established the optimum conditions for producing and characterizing BC films from local coconut water and reported that cultivation time, carbon source percentage, and pH have a major impact on BC. *A. xylinum* was cultured using coconut water culture media supplemented with 3%, 5%, and 7% sugar, whereas the cultivation time varied from 3 days to 5 days to 7 days. The pH was maintained to the value of 3, 5, and 7. BC was dried (until 100 °C) to reduce the moisture content by 4%–5%. This study demonstrated that 5% of carbon source, an incubation period of 7 days, and pH 5 are the optimal conditions for BC film production from local coconut water. Moreover, using electron microscopy, the dried BC film surface showed pores shielded by fibrils (37). One study used *Gluconacetobacter* sp. sju-1 to produce BC from cashew apple juice and the mature coconut water. The optimum parameters for maximum cellulose production were as follow: incubation time = 20 days (480 h); pH = 6.5; temperature = 31 °C; carbon source = fructose; nitrogen source = yeast extract; and growth hormone = indole acetic acid. Under static culture condition, regarding aeration or fermentation condition, dry weight of BC was 14.92 ± 0.30 g L^{-1} with a biosynthetic yield of 74.7 ± 0.32%. However, in case of shaken culture, the dry weight of BC was only 9.63 ± 0.20 g L^{-1}, and the biosynthetic yield was 48.15 ± 0.98. Afterwards the BC was converted to other useful products such as nata/vinegar/reinforced paper, and all using optimized conditions (38). In an attempt to subside the production cost for cellulose through low-cost carbon sources, an effective culture method by *Gluconacetobacter persimmonis* was examined. Several fruit juices, including coconut water, were used as alternative carbon sources for BC production. Carbohydrate analysis of coconut water showed the presence of 1.6% of total sugar. The production of BC provided at 2% in HS medium with 2.0% peptone, 0.5% yeast extract, 0.115% citric acid, and adjusted to pH 6 was observed. The cellulose yield was calculated after 14 days of incubation. The BC produced by the organism from coconut water was almost the same as produced from orange juice (6.18 g/L) (39).

5.4.2 BC FROM SUGAR CANE MOLASSES

Sugar manufacturing and refining units produce molasses (viscous liquid) as a by-product. Molasses is widely used as a microbial fermentation substrate that has different levels of readily biodegradable sugars like sucrose/fructose/glucose in addition to nitrogen/iron/calcium/potassium/magnesium and vitamins (40, 41). It has been reported that in Brazil, molasses is being produced in the proportion of 40–60 Kg/ton of processed cane. Molasses is mainly utilized for manufacturing ethyl alcohol, protein, and animal feed (42, 43), and due to the high content of total reducing sugars this industrial by-product has attracted attention for producing BC (44, 45). A study compared the production of BC from different carbon sources, including molasses, by *G. intermedius* SNT-1 under static conditions where 45.8 g/L of diluted molasses (1:4 v/v) after heat pretreatment with H_2SO_4 resulted in a yield of 12.6 g/L (on a dry basis) of BC. BC pellicle yields were comparable to the nitrogen sources using yeast extract or corn steep liquor along with the heat-pretreated molasses (44). BC obtained from pretreated molasses showed high tensile strength (44). The standard HS medium represents a high cost because it is formulated with synthetic compounds. Thereupon, a study articulated 36 different alternative culture media by exploiting molasses (rawhide sugar cane) inverted by high temperature and by both acid/high temperature as substrates, binded using nutrients like glucose/peptone/yeast extract/Na_2HPO_4, and citric acid. It was found that compared to standard HS medium, alternative medium (formulated with 15 g/L of rawhide molasses, 5 g/L glucose, 1.5 g/L acid citric, and 2.7 g/L Na_2HPO_4, without any nitrogen source addition) resulted in the best cost and yield with 52%, 59%, and 65% in dry, hydrated mass and yield, respectively (45). These studies elucidate the suitability of using sugar cane molasses to reduce the costs for producing BC for industrial use. Additionally, BC production from sugar cane molasses also has lower impact on environment in terms of energy intake and pollution load.

5.4.3 BC FROM FRUIT JUICES

A study investigated an effective culture method to produce BC from fruit juices (grapes, Japanese pear, apple, orange, and pineapple) by *Acetobacter xylinum* NBRC 13693, and the possibility of producing BC from those juices was evaluated. It was observed that BC yields increased by the addition of the nitrogen source to the fruit juices. Additionally, orange and Japanese pear juices were observed to be suitable media for BC production. Various components of orange-like peel and squeeze residue were used to produce BC. BC of 0.65 g (dry weight) was obtained from the orange of 100 g, and the solid residue from the orange was about 17.2 g (46). Another study used watermelon juice (70% v/v) and mandarin (80% v/v) to produce BC by the *Gluconoacetobacter xylinus* CICC10529 strain. The fruit juice media contained ethanol (1% v/v), $MgSO_4.7H_2O$ (1.5% w/v), and K_2HPO_4 (0.1% w/v). Different modes of operation, that is, static and dynamic cultivations of the bacteria, were carried out at 30 °C, and the BC production was observed over 7–10 days. After 2 days, the BC ribbon width produced in the shaker was 25–37 µm; however, the larger width (~40–50 µm) was found in the BC produced in the incubator. Moreover, 10% higher thermal stability was exhibited by the BC produced via the dynamic mode (47). A group of researchers isolated *Gluconacetobacter* sp. gel_SEA623–2 from citrus fruit juice fungus and obtained maximum BC from this bacterium by providing the following optimized conditions: citrus juice: 10%, sucrose: 10%, acetic acid: 1%, ethanol: 1%, temperature 30 °C, and pH 3.5. BC produced by this bacterium exhibited soft physical properties, high tensile strength, and high water retention potential and was considered suitable as a medical and cosmetic material (48). In another study, sisal juice was used in a static cultivation for optimized production of BC by *Komagataeibacter hansenii* ATCC 23769. The study reported 3.38 g/L of BC (highest production) after 10 days of cultivation, using sisal juice (pH 5) at 15 g/L of sugars and supplemented with 7.5 g/L of extract yeast. The BC production from sisal culture was three times higher than the yield in synthetic medium, which indicates that sisal juice is a suitable substrate for BC production (49).

5.4.4 BC PRODUCTION FROM WHEAT STRAW

Wheat straw is an abundant biomass resource across the world, particularly in China. But most of the wheat straw is generally wasted by setting fire to it after kernel harvest, which subsequently causes heavy air pollution. It has been demonstrated that, upon acid hydrolysis followed by bacterial fermentation, wheat straw could be a potentially cheap starting feedstock for BC production (32). However, acid hydrolysis presents certain limitations such as erosion of equipment, high consumption of energy, and production of excessive inhibitory compounds. Enzymatic hydrolysis is an alternative technology for straw saccharification, but its efficiency is also compromised due to wheat straw's complicated structure and resistance to enzymatic hydrolysis (50, 51). In order to enhance enzymatic saccharification of straw and yield of fermentable sugars, a study pretreated the straw with an ionic liquid 1-Allyl-3-methylimidazolium chloride ([AMIM]Cl). It was noted that following optimization, the hydrolytic potential of regenerated straw enhanced significantly as compared to control materials. The sugar yield of straw after pretreatment was 71.2% at 110 °C for 1.5 h with a 3% w/w straw dosage, 3.6 times higher than that of untreated straw (19.6%). Additionally, the BC yield obtained in straw hydrolysates was recorded to be higher than that in glucose-based media. This may be due to the availability of other complex components in the hydrolysate that would enhance the formation of BC (52). For the sake of decreasing the production cost, Hong et al. produced BC by *Acetobacter aceti* subsp. *xylinus* using wheat straw hydrolysates prepared by dilute acid hydrolysis instead of the usual carbon sources. Microbial growth inhibitors were removed by detoxifying wheat straw hydrolysates using various alkalis like sodium hydroxide, calcium hydroxide, and ammonia, in combination with laccase or activated charcoal. The study showed that $Ca(OH)_2$ and activated charcoal-treated hydrolysate resulted in 50% higher BC yield than that from routine carbon sources (32).

5.4.5 BC FROM CORN STEEP LIQUOR (CSL)

CSL, a by-product of corn wet-milling, is a rich source of important nutrient ingredients for supporting microbial growth and fermentation (53). As a fermentation medium supplement, CSL presents a complete source of nitrogen as well as vitamins and other important elements (53). A study investigated the production of BC by *Acetobacter* sp. V6 in shaking culture using CSL (nitrogen source) and molasses (carbon source). BC production was observed after 1 day of incubation in the medium containing molasses and CSL and reached a maximum yield of 3.12 ± 0.03 g/L at 8 days, which was about two-fold higher than the yield in the conventional medium (54). The study further evaluated the physical properties using various techniques and showed the relative crystallinity of BC produced in the complex and molasses media to be 83.02% and 67.27%, respectively. It was finally concluded that molasses and CSL could be a suitable feedstock for the production of BC by *Acetobacter* sp. V6, which will lead to a reduction in production costs (54). Costa et al. analyzed the industrial residues as feedstock for BC production by *Gluconacetobacter hansenii* UCP1619. They synthesized BC pellicles using the HS medium and alternative media composed of various carbon and nitrogen sources, including CSL. Interestingly, none of the tested sources worked except the CSL. The medium formulated with 2.5% CSL and 1.5% glucose produced the highest yield of BC (dry and hydrated mass) and corresponded to 73% of that achieved with the HS medium. The BC produced from CSL containing medium also exhibited greater thermal stability, high crystallinity, and higher tensile strength (55). Moreover, CSL as a feedstock for BC production aids in minimizing the environmental pollution resulting from its disposal, as well as helps in reducing deforestation caused by the use of vegetal cellulose.

5.4.6 BC FROM COFFEE CHERRY HUSK

Coffee cherry husk (CCH) is produced after the processing of coffee cherry by dry process. It was shown that for 1 ton of coffee cherries, about 0.18 ton of CCH is produced (56). A high amount of polyphenols, proteins, carbohydrates, and minerals are present in CCH. The use of CCH in agriculture has been largely restricted due to the presence of undesirable substances like tannins, caffeine, and other polyphenols, and the disposal of coffee waste creates a huge pollution problem (57). In order to investigate the feasibility of CCH for the production of BC by *Gluconacetobacter hansenii* UAC09, Rani and Appaiah used different concentrations of CCH as a carbon source along with various other nutritional constituents such as nitrogen (CSL, urea) and additives (acetic acid, ethyl alcohol). 5.6–8.2 g/L of BC was produced upon the following concentrations of the constituents: CCH extract 1:1 (w/v); CSL 8% (v/v); urea 0.2% (w/v); ethyl alcohol 1.5%; and acetic acid 1.0% (v/v). This yield of BC was three times the yield obtained in a control medium (1.5 g/L). FT-IR analysis showed that BC produced with CCH and HS medium exhibited similar structures. The tensile strength of BC from CCL varied between 28.5 and 42.4 MPa (58). The optimum conditions for maximum BC production reported by another study were water to CCH ratio (1:1), CSL (10%), alcohol (0.5%), pH (6.64), and acetic acid (1.13%). After 14 days of fermentation, the amount of BC obtained was 6.24 g/L, which was

TABLE 5.1
BC from Various Cheap Sources.

BC source	Method	Produced BC	Reference*
Coconut water	Shaking and static culture	2.235 g BC/g biomass	(36, 39)
	Static culture	6.18 g/L	
Pineapple juice	Shaking and static culture	2.452 g BC/g biomass	(36)
Mature coconut water and cashew apple juice	Static culture	14.92±0.30 g/L	(38)
	Shaking culture	9.63±0.20 g/L	
Sugarcane molasses	Static culture	12.6 g/L	(44)
Orange juice	Static culture	0.65 g	(46)
Sisal juice	Static culture	3.38 g/L	(49)
Corn steep liquor (CSL)	Static culture	3.12±0.03 g/L	(54)
Coffee cherry husk	Shaking and static culture	5.6–8.2 g/L	(59)
	Shaking and static culture	6.24 g/L	(59)

*The numbers refer to the numbered references in the text.

comparable to the predicted value of 6.09 g/L (59). Utilization of CCH in bioprocess provides cheaper alternative feedstock as well as helping to reduce the environmental pollution load.

5.5 CONCLUSIONS AND FUTURE PROSPECTS

The BC production process, synthetic media, and capital investment along with operating cost present the main economic restrictions to the commercialization of BC. The common carbon sources, including glucose, fructose, and galactose, also add to its high production costs. To cope with the situation, efforts have been devoted to exploring novel, cheap, natural, and cost-effective carbon sources as well as production strategies. In this regard, several waste carbon sources, industrial effluents, cheap natural products, municipal and industrial wastes, and fruits and vegetables and their wastes have been explored for BC production. Besides lowering the BC production costs, reutilization of these wastes also leads to a reduction in environmental pollution. It has been observed that the use of feedstocks for BC production could be an alternative strategy, as these food resources are rich in glucose/fructose and have high BC production potentials. It is also a fact that as primary food resources, these cannot be applied as alternative sources for BC production. On the other hand, the use of food and other industrial waste for BC production is a wise strategy, where besides reducing BC production costs, recycling these wastes can mitigate environmental pollution issues.

5.6 ACKNOWLEDGMENT

This research was supported by the "The Research Council (TRC)" Oman through Block Research Funding Program (BFP/RGP/EBR/20/261). The work has been partially supported by "The Research Council (TRC)" Oman through Undergraduate Research Grant (BFP/URG/EBR/20/055).

5.7 CONFLICT OF INTEREST

The authors declare no conflict of interest.

REFERENCES

(1) Ul-Islam M, Khan S, Ullah MW, Park JK. Bacterial cellulose composites: Synthetic strategies and multiple applications in bio-medical and electro-conductive fields. *Biotechnology Journal*. 2015;10(12):1847–61.
(2) Ul-Islam M, Ullah MW, Khan S, Park JK. Production of bacterial cellulose from alternative cheap and waste resources: A step for cost reduction with positive environmental aspects. *Korean Journal of Chemical Engineering*. 2020;37:925–37.
(3) Ul-Islam M, Ullah MW, Khan T, Park JK. Bacterial cellulose: Trends in synthesis, characterization, and applications. In *Handbook of hydrocolloids*. Elsevier; 2020. pp. 923–74.
(4) Fatima A, Yasir S, Khan MS, Manan S, Ullah MW, Ul-Islam M. Plant extract-loaded bacterial cellulose composite membrane for potential biomedical applications. *Journal of Bioresources and Bioproducts*. 2020;6:26–32.
(5) Ul-Islam M, Khan S, Ullah MW, Park JK. Comparative study of plant and bacterial cellulose pellicles regenerated from dissolved states. *International Journal of Biological Macromolecules*. 2019;137:247–52.
(6) Sajjad W, Khan T, Ul-Islam M, Khan R, Hussain Z, Khalid A, et al. Development of modified montmorillonite-bacterial cellulose nanocomposites as a novel substitute for burn skin and tissue regeneration. *Carbohydrate Polymers*. 2019;206:548–56.
(7) Ul-Islam M, Ul-Islam S, Yasir S, Fatima A, Ahmed M, Lee YS, et al. Potential applications of bacterial cellulose in environmental and pharmaceutical sectors. *Current Pharmaceutical Design*. 2020;26(45):5793–806.
(8) Kamal T, Ahmad I, Khan SB, Ul-Islam M, Asiri AM. Microwave assisted synthesis and carboxymethyl cellulose stabilized copper nanoparticles on bacterial cellulose nanofibers support for pollutants degradation. *Journal of Polymers and the Environment*. 2019;27(12):2867–77.
(9) Ullah MW, Manan S, Kiprono SJ, Ul-Islam M, Yang G. Synthesis, structure, and properties of bacterial cellulose. *Nanocellulose: From Fundamentals to Advanced Materials*. 2019:81–113.
(10) Hussain Z, Sajjad W, Khan T, Wahid F. Production of bacterial cellulose from industrial wastes: A review. *Cellulose*. 2019;26(5):2895–911.
(11) Islam MU, Ullah MW, Khan S, Shah N, Park JK. Strategies for cost-effective and enhanced production of bacterial cellulose. *International Journal of Biological Macromolecules*. 2017;102:1166–73.
(12) Ali N, Kamal T, Ul-Islam M, Khan A, Shah SJ, Zada A. Chitosan-coated cotton cloth supported copper nanoparticles for toxic dye reduction. *International Journal of Biological Macromolecules*. 2018;111:832–8.
(13) Islam MU, Khan S, Ullah MW, Park JK. Recent Advances in Biopolymer Composites for Environmental Issues. *Handbook of Composites from Renewable Materials*. 2017:673–91.

(14) Ul-Islam M, Wajid Ullah M, Khan S, Kamal T, Ul-Islam S, Shah N, et al. Recent advancement in cellulose based nanocomposite for addressing environmental challenges. *Recent Patents on Nanotechnology.* 2016;10(3):169–80.

(15) Ul-Islam M, Khan S, Ullah MW, Park JK. Structure, chemistry and pharmaceutical applications of biodegradable polymers. *Handbook of Polymers for Pharmaceutical Technologies.* 2015;3:517–40.

(16) Ul-Islam M, Khan S, Khattak WA, Ullah MW, Park JK. Synthesis, chemistry, and medical application of bacterial cellulose nanocomposites. In *Eco-friendly polymer nanocomposites.* Springer; 2015. pp. 399–437.

(17) Cheng Z, Yang R, Liu X, Liu X, Chen H. Green synthesis of bacterial cellulose via acetic acid pre-hydrolysis liquor of agricultural corn stalk used as carbon source. *Bioresource Technology.* 2017;234:8–14.

(18) Akinsemolu AA. The role of microorganisms in achieving the sustainable development goals. *Journal of Cleaner Production.* 2018;182:139–55.

(19) Khattak WA, Khan T, Ul-Islam M, Wahid F, Park JK. Production, characterization and physico-mechanical properties of bacterial cellulose from industrial wastes. *Journal of Polymers and the Environment.* 2015;23(1):45–53.

(20) Shah N, Ul-Islam M, Khattak WA, Park JK. Overview of bacterial cellulose composites: A multipurpose advanced material. *Carbohydrate Polymers.* 2013;98(2):1585–98.

(21) Ul-Islam M, Subhan F, Islam SU, Khan S, Shah N, Manan S, et al. Development of three-dimensional bacterial cellulose/chitosan scaffolds: Analysis of cell-scaffold interaction for potential application in the diagnosis of ovarian cancer. *International Journal of Biological Macromolecules.* 2019;137:1050–9.

(22) Ha JH, Shah N, Ul-Islam M, Khan T, Park JK. Bacterial cellulose production from a single sugar α-linked glucuronic acid-based oligosaccharide. *Process Biochemistry.* 2011;46(9):1717–23.

(23) Tan J, Xern C. *Production of crude palm oil through microwave pretreatment cum solvent extraction.* University of Malaya; 2017.

(24) Cavka A, Guo X, Tang S-J, Winestrand S, Jönsson LJ, Hong F. Production of bacterial cellulose and enzyme from waste fiber sludge. *Biotechnology for Biofuels.* 2013;6(1):25.

(25) Ha JH, Shehzad O, Khan S, Lee SY, Park JW, Khan T, et al. Production of bacterial cellulose by a static cultivation using the waste from beer culture broth. *Korean Journal of Chemical Engineering.* 2008;25(4):812.

(26) Ha JH, Shah N, Ul-Islam M, Park JK. Potential of the waste from beer fermentation broth for bio-ethanol production without any additional enzyme, microbial cells and carbohydrates. *Enzyme and Microbial Technology.* 2011;49(3):298–304.

(27) Li Z, Wang L, Hua J, Jia S, Zhang J, Liu H. Production of nano bacterial cellulose from waste water of candied jujube-processing industry using Acetobacter xylinum. *Carbohydrate Polymers.* 2015;120:115–9.

(28) Huang C, Yang XY, Xiong L, Guo HJ, Luo J, Wang B, et al. Evaluating the possibility of using acetone-butanol-ethanol (ABE) fermentation wastewater for bacterial cellulose production by G luconacetobacter xylinus. *Letters in Applied Microbiology.* 2015;60(5):491–6.

(29) Revin V, Liyaskina E, Nazarkina M, Bogatyreva A, Shchankin M. Cost-effective production of bacterial cellulose using acidic food industry by-products. *Brazilian Journal of Microbiology.* 2018;49:151–9.

(30) Tsouko E, Kourmentza C, Ladakis D, Kopsahelis N, Mandala I, Papanikolaou S, et al. Bacterial cellulose production from industrial waste and by-product streams. *International Journal of Molecular Sciences.* 2015;16(7):14832–49.

(31) Goelzer F, Faria-Tischer P, Vitorino J, Sierakowski M-R, Tischer C. Production and characterization of nanospheres of bacterial cellulose from Acetobacter xylinum from processed rice bark. *Materials Science and Engineering: C.* 2009;29(2):546–51.

(32) Hong F, Zhu YX, Yang G, Yang XX. Wheat straw acid hydrolysate as a potential cost-effective feedstock for production of bacterial cellulose. *Journal of chemical technology & biotechnology.* 2011;86(5):675–80.

(33) Hungund BS, Gupta S. Production of bacterial cellulose from Enterobacter amnigenus GH-1 isolated from rotten apple. *World Journal of Microbiology and Biotechnology.* 2010;26(10):1823–8.

(34) Thompson DN, Hamilton MA, editors. *Production of bacterial cellulose from alternate feedstocks.* Twenty-Second Symposium on Biotechnology for Fuels and Chemicals; 2001: Springer.

(35) Kongruang S. Bacterial cellulose production by Acetobacter xylinum strains from agricultural waste products. *Applied Biochemistry and Biotechnology.* 2008;148(1–3): 245–56. Epub 2008/04/18. doi: 10.1007/s12010-007-8119-6. PubMed PMID: 18418756.

(36) Lestari P, Elfrida N, Suryani A, Suryadi Y. Study on the production of bacterial cellulose from Acetobacter xylinum using agro-waste. *Jordan Journal of Biological Sciences.* 2014;147(1570):1–6.

(37) Indrianingsih A, Rosyida V, Jatmiko T, Prasetyo D, Poeloengasih C, Apriyana W, et al., editors. *Preliminary study on biosynthesis and characterization of bacteria cellulose films from coconut water.* IOP Conference Series: Earth and Environmental Science; 2017: IOP Publishing.

(38) Gayathry G. *Bioconversion and value addition of mature coconut water and cashew apple juice into bacterial cellulose using gluconacetobacter sp.* Tamil Nadu Agricultural University Coimbatore; 2013.

(39) Hungund B, Prabhu S, Shetty C, Acharya S, Prabhu V, Gupta S. Production of bacterial cellulose from Gluconacetobacter persimmonis GH-2 using dual and cheaper carbon sources. *Journal of Microbial and Biochemical Technology.* 2013;5(2).

(40) Bae S, Shoda M. Production of bacterial cellulose by Acetobacter xylinum BPR2001 using molasses medium in a jar fermenter. *Applied Microbiology and Biotechnology.* 2005;67(1):45–51.

(41) Sanadi NA, Fan Y, Leow C, Wong J, Koay YS, Lee C, et al. Growth of bacillus coagulans using molasses as a nutrient source. *Chemical Engineering Transactions.* 2017;56:511–6.

(42) Soccol CR, de Souza Vandenberghe LP, Medeiros ABP, Karp SG, Buckeridge M, Ramos LP, et al. Bioethanol from lignocelluloses: Status and perspectives in Brazil. *Bioresource Technology.* 2010;101(13):4820–5.

(43) Dien B, Cotta M, Jeffries T. Bacteria engineered for fuel ethanol production: Current status. *Applied Microbiology and Biotechnology.* 2003;63(3):258–66.

(44) Tyagi N, Suresh S. Production of cellulose from sugarcane molasses using Gluconacetobacter intermedius SNT-1: Optimization & characterization. *Journal of Cleaner Production.* 2016;112:71–80.

(45) Costa A, Nascimento V, De Amorim J, Gomes E, Araujo L, Sarubbo L. Residue from the production of sugar cane: An alternative nutrient used in biocellulose production by Gluconacetobacter Hansenii. *Chemical Engineering Transactions.* 2018;64:7–12.

(46) Kurosumi A, Sasaki C, Yamashita Y, Nakamura Y. Utilization of various fruit juices as carbon source for production of bacterial cellulose by Acetobacter xylinum NBRC 13693. *Carbohydrate Polymers.* 2009;76(2):333–5.

(47) Kosseva MR, Li M, Zhang J, He Y, Tjutju NA, editors. *Study on the bacterial cellulose production from fruit juices.* Proceeding of the 2nd International Conference on Bioscience and Biotechnology; 2017.

(48) Kim S, Lee S, Park K, Park S, An H, Hyun J, et al. Gluconacetobacter sp. gel_SEA-623–2, bacterial cellulose producing bacterium isolated from citrus fruit juice. *Saudi Journal of Biological Sciences.* 2017;24(2):314–9.

(49) Lima H, Nascimento E, Andrade F, Brígida A, Borges MdF, Cassales A, et al. Bacterial cellulose production by Komagataeibacter hansenii ATCC 23769 using sisal juice-an agroindustry waste. *Brazilian Journal of Chemical Engineering.* 2017;34(3):671–80.

(50) Chandra RP, Bura R, Mabee W, Berlin dA, Pan X, Saddler J. *Substrate pretreatment: The key to effective enzymatic hydrolysis of lignocellulosics? Biofuels.* Springer; 2007. pp. 67–93.

(51) Kumar P, Barrett DM, Delwiche MJ, Stroeve P. Methods for pretreatment of lignocellulosic biomass for efficient hydrolysis and biofuel production. *Industrial & Engineering Chemistry Research.* 2009;48(8):3713–29.

(52) Chen L, Hong F, Yang X-x, Han S-f. Biotransformation of wheat straw to bacterial cellulose and its mechanism. *Bioresource Technology.* 2013;135:464–8.

(53) Kona R, Qureshi N, Pai J. Production of glucose oxidase using Aspergillus niger and corn steep liquor. *Bioresource Technology.* 2001;78(2):123–6.

(54) Jung H-I, Lee O-M, Jeong J-H, Jeon Y-D, Park K-H, Kim H-S, et al. Production and characterization of cellulose by Acetobacter sp. V6 using a cost-effective molasses—corn steep liquor medium. *Applied Biochemistry and Biotechnology.* 2010;162(2):486–97.

(55) Costa AF, Almeida FC, Vinhas GM, Sarubbo LA. Production of bacterial cellulose by Gluconacetobacter hansenii using corn steep liquor as nutrient sources. *Frontiers in Microbiology.* 2017;8:2027.

(56) Adams M, Dougan J. Biological management of coffee processing wastes. *Tropical Science.* 1981;23(3):177–96.

(57) Venugopal C, Rai MR, Appaiah KA. Mycotypha sps strain no. AKM 1801-Novel thermophilic fungi for alkalization of coffee husk effluent. *Asian Journal of Microbiology, Biotechnology and Environmental Sciences.* 2004;6:525–7.

(58) Rani MU, Appaiah KA. Production of bacterial cellulose by Gluconacetobacter hansenii UAC09 using coffee cherry husk. *Journal of Food Science and Technology.* 2013;50(4):755–62.

(59) Rani MU, Rastogi NK, Appaiah KA. Statistical optimization of medium composition for bacterial cellulose production by Gluconacetobacter hansenii UAC09 using coffee cherry husk extract—an agro-industry waste. *Journal of Microbiology and Biotechnology.* 2011;21(7):739–45.

(60) Fernandes M, Gama M, Dourado F, Souto AP. Development of novel bacterial cellulose composites for the textile and shoe industry. *Microbial Biotechnology.* 2019;12:650–61.

(61) Khan S, Ul-Islam M, Ullah MW, Kim Y, Park JK. Synthesis and characterization of a novel bacterial cellulose–poly(3,4-ethylenedioxythiophene)–poly(styrene sulfonate) composite for use in biomedical applications. *Cellulose.* 2015;22:2141–48.

(62) Ullah MW, Ul Islam M, Khan S, Shah N, Park JK. Recent advancements in bioreactions of cellular and cell-free systems: A study of bacterial cellulose as a model. *Korean Journal of Chemical Engineering.* 2017;34:1591–99.

6 Synthesis Routes and Applications of Cellulose in Food Industry

Sehrish Manan, Ajmal Shahzad, Mazhar Ul-Islam, Muhammad Wajid Ullah, and Guang Yang

CONTENTS

6.1 INTRODUCTION

Cellulose microfibrils are insoluble rod-like structures comprised of β-1,4-linked glucose molecules linked through hydrogen bonding. The cellulose-producing organisms, including bacteria, plants, and algae, contain the cellulose synthase (CS) enzyme that catalyzes the polymerization of glucan chains (1). In plants, cellulose is the main component of the cell wall, where it accounts for approximately one-third of the total biomass. Based on the chemical composition and structural organization, the plant cell wall is divided into the primary and secondary cell wall (2). The primary cell wall contains cellulose, hemicellulose, and pectin, while the secondary cell wall is comprised of cellulose, hemicellulose, and lignin and is more rigid and denser than the primary cell wall (3). Bacterial cellulose (BC) is produced by several bacterial genera (4) and some algae, as well as synthesized enzymatically by the cell-free enzyme systems (5, 6). Unlike plant cellulose (PC), BC does not contain lignin, pectin, and hemicellulose and represents the purest form of cellulose (7).

DOI: 10.1201/9781003118756-6

Typically, cellulose consists of β-1,4-linked glucan chains. Hydrogen bonding is formed between the oxygen and the hydroxyl group, both within an individual glucose chain and among the adjacent chains. The hydrogen bonding and van der Waals forces favor the aggregation of glucan chains and result in the formation of crystalline cellulose through the stacking of the cellulose microfibrils (8, 9). The natural crystalline form of cellulose is cellulose I, where cellulose Iα exists as a single-chain triclinic unit cell and cellulose Iβ possesses the two-chain monoclinic unit cell. Cellulose I could be conclusively converted into a more stable form known as cellulose II (8). In higher plants, the percentage of cellulose Iα and Iβ depends upon the species and the nature of the cell wall (3, 10). Cellulose produced by the microbial cell and cell-free enzyme systems possesses cellulose I and cellulose II polymorphic structures, respectively (7, 11).

Cellulose nanofibers have many interesting properties like high crystallinity, excellent mechanical properties, high aspect ratio, and remarkable surface chemistry (12). The characteristics of cellulose depend upon its source, synthesis methods, structural organization, culture time and condition, and carbon source. The nanofibers of cellulose are loosely arranged and have many empty spaces, hence forming a porous matrix with high surface area. Any structural changes in cellulose affect its physiochemical, mechanical, thermal, and biological properties. For example, the highly porous structure of cellulose contributes to its high water-holding capacity (WHC), usually 100–200 times more than its dry weight, as well as controlled water release rate (WRR), while the 3D network structure and compact fiber arrangement give high density and better resistance to the applied force (13, 14). Similarly, the porous morphology of cellulose nanofibers allow the impregnation of a variety of materials of different shapes and sizes, thus showing the potential to form composites (14, 15). The use of cellulosic materials is a topic of immense research interest in different fields such as cosmetics (16, 17), medical (18–21), pharmaceutics (22), energy (23), environmental (24–26), food industry (27), and others (28–32), as revealed by the growing number of publications in last decade (Table 6.1). For food applications, the selection of a certain product should demonstrate certain features such as nontoxicity, low cost, simple synthesis, availability, renewability, and biodegradability. The toxicological test, including acute and chronic oral toxicity assays for cellulose, show no reproductive toxicity and inflammatory reaction in mice (33–36). Besides, several pieces of research reported that cellulose is not genotoxic. In view of the long-lasting history of cellulose as a food ingredient in the Asian region and its use in biomedical supplies, it can be determined that it is harmless and could be safely used for food applications (34, 37). Thus, cellulose has been categorized as 'generally recognized as safe' (GRAS) by the USA Food and Drug Administration since 1992 (38). The digestion and absorption of cellulose in human gut is illustrated in Figure 6.1. Although both BC and PC have been extensively utilized for different food applications, the comparatively low production cost makes PC a preferred choice over BC, as indicated by the relatively high number of publications in the last 10 years (Table 6.1).

TABLE 6.1
A Comparison of the Total Number of Publications of Plant Cellulose and Bacterial Cellulose and the Number of Publications in Food Applications for Both Sources of Cellulose in a Decade

Year	Number of publications		Number of publications on food applications	
	Plant cellulose	Bacterial cellulose	Plant cellulose	Bacterial cellulose
2020	2,456	1,420	347	144
2019	4,455	2,236	748	264
2018	4,770	2,053	708	244
2017	4,252	1,753	619	188
2016	3,946	1,583	560	179
2015	3,700	1,320	528	139
2014	3,474	1,188	489	124
2013	3,237	1,133	489	119
2012	2,840	951	442	118
2011	2,770	879	353	93

Source: web of science; search key word (title): 'Plant cellulose' for plant cellulose; and 'Bacterial cellulose' and 'Bacterial nanocellulose' for bacterial cellulose). Data accessed on September 16, 2020.

This chapter provides an overview of the general structure and biosynthesis of cellulose by plants and microorganisms and their potential applications in the food industry. It further describes the utilization of cellulose and cellulose-based composites as the raw materials, which could be used as raw food material, food packaging material, additive food ingredients, and as a delivery system for active food ingredients. It provides insights into the development of cellulose-based intelligent pH and antimicrobial sensors for application in the food industry.

6.2 CELLULOSE SYNTHESIS

6.2.1 PLANTS

In higher plants, typically each cell (protoplast) is enclosed in an extracellular matrix known as the cell wall. During growth and development and in response to external stimuli, the plant cell wall is continuously modified. Generally, the plant cell wall is comprised of three main types of polysaccharides, including cellulose, hemicellulose, and pectin, along with some proteins and other chemical compounds (40). In higher plants, the cellulose synthase complex (CSC), which probably regulates the cellulose synthesis, was first explained in freeze-fracture experiments on maize (41, 42). When plant cells split at the plasma membrane and the cell wall, the globular

Positive charged CNC binds with salivary protein and form "protein corona"

CNC mixed with saliva

Negative charged CNC repelled by the proteins

CNC moves down by the peristalsis through the esophagus into stomach

Uncharged CNC is inactive, but experiences the shear forces

Gastric acid and enzymes cannot trigger CNC's breaking

Penetrates the mucus layer and reach the epithelium in small intestine

Through the M-cells in the Peyer's patches

Through enterocytes by passive diffusion

Through enterocytes by transcytosis

Through the paracellular space

Mouth

Esophagus

Stomach

Small Intestine

Peyer's patches

Enterocytes

Mucus layer

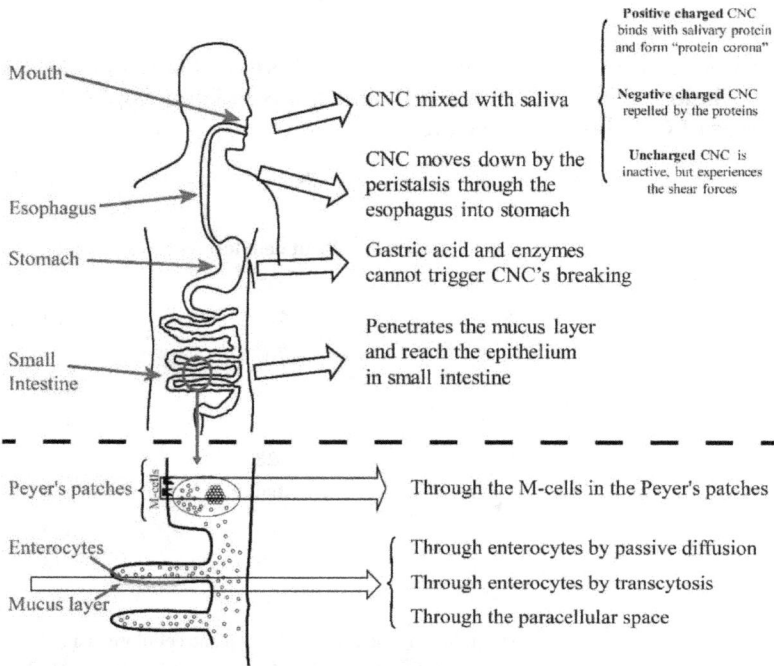

FIGURE 6.1 Digestion and absorption of CNC in the human body. Figure reproduced from (39).

Source: El Miri, N., Abdelouahdi, K., Barakat, A., Zahouily, M., Fihri, A., Solhy, A., et al. (2015). Bio-nanocomposite films reinforced with cellulose nanocrystals: Rheology of film-forming solutions, transparency, water vapor barrier and tensile properties of films. *Carbohydr. Polym.* doi:10.1016/j. carbpol.2015.04.051.

plasma membrane-associated complexes are present at the edges of cellulose microfibrils, currently known as CSC. A rosette six-fold structure was observed among the two leaflets of the plasma membrane, which was later suggested as the functional unit of cellulose synthesis (41, 42). The first plant cellulose synthase A (*CesA*) gene was identified in cotton based on its sequence similarity with the bacterial cellulose synthase gene (43). The UDP-glucose is the most likely possible substrate of CESA. Furthermore, a deletion of a region from a cotton *CesA* blocks the UDP-glucose binding (43), while a point mutation in CS, as reported in *Arabidopsis CesA*, affects the cellulose production. The predicted structure of the presumed cytosolic domain of the cotton CESA aligned relatively well with the crystal structure of the bacterial cellulose synthase, suggesting a common catalytic mechanism during the synthesis of BC and PC (44, 45). In addition to the conserved sequence similarity to bacterial cellulose synthase, the plant *CesA* contains the plant reserved regions, which form the rosette structure and interact with the cellulose biosynthesis regulatory elements (45). *Arabidopsis* possess ten *CesA* genes, among which *CesA1–3*, *CesA5*, *6*, and *9* are required for primary cell wall synthesis while *CesA4*, *7*, and *8* are responsible for production of secondary cell wall components (46, 47). Moreover, plants also contain several cellulose synthase-like (*CLS*) genes (46, 48), which are classified into ten families: *CLSA, CLSB, CLSC, CLSD, CLSE, CLSF, CLSG, CLSH, CLSI,* and *CLSJ* (48), among which the members of *CLSD* are involved in cellulose synthesis (49). Furthermore, the interaction between CESA and KORRIGAN (KOR), a putative (1→4)-β-glucanase, was studied (46, 50). Although the specific function of KOR in cellulose synthesis remains unknown, it is suggested to be involved in the release of glucan chains after their synthesis (51). Of the cell wall components, cellulose is produced at the plasma membrane (52); while similar to other secretory proteins, the cell wall proteins are also synthesized in the endoplasmic reticulum (ER) and modified in the Golgi apparatus (GA) and ER (53). The soluble cell wall polysaccharides are manufactured in the GA and secreted via the trans-Golgi network (54, 55). These findings suggest that CESAs are assembled in the CSCs, ER, or GA and subsequently moved to the plasma membrane through tiny compartments in a cytoskeleton-dependent manner (40, 55). The trafficking of CSCs through the plasma membrane is likely to initiate randomly; nevertheless, the cortical microtubules are generally known for directing the movement of CSCs (Figure 6.2). The microtubule-based regulation of CSC is reinforced by the cellulose synthase interactive (CSI) family (54, 56). The CSI1 interacts with both the CESA subunits and microtubule, while its disruption causes misalignment between the CSC and microtubules routes (59). Once the CSC stops the cellulose production, it is drawn back from the plasma membrane. The CSC is either recycled to the plasma membrane or retrieved to the trans-Golgi network or moved to the vacuole for degradation through endosomes (58).

6.2.2 Bacteria

Bacterial cellulose, also known as bacterial nanocellulose (BNC), is a biopolymer produced by several bacterial genera, including *Acetobacter, Rhizobium, Agrobacterium, Aerobacter, Achromobacter, Azotobacter, Salmonella, Escherichia,* and *Sarcina* (4, 60), a few algae such as *Valonia* and *Chaetamorpha* spp., fungi (61–63), and cell-free systems (5, 6) by utilizing glucose and other carbon sources. The carbon source is converted to cellulose via four enzymatic reactions: phosphorylation of glucose to glucose-6-phosphate (Glc-6-P) by glucokinase, conversion of Glc-6-P to glucose-1-phosphate (Glc-1-P) by phosphoglucomutase, conversion of Glc-1-P to UDP-glucose by UDP-glucose pyrophosphorylase, and finally the conversion of UDP-glucose to cellulose by cellulose synthase. BC is synthesized within the microbial cells in the form of β-1,4-glucan chains and subsequently extruded across the cell membrane through a four-step process, including the synthesis of sugar nucleotides via activation of monosaccharides, assembly of glucose monosaccharides through progressive addition of sugar nucleotides via polymerization, acylation of glucan units, and finally the extracellular transport of glucan chains. The biosynthesis of BC by microbial cells is an aerobic process, whereas it is produced anaerobically by the cell-free enzyme systems (11, 64). BC is produced as a dense gelatinous sheath or a membrane at the air-medium interface under static cultivation (65), pellets under shaking cultivation (66), and granules or sphere-like particles under agitation cultivation (67).

The molecular studies of biosynthesis of BC synthesis reveal the involvement of at least four different genes, including *axcess A, axcess B, axcess C,* and *axcess D,* which are organized in the form of an operon (Figure 6.3) and encode specific

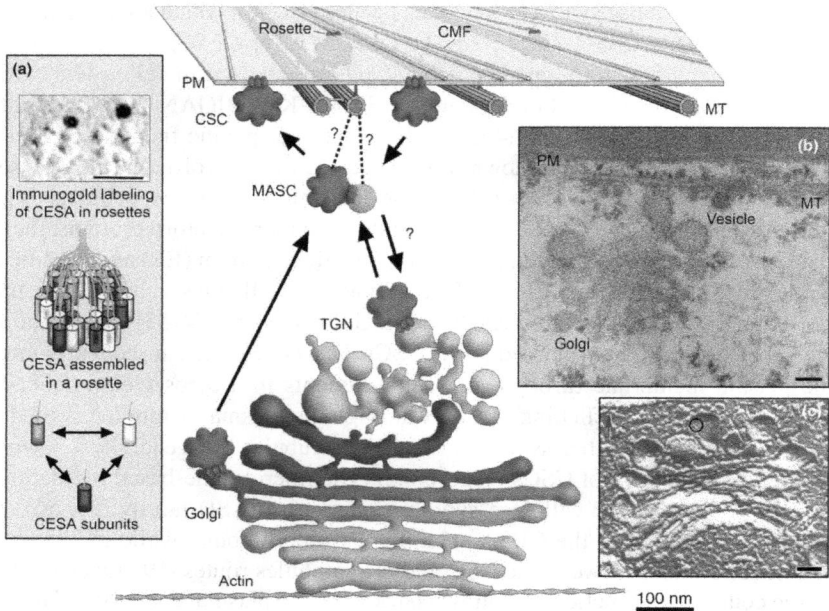

FIGURE 6.2 CESA trafficking pathways elucidated by electron microscopy data and live cell imaging. Hexameric rosette complexes containing CESA proteins and their binding partners are present in the plasma membrane (PM), where they migrate along cortical microtubules as they extrude cellulose microfibrils (CMFs) into the periplasmic space. The rosette consists of homodimers and heterodimers of three unique CESA isoforms (inset A). The micrograph shows immunogold labeling of cotton CESA proteins in rosettes from freeze fractures of *Vigna angularis* plasma membranes. Preassembled rosettes can be detected in the endoplasmic reticulum and in the Golgi apparatus (circled in inset C). Live cell imaging of GFP:CESA3 has demonstrated that Golgi bodies pause transiently on cortical microtubules during the secretion of CSCs to the plasma membrane. Interactions between the Golgi apparatus and cortical microtubules can also be visualized by electron microscopy: inset B shows a Golgi body associated with a microtubule in a hypocotyl epidermis cell upon treatment with the cellulose synthesis inhibitor CGA615′315 (5 nM). Finally, CSCs are internalized into microtubule-associated cellulose synthase compartments (MASCs), which are tethered to cortical microtubules via interactions with the CSC itself or with a component of MASC vesicles. MASC-like vesicles are also likely intermediates in the secretion of CSCs to the plasma membrane. It is currently unclear whether the *trans*-Golgi network (TGN) is involved in the secretion of CSCs, in their internalization, or in both of these processes. Scale bar in inset A = 30 nm, all other scale bars = 100 nm. Figure reproduced from (59).

Source: Hagiwara, A., Imai, N., Sano, M., Kawabe, M., Tamano, S., Kitamura, S., et al. (2010). A 28-day oral toxicity study of fermentation-derived cellulose, produced by Acetobacter aceti subspecies xylinum, in F344 rats. *J. Toxicol. Sci.* doi:10.2131/jts.35.317.

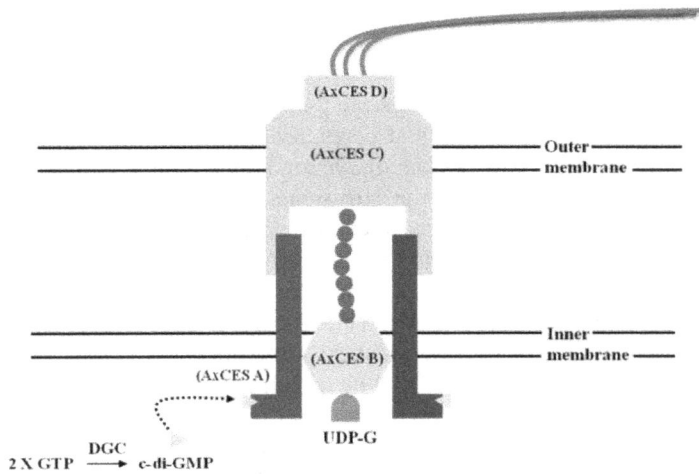

FIGURE 6.3 Illustration of schematic model of cellulose-synthesizing machinery in microbial cell system. Figure reproduced from (5).

Source: Aljohani, W., Ullah, M. W., Zhang, X., and Yang, G. (2018). Bioprinting and its applications in tissue engineering and regenerative medicine. *Int. J. Biol. Macromol.* 107, 261–275. doi:10.1016/j.ijbiomac.2017.08.171.

TABLE 6.2

A Comparison of Structural, Physicomechanical, and Thermal Properties of Plant and Bacterial Cellulose.

Properties	Plant cellulose		Bacterial cellulose	Ref.
	Cellulose nanocrystals	Cellulose nanofibers		
Crystallinity (%)	54–88	59–64	65–79	(77)
Degree of polymerization (DP)	500–15,000	≥500	800–10,000	(78)
Fiber length (nm)	150–300	85–225	70–80	(79)
Young's modulus (GPa)	50–100	39–78	15–30	(79)
Density (cm^{-3})	1.6	1.566	1.5	(79)
Purity	Low	Low	High	(80)
Particle length (μm)	0.05–0.5	0.5–2	>1	(81)
Particle width (nm)	3–10	4–20	30–50	(81)
Particle height (nm)	3–10	4–20	6–10	(81)
Surface area (m^2/g)	---	5–10	24–40	(82)
Post-synthesis processing	Yes	Yes	No	(83)
Porosity	Low	Medium	High	(84)
Thermal stability	High	High	High	(82)
Hydrophilicity	Comparatively low	Moderate	High	(82)

proteins AxCESA, AxCESB, AxCESC, and AxCESD, respectively, which perform specific functions (68). The operon is involved in the intracellular biosynthesis, extracellular transport, and in vitro assembly of cellulose fibrils into the high ordered structures (69). The intracellular synthesized β-1,4-glucan chains are extruded across the external barrier through a series of pores, termed as terminal complexes (TCs) (11, 68). In bacteria, for example *G. xylinum*, among the four subunits (68), the last two are proposed to be involved in the extrusion and crystallization of β-1,4-glucan chains (70); however, they are yet to be characterized for their possible interaction and action mechanism. The extruded chains self-aggregate via hydrogen bonding and form twisted subelementary fibrils, which further crystallize and form compact and well-arranged ribbons and bundles (71). As BC is produced as a hydrogel at the air-medium interface by the microbial cells, the thickness of hydrogel increases downward with incubation time until all microbial cells entrapped become inactive or die due to the exhaustion of nutrients and oxygen (71, 72). The unique molecular configuration endows BC with structural, physicomechanical, thermal, and biological features that are superior to PC (71). A comparative analysis of structural, physicomechanical, and thermal properties of BC and PC is given in Table 6.2. The high production cost and issues associated with its production such as low yield and productivity by the microbial cells restrict the commercialization of BC-based products. Some of these issues have been addressed by identifying novel bacterial strains, developing advanced reactors, supplementing carbon sources, and utilizing low-cost substrates (64, 73–76).

6.3 FOOD APPLICATIONS

6.3.1 RAW FOOD MATERIAL

The use of cellulose traditionally emerges from the manufacturing of kombucha tea and *nata de coco* dessert (83, 85). BC is produced and chopped into small pieces followed by dipping in sugar syrup, and is presented as a dessert known as nata, which originated from the Philippines and presently is a traditional dessert in Southeast Asian regions. *Nata de coco* has become famous globally due to its simple manufacturing process, smooth texture, and pleasant mouth feel (83). A variety of fruit flavors of nata are prepared by using different fruit juices as the culture medium. For example, *nata de coco* is manufactured from pineapple juice by bacteria, and this type of nata is rich in pineapple flavor (83). Likewise, nata with coconut flavor is also prepared by using coconut water as the culture medium. Furthermore, BC is also prepared as kombucha tea by fermenting the sweet tea using a symbiotic culture of lactic acid bacteria, yeast, and *Komagataeibacter* bacteria (86, 87). Kombucha tea is carbonated and slightly acidic in taste, which makes it more acceptable to consumers. A few metabolic products of associated yeast and bacteria, like acetic acid and other organic acids, impart antibacterial properties and inhibit the contamination of the beverages by different pathogenic microbes (88). A study demonstrated that kombucha tea consumption reduces lipid accumulation and further minimizes the risk of liver damage; for example, it promoted liver restoration in mice because of its high fiber content, low calories, and antioxidant ingredients (89). Kombucha tea also has promising therapeutic effects and immunomodulatory properties, which reportedly improved the autoimmune encephalomyelitis in experimental mouse models (89).

6.3.2 FOOD ADDITIVE AND STABILIZER

The consistency of processed food products always remains the main concern in the food business. Similarly, the thinning and phase separation of freshly made beverages and other fluid products with the elapse of time are other main issues in the food industry (39). Long-term storage makes the beverages thin, which forms a water layer at the bottom or top of the container, although the product is safe for consumption. Therefore, there is great demand for developing a product with a long shelf life and better thickening and emulsifying properties. To meet these criteria, many kinds of food stabilizers have been designed and tested to improve the features of processed foods. For example, several types of gums such as gellan gum (90), locust bean gum (91), carrageenan (92), and xanthan gum (93) have been used as food stabilizers due to their excellent stabilization and emulsification properties. However, these gums have high production cost, laborious purification protocols, and indigestion and safety issues (94, 95).

Cellulose has been explored in the food industry for its potential emulsifying properties and as a stabilizer due to its thick structure, high liquid-holding capacity, and suspending nature. BC is used as the reinforcement material to strengthen the fragile food hydrogels as well as the heat-stable suspending agent (96). Moreover, the addition of cellulose improves the quality of pasty foods by reducing its stickiness, thus could be added to jams as the non-caloric agent (Table 6.3). For example, in addition to preventing the clumping of chocolate drinks, the addition of BC to tofu has remarkably increased its strength and increased its stiffness and texture (97). In a Japanese processed seafood kamaboko, the addition of BC improved the springiness and sensory quality because of its better enduring aging process property (98). Another study reported that the addition of cellulose to ice cream as a typical cream alternative improves its melting resistance, gives smooth texture, and decreases its total calorie content (99). BC is capable of maintaining the shape of ice cream for a minimum of 60 minutes at room temperature after its removal from the freezer. Additionally, BC has been used in the production of vegetarian meat along with Monascus, the extract of a naturally pigmented red fungus (100). The obtained product not only possessed natural meat-like flavor but also showed color and morphological stability for a longer time (101). This BC-Monascus product could be a better replacement for animal meat-based products for consumers with dietary restrictions due to its non-animal source.

6.3.3 FUNCTIONAL FOOD INGREDIENT

The functional food industry, comprised of beverages, food, and supplements, is among one of the fastest growing sectors of the food industry in recent years. Functional food either contains new ingredients or increases the quality of pre-existing ingredients, and it possesses a disease prevention function or boosts natural health. Several

TABLE 6.3

Application of Cellulose as Food Additive/Ingredient. Table reproduced from (102). FMC Corporation (2014) (division formerly "FMC Biopolymer" now FMC Health and Nutrition) cellulose gel-product description, application guide, and recipes. www.fmcbiopolymer.com/Food/Ingredients/CelluloseGel/Introduction.aspx (accessed on March 19, 2014).

Type	Concentration (%)	Application	Functions
Colloidal	3–5	Mixes for power and candy bars	Stabilizes the emulsion; suspends the solids; improves creaminess and pulpiness; adds opacity
	0.5–3	Batters and breading	Improves cling; reduces drying time; reduces fat absorption during frying; reduces sogginess if finished product is stored under heat lamps
	0.25–0.7	Chocolate drinks	Adds creaminess; suspends the solids; stable under high-temperature processing; adds opacity
	1–3	Dressings	Enhances the mouthfeel characteristics; mimics the mouthfeel of oil; stabilizes emulsions; suspends the solids; improves cling; opacifier
	0.8–2	Fillings	Prevents boil-out during baking/heating (no leakage or rupture); improves texture and flavor release
	0.5–2	Food service	Stabilizes microwave sauces; reduces skinning on sauces held on steam table; helps keep fried foods crisp under heat lamps; reduces fat pick-up during frying
	0.5–1	High-fiber drinks	Increases dietary fiber; adds body and creaminess; suspends solids
	0.2–1	Icings	Controls flow and moisture migration; imparts stability; increases creaminess
	0.3–1.3	Sauces	Shear stability allows pumping without viscosity loss; stabilizes emulsions; improves cling; adds body and creaminess; prevents boil-out—imparts heat stability; adds opacity
	0.1–1	Ice cream, frozen desserts	Controls ice crystal growth: leads to smaller ice crystals; enhances the mouthfeel; improves creaminess and meltaway; replaces fat in low-fat recipes
Powder	0.5–2.5	Confectionary	Controls moisture absorption; nonnutritive bulk filler
	0.5–1	High-fiber drinks	Increases dietary fiber; adds body and creaminess; suspends solids

supplementary food products have already been introduced to the market to fulfill consumer demands. However, the addition of these dietary fibers could adversely affect the flavor, texture, color, and taste of the particular food product.

Cellulose has been actively explored as a food additive due to its natural origin, fibrous nature, and low calories. The use of cellulose increases the amount of dietary fiber in modified food products, as well as reduces the adverse effects of conventional dietary fibers. Currently, the production of low-calorie foods addresses one of the major public health concerns. Nanocellulose is used in the manufacturing of foods with low-fat content. A Turbak patent (US 4378381) suggested that nanocellulose could be a substitute for oil to formulate low-calorie salad dressing. The addition of 10% cellulose of bacterial origin in meatballs gave similar properties to those of control in terms of shelf life. These meatballs are chewy and juicy too, thus they are considered to be the potential replacement for fat in emulsified meat products (103). Robson proposed that cellulose could be helpful in reducing the energy density of several processed food products to <1.6 kcal/g (104). According to a study on mice, the dietary supplementation of BC significantly reduced the obesity induced by the fat-rich diet by reducing liver inflammation and damage. It further improved the antioxidant defense system and regulated adipogenesis (105). Moreover, the addition of BC to food supplements improved the mucosa and the thickness of colon muscles and increased the length of villus cells; hence, it could provide protection to colonic muscle cells from apoptosis (106, 107). These studies recommend that cellulose could be used as a promising dietary fiber to regulate gut microbes and for treating constipation. Microcrystalline cellulose (MCC)-potato starch composites showed excellent antilipidemic activity in mice compared to the traditional fiber composites (108). Likewise, the presence of crystalline cellulose in the feed of pigs efficiently lowered the total and LDL cholesterol compared to the control groups (109).

6.3.4 DELIVERY OF BIOACTIVE AGENTS

Recently, the demand for foods with functional food ingredients has been rising significantly. The addition of healthy but unstable additives such as probiotics and antioxidants in food is in great demand. The production of functional food requires the stability of bioactive compounds during the manufacturing, storage, and ingestion in the human gut. Besides stable functional food ingredients, the bioavailability of these beneficial items at the site of nutrient absorption is also important. The microencapsulation of such ingredients is a prerequisite to control the stability and ensure their targeted delivery. Biopolymers are capable of safely delivering the bioactive compounds and therapeutics in food systems.

Cellulose derivatives have key roles in the development of microencapsulated food systems (110). A variety of microencapsulated fish oils for dietary applications were produced by using the cellulose derivatives as the stabilizers to make oil/water emulsion (111, 112). Ethyl-cellulose (EC) microcapsules containing folic acid protect the degradation of nutrients at low pH in the stomach while allowing its release at high pH in the intestine, the site of its physiological absorption (113). EC is also helpful in the microencapsulation of probiotic bacteria in beverages with high moisture content at ambient temperatures (114). According to another study, the EC-coated limonene and vanillin microcapsules showed sustained drug release (115). In another study, BC aerogels loaded with L-ascorbic acid showed great potential to hold their initial water content, which can be 100% of its initial water content. The high water-holding capacity with large pore volume supports the use of BC aerogels as the sustained release medium (17). Many researchers have investigated the use of BC as a carrier for delivering nanoparticles such as those of Ag, ZnO, and TiO_2 (116–120) and other nanomaterials (121) with antimicrobial activities. These nanoparticles are widely used in food packaging and for biomedical applications such as wound dressing. BC as an efficient transport mechanism for proteins with serum albumin was studied by Müller et al., who observed that freeze-dried BC displayed low protein loading compared to the BC materials dehydrated with other approaches (122). Several researchers have designed BC films loaded with nisin and utilized them to control *Listeria* monocyte genus and total aerobic bacteria on the surface of vacuum-packaged frankfurters, where BC acted as the matrix to deliver nisin to the target site (123, 124). Currently, different polysaccharides and proteins are receiving immense consideration in developing suitable encapsulants that could be electrosprayed (125). During the electrospraying process, the polymer solution is turned to particles by applying high voltage (126). The high structural organization of pure proteins is a presently a major hindrance to ensure their sole availability for electrospraying (125). This issue could be resolved by blending the protein with polysaccharides such as starch (127), pullulan (128), and cellulose (129), which effectively enhance the sprayability of natural polymers. For example, whey and soy proteins encapsulated in BC were studied for their electrosprayability. The findings of this study showed the suitability of BC-whey composite as a suitable material for encapsulation of bioactive and volatile compounds (129).

6.3.5 Food Packaging

Packaging protects food from environmental factors and microbial contamination. The awareness of foodborne diseases encouraged the importance of avoiding microbial contamination of raw food materials as well as processed food items. Food contamination is a major cause of food spoilage and wastage; it also threatens the life of the consumer (130). Currently, the food-packing sector is dominated by the nanotechnology-derived food packaging materials (131). The increased knowledge of the environmental effects of plastic-based packaging materials encouraged the growing commercialization of nanomaterials for food packaging (132, 133). Nanomaterial-based food packaging can be classified into two types: active packaging and improved packaging. Active packaging is achieved through direct interaction of the nanomaterial with the food product, whereas improved packaging involves the incorporation of nanomaterials into a polymer matrix for enhanced barrier properties (132). Active packaging can function as flavoring, antioxidant, moisture absorber, UV barrier, oxygen scavenger, and antimicrobial.

Although pristine cellulose lacks antimicrobial activity, nanocellulose-based antimicrobial materials can be prepared by conjugating nanocellulose and antimicrobial agents through chemical or physical modification techniques (134, 135). The cellulose-derived edible films and coatings for wrapping/packaging are in great demand due to their compatibility with a variety of food products (12). The coatings and films incorporated with BC not only allow the regulation of gas (CO_2 and O_2) and water permeability but also provide strength to the packaging material (136). The barrier properties of corn starch thermoplastic were improved by adding BC nanowhiskers for its food packaging application (137). In a study, BC-based food packaging film with high hydrophobicity and controlled gas (N_2, CO_2, and O_2) permeability was designed (136). Similarly, a biodegradable mono- and multilayer BC/sorbic acid (BC/SA) film with antimicrobial features was prepared (138). The BC/SA films tested against *E. coli* showed promising antimicrobial activity (139, 140). A recent study reported the development of transparent antimicrobial BC films through the impregnation of ginger nanofibers, which not only served as the antimicrobial agent but also a rich source of cellulose (141). Green bionanocomposites for food packaging are developed by incorporating nanocellulose into polylactic acid (PLA). The PLA/nanocellulose composite showed enhanced antimicrobial properties (142). Furthermore, composite materials with antimicrobial features were prepared, where ground BC served as the reinforcing material while polyvinyl alcohol (PVA) was used as the polymeric matrix and SA as the antimicrobial agent (139). These studies highlight the use of cellulose-based antimicrobial composites for food packaging and other applications. In another study, Yang et al. developed a BC/silver nanoparticle composite and explored its properties as antimicrobial food packaging material (143). In a study, tough, UV-resistant, and thermally stable BC-based composite films were prepared by combining BC, PVA, and chitosan: PVA improved the water solubility of films and increased the water-swelling degree (144). BC can also be used as a preservative, where the coating of BC nanofibers suspension keeps cut apples fresh for a long time, indicating that it could be applied to coat fresh-cut vegetables and fruits (139). Antioxidant packaging films were prepared by utilizing BC, epigallocatechin-3-gallate (EGCG), and chitosan, where BC acted as the reinforcement material (145). In a study, the nanocrystal BC was added to formulate high-performance, edible, and biodegradable gel-like covers for food wrapping (146). The aerogels of rice and oat husk cellulose nanocrystals could be applied as the water absorbers in food packages. In a study, 402.8% of water absorption capacity was noticed in rice cellulose aerogels (27). Table 6.4 summarizes the different food packaging applications of cellulose obtained from different origins.

6.3.6 PICKERING EMULSION

Presently, cellulose is widely used as a green thickener and stabilizer in the food industry. Especially, cellulose nanowhiskers or CNCs effectively stabilize the oil/fat in food products by developing a stable emulsion that acts as the biocompatible stabilizers in the food system (110). Usually, the emulsions contain tiny spherical droplets of two fluids stabilized by the surface active compounds, for example surface active polymers (175). Pickering emulsions are produced as the substitution for conventional surfactants with solid particles (176). Compared to the traditional emulsions, Pickering emulsions have better features such as their adjustable permeability, high elastic properties, and high stability against coalescence (177, 178). The morphological properties of particles used in Pickering emulsions govern the stability of

TABLE 6.4

Examples of Cellulose Isolated from Various Plant and Bacterial Sources That Are Applied to Food Packaging.

Cellulose source	Reinforcement material	Properties	Ref.
Sugarcane bagasse	Starch film	Reduced hydrophilicity, improved water barrier property	(148)
Sugarcane bagasse	PVA/CMC	Increased tensile strength, improved water permeability	(149)
Sugarcane bagasse	Linear or cross-linked PVA	Improvement in tensile strength, elongation at break, yield force and toughness, enhanced thermal stability	(150)
Sugarcane bagasse/ graphene oxide (hybrid nanofiller)	PVA	Improved Young's modulus, tensile strength and toughness, moisture absorption	(151)
Sugarcane bagasse	CMC/starch	Improved optical transparency and tensile properties, reduced water vapor permeability	(152)
Empty fruit bunches	Starch-chitosan	Increased tensile strength	(153)
Phormium tenax leaves	PLA	Enhanced thermal stability	(154)
Linum usitatissimum	Chitosan	Reduced film transparency	(155)
Ramie	PVA/chitosan	Improved mechanical properties, oxygen barrier properties, and antimicrobial activities	(156)
Sweet potato	Cassava starch	Enhanced tensile strength and elongation at break, good moisture absorption	(157)
Mulberry	Agar	Enhanced mechanical and water vapor barrier properties	(158)
Mulberry	Alginate	Enhanced water vapor permeability and mechanical properties	(159)
Actinidia deliciosa pruning residue	PVA/chitosan	Enhanced mechanical properties	(160)
Coffee silverskin	PLA	Excellent oxygen barrier and water permeability property, increased Young's modulus and tensile strength	(161)
Red algae waste	PVA	Enhanced Young's modulus, tensile strength, and toughness	(162)
Seaweed	PVA	Improved tensile strength, water vapor permeability, and thermal stability	(163)

Source: Table adapted from Sethaphong et al. (2013). (147) and modified.

Cellulose source	Reinforcement material	Properties	Ref.
Bleached kraft bagasse pulp	Wheat gluten	Significant improvement in tensile strength and water resistance	(164)
Softwood kraft pulp	poly(3-hydro xybutyrate-co-3-hydroxyvalerate)	Improved tensile strength and Young's modulus, reduction in elongation at break	(165)
Wheat straw	CMC	Increased tensile strength and reduced water vapor permeability	(166)
Wheat straw	Wheat straw hemicelluloses	Reduced water sensitivity and permeability, enhanced modulus, water resistance, and elongation	(167)
Rice straw	CMC	Increased tensile strength of the composite by reduced water vapor permeability	(168)
Rice husk	Starch	Enhanced thermal stability, tensile properties, and storage modulus and reduced water uptake of the starch biocomposite	(168)
Banana peels	Pectin	Improved tensile properties, water vapor, and water resistance properties	(169)
Barley straw and husk	PVA blended with natural chitosan	Reduced film transparency, enhanced thermal and mechanical properties, inhibited fungal and bacterial growth	(170)
Oat husk	Whey protein isolate	Reduced film transparency and water vapor permeability, high Young's modulus and tensile strength	(171)
Corn cob	PVA	Enhanced tensile strength, reduced water vapor permeability	(172)
Industrial waste cotton	PVA	High transparency and flexibility	(173)
BC	Chitosan and epigallocatechin-3-gallate (EGCG)	Antioxidant activity	(174)
BC	Silver nanoparticles	Antimicrobial activity	(143)
BC	PVA and chitosan	Enhanced thermal stability, UV-resistant, high toughness	(144)
BC	PVA and SA	Antimicrobial activity	(156)

the respective emulsion (179). In earlier studies, mostly spherical particles such as protein-based nanoparticles were used to prepare Pickering emulsions. In contrast to the spherical particles, the CNC particles give high stability to the emulsions due to their rod-like structure with high aspect ratio; thus, they are capable of interacting with each other and form a stable bridge-like structure at the edges (178). Researchers have prepared three highly stabilized oil/water Pickering emulsions by using CNCs from BC (BCN), cotton (CCN), and *Cladophora* (ClaCN) with an aspect ratio of 47, 13, and 160, respectively (175). The highly hydrolyzed CNCs from the stem of *Asparagus officinalis L.* showed high emulsification efficiency (177). Another study reported that smaller CNCs possess higher creamy index, which could be attributed to their better emulsification efficiency (180). The hydrolysis of cellulose, such as by using H_2SO_4 or H_3PO_4, can produce negatively charged surfaces of CNCs; thus, these provide better colloidal stability through powerful electrostatic repulsion between the negatively charged CNCs (180, 181). The emulsions stabilized by the acid-hydrolyzed CNCs commonly use basic salts such as sodium chloride for electrostatic screening (180).

The stability of a Pickering emulsion is defined by the formation of a thick and dense layer generated by the conclusive particle adsorption at the water/oil interface, consequently preventing the amalgamation and ripening of emulsion (176). In emulsification, the wettability of particles is an important factor that contributes to the stability of emulsions, as the particles could be partly wetted by the continuous/dispersed phase during processing. Generally, wettability is characterized by a three-phase contact angle (θ) present in the water-particle-oil interface (179, 182, 183). The (200) β/(220) α crystalline edge of cellulose chains forms the "hydrophobic face" of cellulose, where the acquired amphiphilic property assists the adsorption at oil/water interface (175). The use of cellulose as the Pickering stabilizer, however, is restricted due to its high hydrophilicity (184); in recent years, several hydrophobic surface modification strategies, including oxidation, graft polymerization, and esterification, are designed to produce cellulose with a hydrophobic surface (185, 186). The surface modification of nanocellulose, such as through ocetnyl succinic anhydride (OSA) increased the contact angle from 51.7°–56° (untreated CNCs) to 82.1°–85.0° (OSA-treated CNCs), which allows nanocellulose to be partially wetted by both the phases and thus stabilizes the oil/water emulsion (185, 186). The resultant emulsions showed droplet flocculation at pH < 4.0 (without NaCl) or ionic strength ≥ 20 mM NaCl (pH 7.0), while the elastic gel-like behavior was detected at ≥20 mM NaCl (186). The surface modification of cellulose with enhanced hydrophobicity could demonstrate the preparation of gel-like water/oil Pickering high-internal-phase-emulsions (HIPEs) (187). In a study, phosphorylated CNCs (P-CNCs) were prepared through H_3PO_4 hydrolysis, and cellulose was further modified by using the glycidyl trimethyl ammonium chloride modified chitosan (GCh) to produce GChP-CNCs nanoparticles. These P-CNC particles showed characteristics of both the emulsifier and the nanoparticles (188). The property of cellulose as the oil/water Pickering emulsion was further enhanced by adding chitosan to cellulose (189). The bovine serum albumin (BSA)-covered CNCs enriched the gel-like behavior of HIPEs with a reduced drop size (187). Similarly, another study investigated the suspensions stabilized by cellulose and sodium caseinate (CAS) at pH 3 (190). The combined effect of CNCs-CAS enhanced the emulsification properties compared to the CNCs or CAS alone. By using the ultrasound-assisted *in situ* coprecipitation method, Low et al. prepared Fe_3O_4-CNC nanoparticles. The obtained emulsion showed responses toward both altered pH and magnetic field (191). Further, it was noticed that Fe_3O_4-CNC emulsion containing curcumin showed a sustained release in response to the external magnetic field. The curcumin release in response to the magnetic field increased to 53.30% in comparison to the absence of an external magnetic field where curcumin release was only 14.59%. The curcumin-loaded suspensions efficiently suppressed the development of colon tumor to 18% in the presence of an external magnetic field (192). This approach provides a platform for developing the cellulose-based Pickering emulsions for the release of therapeutic and bioactive compounds upon the application of external magnetic field. CNCs and tannic acid were used as stabilizers to make avocado oil-based HIPEs (140). This HIPE system gives oxidative stability to avocado oil owing a low lipid hydroperoxides, malondialdehyde, and hexanal levels after thermally accelerated storage. The oregano and CNC-based Pickering emulsion showed antimicrobial behavior (193). Many other oils such as eugenol, cinnamaldehyde, white essential oil, and limonene, with antimicrobial properties, have been developed by using CNC-based stabilizers (194–196).

6.4 CONCLUSIONS

Cellulose has great potential in the food industry and other fields due to its unique structural, physicochemical, thermal, mechanical, and biological features. The biodegradability, nontoxic nature, and production from renewable resources precisely meet the needs of modern society for food safety and environmental protection. In the food industry, cellulose and its derivatives are receiving tremendous attention for application in packaging, delivery of bioactive compounds, and as an emulsifier, where these are used as the reinforcement material. Due to its highly fibrous network and cohesive energy density, cellulose has the capability to decrease O_2, CO_2, and water permeability. Furthermore, the large surface area of cellulose aids the absorption and doping of antioxidant and antimicrobial agents into its matrix. The enhanced permeability and mechanical properties make cellulose a promising candidate for the sustainable release of therapeutics and bioactive compounds through its matrix. The addition of molecules to the cellulose matrix and their subsequent release at the target site inhibit the microbial growth for improvement of food matrices such as mixed vegetables and cooked ham and reduce the chances of weight loss and freshness of fragile fruits like strawberries. The presence of cellulose and cellulose derivate in food products not only influences the rheological and mechanical properties of the final product for improved sensory, textural, and organoleptic attributes, but also cuts the calorie level. As far as toxicity is concerned, under known conditions, cellulose is considered safe upon oral consumption, dermal applications, and inhalation exposure. Further studies should be carried out to understand the movement of particles from cellulosic films into the food and their effect on food safety, texture, and flavor.

ACKNOWLEDGMENT

This work was supported by the National Natural Science Foundation of China (21774039, 51973076), BRICS STI Framework Programme 3rd call 2019 (2018YFE0123700), China Postdoctoral Science Foundation (2016M602291), and the Fundamental Research Funds for Central Universities, Open Research Fund of State Key Laboratory of Polymer Physics and Chemistry, and Changchun Institute of Applied Chemistry, Chinese Academy of Sciences.

CONFLICT OF INTEREST

All authors declare no competing financial conflict of interest associated with the publication of this work.

REFERENCES

(1) Abral, H., Ariksa, J., Mahardika, M., Handayani, D., Aminah, I., Sandrawati, N., et al. (2020). Transparent and antimicrobial cellulose film from ginger nanofiber. *Food Hydrocoll.* 98, 105266.

(2) Aceituno-Medina, M., Mendoza, S., Lagaron, J. M., and López-Rubio, A. (2013). Development and characterization of food-grade electrospun fibers from amaranth protein and pullulan blends. *Food Res. Int.* doi:10.1016/j.foodres.2013.07.055.

(3) Adel, A. M., and El-Shinnawy, N. A. (2012). Hypolipidemic applications of microcrystalline cellulose composite synthesized from different agricultural residues. *Int. J. Biol. Macromol.* doi:10.1016/j.ijbiomac.2012.08.003.

(4) Ahrem, H., Pretzel, D., Endres, M., Conrad, D., Courseau, J., Müller, H., et al. (2014). Laser-structured bacterial nanocellulose hydrogels support ingrowth and differentiation of chondrocytes and show potential as cartilage implants. *Acta Biomater.* 10, 1341–1353. doi:10.1016/j.actbio.2013.12.004.

(5) Aljohani, W., Ullah, M. W., Zhang, X., and Yang, G. (2018). Bioprinting and its applications in tissue engineering and regenerative medicine. *Int. J. Biol. Macromol.* 107, 261–275. doi:10.1016/j.ijbiomac.2017.08.171.

(6) Amarasinghe, H., Weerakkody, N. S., and Waisundara, V. Y. (2018). Evaluation of physicochemical properties and antioxidant activities of kombucha "Tea Fungus" during extended periods of fermentation. *Food Sci. Nutr.* doi:10.1002/fsn3.605.

(7) Araki, J. (2013). Electrostatic or steric?-preparations and characterizations of well-dispersed systems containing rod-like nanowhiskers of crystalline polysaccharides. *Soft Matter.* doi:10.1039/c3sm27514k.

(8) Atalla, R. H., and VanderHart, D. L. (1984). Native cellulose: A composite of two distinct crystalline forms. *Science* 80. doi:10.1126/science.223.4633.283.

(9) Baek, J., Wahid-Pedro, F., Kim, K., Kim, K., and Tam, K. C. (2019). Phosphorylated-CNC/modified-chitosan nanocomplexes for the stabilization of Pickering emulsions. *Carbohydr. Polym.* doi:10.1016/j.carbpol.2018.11.006.

(10) BahramParvar, M., Tehrani, M. M., and Razavi, S. M. A. (2013). Effects of a novel stabilizer blend and presence of κ-carrageenan on some properties of vanilla ice cream during storage. *Food Biosci.* doi:10.1016/j.fbio.2013.05.001.

(11) Bera, B. (2016). Literature review on electrospinning process (A fascinating fiber fabrication technique). *Imp. J. Interdiscip. Res. IJIR.* 2(8), 972–984.

(12) Berton-Carabin, C. C., and Schroën, K. (2015). Pickering emulsions for food applications: Background, trends, and challenges. *Annu. Rev. Food Sci. Technol.* doi:10.1146/annurev-food-081114-110822.

(13) Bhardwaj, N., and Kundu, S. C. (2010). Electrospinning: A fascinating fiber fabrication technique. *Biotechnol. Adv.* doi:10.1016/j.biotechadv.2010.01.004.
(14) Brett, C. T. (2000). Cellulose mlicrofibrils in plants: Biosynthesis, deposition, and integration into the cell wall. *Int. Rev. Cytol.* doi:10.1016/s0074-7696(00)99004-1.
(15) Bringmann, M., Landrein, B., Schudoma, C., Hamant, O., Hauser, M. T., and Persson, S. (2012). Cracking the elusive alignment hypothesis: The microtubule-cellulose synthase nexus unraveled. *Trends Plant Sci.* doi:10.1016/j.tplants.2012.06.003.
(16) Cazón, P., Vázquez, M., and Velazquez, G. (2019). Composite films with UV-barrier properties based on bacterial cellulose combined with chitosan and poly(vinyl alcohol): Study of puncture and water interaction properties. *Biomacromolecules.* doi:10.1021/acs. biomac.9b00317.
(17) Chang, C., and Zhang, L. (2011). Cellulose-based hydrogels: Present status and application prospects. *Carbohydr. Polym.* doi:10.1016/j.carbpol.2010.12.023.
(18) Chau, C. F., Yang, P., Yu, C. M., and Yen, G. C. (2008). Investigation on the lipid- and cholesterol-lowering abilities of biocellulose. *J. Agric. Food Chem.* doi:10.1021/ jf7035802.
(19) Chen, Q. H., Zheng, J., Xu, Y. T., Yin, S. W., Liu, F., and Tang, C. H. (2018). Surface modification improves fabrication of Pickering high internal phase emulsions stabilized by cellulose nanocrystals. *Food Hydrocoll.* doi:10.1016/j.foodhyd.2017.09.005.
(20) Cosgrove, D. J. (2005). Growth of the plant cell wall. *Nat. Rev. Mol. Cell Biol.* doi:10.1038/ nrm1746.
(21) Cropotova, J., and Popel, S. (2013). A way to prevent syneresis in fruit fillings prepared with gellan gum. *Sci. Pap. Ser. D. Anim. Sci.* 56, 326–332.
(22) Crowell, E. F., Gonneau, M., Stierhof, Y. D., Höfte, H., and Vernhettes, S. (2010). Regulated trafficking of cellulose synthases. *Curr. Opin. Plant Biol.* doi:10.1016/j. pbi.2010.07.005.
(23) Cullen, R. T., Searl, A., Miller, B. G., Davis, J. M. G., and Jones, A. D. (2000). Pulmonary and intraperitoneal inflammation induced by cellulose fibres. *J. Appl. Toxicol.* doi:10.1002/(SICI)1099-1263(200001/02)20:1<49::AID-JAT627>3.0.CO;2-L.
(24) Curvello, R., Raghuwanshi, V. S., and Garnier, G. (2019). Engineering nanocellulose hydrogels for biomedical applications. *Adv. Colloid Interface Sci.* 267, 47–61. doi:https:// doi.org/10.1016/j.cis.2019.03.002.
(25) Dai, H., Wu, J., Zhang, H., Chen, Y., Ma, L., Huang, H., et al. (2020). Recent advances on cellulose nanocrystals for Pickering emulsions: Development and challenge. *Trends Food Sci. Technol.* doi:10.1016/j.tifs.2020.05.016.
(26) Davidov-Pardo, G., Roccia, P., Salgado, D., León, A. E., and Pedroza-Islas, R. (2008). Utilization of different wall materials to microencapsulate fish oil evaluation of its behavior in bread products. *Am. J. Food Technol.* doi:10.3923/ajft.2008.384.393.
(27) Deng, Z., Jung, J., Simonsen, J., and Zhao, Y. (2018). Cellulose nanocrystals Pickering emulsion incorporated chitosan coatings for improving storability of postharvest Bartlett pears (Pyrus communis) during long-term cold storage. *Food Hydrocoll.* doi:10.1016/j. foodhyd.2018.06.012.
(28) de Oliveira, J. P., Bruni, G. P., el Halal, S. L. M., Bertoldi, F. C., Dias, A. R. G., and Zavareze, E. da R. (2019). Cellulose nanocrystals from rice and oat husks and their application in aerogels for food packaging. *Int. J. Biol. Macromol.* doi:10.1016/j. ijbiomac.2018.11.205.
(29) Di, Z., Shi, Z., Ullah, M. W., Li, S., and Yang, G. (2017). A transparent wound dressing based on bacterial cellulose whisker and poly(2-hydroxyethyl methacrylate). *Int. J. Biol. Macromol.* 105, 638–644. doi:10.1016/j.ijbiomac.2017.07.075.
(30) Dobre, L. M., and Stoica-Guzun, A. (2013). Antimicrobial ag-polyvinyl alcohol-bacterial cellulose composite films. *J. Biobased Mater. Bioenergy.* doi:10.1166/ jbmb.2013.1272.

(31) Dobre, L. M., Stoica-Guzun, A., Stroescu, M., Jipa, I. M., Dobre, T., Ferdeş, M., et al. (2012). Modelling of sorbic acid diffusion through bacterial cellulose-based antimicrobial films. *Chem. Pap.* doi:10.2478/s11696-011-0086-2.

(32) Dourado, F., Gama, M., and Rodrigues, A. C. (2017). A Review on the toxicology and dietetic role of bacterial cellulose. *Toxicol. Reports.* doi:10.1016/j.toxrep.2017.09.005.

(33) Dourado, F., Leal, M., Martins, D., Fontão, A., Cristina Rodrigues, A., and Gama, M. (2016). Celluloses as food ingredients/additives: Is there a room for BNC? *Bacterial Nanocellulose: From Biotechnology to Bio-Economy* doi:10.1016/B978-0-444-63458-0.00007-X.

(34) Du Le, H., Loveday, S. M., Singh, H., and Sarkar, A. (2020). Pickering emulsions stabilised by hydrophobically modified cellulose nanocrystals: Responsiveness to pH and ionic strength. *Food Hydrocoll.* doi:10.1016/j.foodhyd.2019.105344.

(35) Echegoyen, Y., and Nerín, C. (2013a). Nanoparticle release from nano-silver antimicrobial food containers. *Food Chem. Toxicol.* doi:10.1016/j.fct.2013.08.014.

(36) Echegoyen, Y., and Nerín, C. (2013b). Nanoparticle release from nano-silver antimicrobial food containers. *Food Chem. Toxicol.* doi:10.1016/j.fct.2013.08.014.

(37) El Achaby, M., El Miri, N., Aboulkas, A., Zahouily, M., Bilal, E., Barakat, A., et al. (2017). Processing and properties of eco-friendly bio-nanocomposite films filled with cellulose nanocrystals from sugarcane bagasse. *Int. J. Biol. Macromol.* doi:10.1016/j.ijbiomac.2016.12.040.

(38) El Achaby, M., Kassab, Z., Aboulkas, A., Gaillard, C., and Barakat, A. (2018). Reuse of red algae waste for the production of cellulose nanocrystals and its application in polymer nanocomposites. *Int. J. Biol. Macromol.* doi:10.1016/j.ijbiomac.2017.08.067.

(39) El Miri, N., Abdelouahdi, K., Barakat, A., Zahouily, M., Fihri, A., Solhy, A., et al. (2015). Bio-nanocomposite films reinforced with cellulose nanocrystals: Rheology of film-forming solutions, transparency, water vapor barrier and tensile properties of films. *Carbohydr. Polym.* doi:10.1016/j.carbpol.2015.04.051.

(40) El Miri, N., El Achaby, M., Fihri, A., Larzek, M., Zahouily, M., Abdelouahdi, K., et al. (2016). Synergistic effect of cellulose nanocrystals/graphene oxide nanosheets as functional hybrid nanofiller for enhancing properties of PVA nanocomposites. *Carbohydr. Polym.* doi:10.1016/j.carbpol.2015.10.072.

(41) El-Wakil, N. A., Hassan, E. A., Abou-Zeid, R. E., and Dufresne, A. (2015). Development of wheat gluten/nanocellulose/titanium dioxide nanocomposites for active food packaging. *Carbohydr. Polym.* doi:10.1016/j.carbpol.2015.01.076.

(42) Endler, A., Sánchez-Rodríguez, C., and Persson, S. (2010). Cellulose squeezes through. *Nat. Chem. Biol.* 6, 883–884. doi:10.1038/nchembio.480.

(43) Fabra, M. J., López-Rubio, A., Ambrosio-Martín, J., and Lagaron, J. M. (2016). Improving the barrier properties of thermoplastic corn starch-based films containing bacterial cellulose nanowhiskers by means of PHA electrospun coatings of interest in food packaging. *Food Hydrocoll.* 61, 261–268.

(44) Farooq, U., Ullah, M. W., Yang, Q., Aziz, A., Xu, J., Zhou, L., et al. (2020). High-density phage particles immobilization in surface-modified bacterial cellulose for ultra-sensitive and selective electrochemical detection of Staphylococcus aureus. *Biosens. Bioelectron.* 157, 112163. doi:10.1016/j.bios.2020.112163.

(45) Farooq, U., Yang, Q., Ullah, M. W., and Wang, S. (2018). Bacterial biosensing: Recent advances in phage-based bioassays and biosensors. *Biosens. Bioelectron.* 118, 204–216. doi:10.1016/j.bios.2018.07.058.

(46) Fatima, A., Yasir, S., Khan, M. S., Manan, S., Ullah, M. W., and Ul-Islam, M. (2021). Plant extract-loaded bacterial cellulose composite membrane for potential biomedical applications. *J. Bioresour. Bioprod.* 6, 26–32. doi:10.1016/j.jobab.2020.11.002.

(47) Fortunati, E., Benincasa, P., Balestra, G. M., Luzi, F., Mazzaglia, A., Del Buono, D., et al. (2016). Revalorization of barley straw and husk as precursors for cellulose nanocrystals extraction and their effect on PVA_CH nanocomposites. *Ind. Crops Prod.* doi:10.1016/j.indcrop.2016.07.047.

(48) Fortunati, E., Luzi, F., Puglia, D., Dominici, F., Santulli, C., Kenny, J. M., et al. (2014). Investigation of thermo-mechanical, chemical and degradative properties of PLA-limonene films reinforced with cellulose nanocrystals extracted from Phormium tenax leaves. *Eur. Polym. J.* doi:10.1016/j.eurpolymj.2014.03.030.

(49) Fu, L., Zhou, P., Zhang, S., and Yang, G. (2013). Evaluation of bacterial nanocellulose-based uniform wound dressing for large area skin transplantation. *Mater. Sci. Eng. C* 33, 2995–3000. doi:10.1016/j.msec.2013.03.026.

(50) Gallegos, A. M. A., Carrera, S. H., Parra, R., Keshavarz, T., and Iqbal, H. M. N. (2016). Bacterial cellulose: A sustainable source to develop value-added products—A review. *BioResources.* doi:10.15376/biores.11.2.Gallegos.

(51) Galway, M. E., Eng, R. C., Schiefelbein, J. W., and Wasteneys, G. O. (2011). Root hair-specific disruption of cellulose and xyloglucan in AtCSLD3 mutants, and factors affecting the post-rupture resumption of mutant root hair growth. *Planta.* doi:10.1007/s00425-011-1355-6.

(52) Gan, I., and Chow, W. S. (2018a). Antimicrobial poly(lactic acid)/cellulose bionanocomposite for food packaging application: A review. *Food Packag. Shelf Life.* doi:10.1016/j.fpsl.2018.06.012.

(53) Gan, I., and Chow, W. S. (2018b). Antimicrobial poly(lactic acid)/cellulose bionanocomposite for food packaging application: A review. *Food Packag. Shelf Life* 17, 150–161. doi:10.1016/j.fpsl.2018.06.012.

(54) George, J., and Siddaramaiah (2012). High performance edible nanocomposite films containing bacterial cellulose nanocrystals. *Carbohydr. Polym.* doi:10.1016/j.carbpol.2011.10.019.

(55) Gómez H., C., Serpa, A., Velásquez-Cock, J., Gañán, P., Castro, C., Vélez, L., et al. (2016). Vegetable nanocellulose in food science: A review. *Food Hydrocoll.* doi:10.1016/j.foodhyd.2016.01.023.

(56) Gu, Y., and Somerville, C. (2010). Cellulose synthase interacting protein: A new factor in cellulose synthesis. *Plant Signal. Behav.* doi:10.4161/psb.5.12.13621.

(57) Guo, Y., Zhang, X., Hao, W., Xie, Y., Chen, L., Li, Z., et al. (2018). Nano-bacterial cellulose/soy protein isolate complex gel as fat substitutes in ice cream model. *Carbohydr. Polym.* doi:10.1016/j.carbpol.2018.06.078.

(58) Habibi, H., and Khosravi-Darani, K. (2017). Effective variables on production and structure of xanthan gum and its food applications: A review. *Biocatal. Agric. Biotechnol.* doi:10.1016/j.bcab.2017.02.013.

(59) Hagiwara, A., Imai, N., Sano, M., Kawabe, M., Tamano, S., Kitamura, S., et al. (2010). A 28-day oral toxicity study of fermentation-derived cellulose, produced by Acetobacter aceti subspecies xylinum, in F344 rats. *J. Toxicol. Sci.* doi:10.2131/jts.35.317.

(60) He, X., and Hwang, H. M. (2016). Nanotechnology in food science: Functionality, applicability, and safety assessment. *J. Food Drug Anal.* doi:10.1016/j.jfda.2016.06.001.

(61) Huang, Y., Zhu, C., Yang, J., Nie, Y., Chen, C., and Sun, D. (2014). Recent advances in bacterial cellulose. *Cellulose.* doi:10.1007/s10570-013-0088-z.

(62) Huber, T., Müssig, J., Curnow, O., Pang, S., Bickerton, S., and Staiger, M. P. (2012). A critical review of all-cellulose composites. *J. Mater. Sci.* doi:10.1007/s10853-011-5774-3.

(63) Hungund, B. (2013). Production of bacterial cellulose from gluconacetobacter persimmonis GH-2 using dual and cheaper carbon sources. *J. Microb. Biochem. Technol.* 5. doi:10.4172/1948-5948.1000095.

(64) Huq, T., Riedl, B., Bouchard, J., Salmieri, S., and Lacroix, M. (2014). Microencapsulation of nisin in alginate-cellulose nanocrystal (CNC) microbeads for prolonged efficacy against Listeria monocytogenes. *Cellulose*. doi:10.1007/s10570-014-0432-y.

(65) Hussain, Z., Sajjad, W., Khan, T., and Wahid, F. (2019). Production of bacterial cellulose from industrial wastes: A review. *Cellulose* 26, 2895–2911. doi:10.1007/s10570-019-02307-1.

(66) Hyun, J., Lee, Y., Wang, S., Kim, J., Kim, J., Cha, J. H., et al. (2016). Kombucha tea prevents obese mice from developing hepatic steatosis and liver damage. *Food Sci. Biotechnol.* doi:10.1007/s10068-016-0142-3.

(67) Jamshaid, A., Hamid, A., Muhammad, N., Naseer, A., Ghauri, M., Iqbal, J., et al. (2017). Cellulose-based materials for the removal of heavy metals from wastewater—an overview. *ChemBioEng Rev.* doi:10.1002/cben.201700002.

(68) Janpetch, N., Saito, N., and Rujiravanit, R. (2016). Fabrication of bacterial cellulose-ZnO composite via solution plasma process for antibacterial applications. *Carbohydr. Polym.* 148, 335–344. doi:10.1016/j.carbpol.2016.04.066.

(69) Jasim, A., Ullah, M. W., Shi, Z., Lin, X., and Yang, G. (2017). Fabrication of bacterial cellulose/polyaniline/single-walled carbon nanotubes membrane for potential application as biosensor. *Carbohydr. Polym.* 163, 62–69. doi:10.1016/j.carbpol.2017.01.056.

(70) Jipa, I. M., Stoica-Guzun, A., and Stroescu, M. (2012). Controlled release of sorbic acid from bacterial cellulose based mono and multilayer antimicrobial films. *LWT—Food Sci. Technol.* doi:10.1016/j.lwt.2012.01.039.

(71) Jorfi, M., and Foster, E. J. (2015). Recent advances in nanocellulose for biomedical applications. *J. Appl. Polym. Sci.* 132. doi:10.1002/app.41719.

(72) Juraniec, M., and Gajda, B. (2020). Cellulose biosynthesis in plants—the concerted action of CESA and non-CESA proteins. *Biol. Plant.* doi:10.32615/bp.2020.065.

(73) Kalashnikova, I., Bizot, H., Bertoncini, P., Cathala, B., and Capron, I. (2013). Cellulosic nanorods of various aspect ratios for oil in water Pickering emulsions. *Soft Matter.* doi:10.1039/c2sm26472b.

(74) Kargarzadeh, H., Johar, N., and Ahmad, I. (2017). Starch biocomposite film reinforced by multiscale rice husk fiber. *Compos. Sci. Technol.* doi:10.1016/j.compscitech.2017.08.018.

(75) Khan, S., Ul-Islam, M., Khattak, W. A., Ullah, M. W., and Park, J. K. (2015). Bacterial cellulose-titanium dioxide nanocomposites: Nanostructural characteristics, antibacterial mechanism, and biocompatibility. *Cellulose*. doi:10.1007/s10570-014-0528-4.

(76) Khattak, W. A., Khan, T., Ha, J. H., Ul-Islam, M., Kang, M. K., and Park, J. K. (2013). Enhanced production of bioethanol from waste of beer fermentation broth at high temperature through consecutive batch strategy by simultaneous saccharification and fermentation. *Enzyme Microb. Technol.* 53, 322–330. doi:10.1016/j.enzmictec.2013.07.004.

(77) Kim, Y., Ullah, M. W., Ul-Islam, M., Khan, S., Jang, J. H., and Park, J. K. (2019). Self-assembly of bio-cellulose nanofibrils through intermediate phase in a cell-free enzyme system. *Biochem. Eng. J.* 142, 135–144. doi:10.1016/j.bej.2018.11.017.

(78) Klemm, D., Heublein, B., Fink, H. P., and Bohn, A. (2005). Cellulose: Fascinating biopolymer and sustainable raw material. *Angew. Chemie—Int. Ed.* 44, 3358–3393. doi:10.1002/anie.200460587.

(79) Kolanowski, W., Laufenberg, G., and Kunz, B. (2004). Fish oil stabilisation by microencapsulation with modified cellulose. *Int. J. Food Sci. Nutr.* doi:10.1080/09637480410001725157.

(80) Kong, L., and Ziegler, G. R. (2014). Formation of starch-guest inclusion complexes in electrospun starch fibers. *Food Hydrocoll.* doi:10.1016/j.foodhyd.2013.12.018.

(81) Kubiak, K., Kurzawa, M., Jedrzejczak-Krzepkowska, M., Ludwicka, K., Krawczyk, M., Migdalski, A., et al. (2014). Complete genome sequence of *Gluconacetobacter xylinus* E25 strain-Valuable and effective producer of bacterial nanocellulose. *J. Biotechnol.* doi:10.1016/j.jbiotec.2014.02.006.

(82) Kunchitwaranont, A., Chiewchan, N., and Devahastin, S. (2019). Use and understanding of the role of spontaneously formed nanocellulosic fiber from lime (citrus aurantifolia swingle) residues to improve stability of sterilized coconut milk. *J. Food Sci.* doi:10.1111/1750-3841.14937.

(83) Lampugnani, E. R., Flores-Sandoval, E., Tan, Q. W., Mutwil, M., Bowman, J. L., and Persson, S. (2019). Cellulose synthesis—central components and their evolutionary relationships. *Trends Plant Sci.* doi:10.1016/j.tplants.2019.02.011.

(84) Lee, K. Y., Buldum, G., Mantalaris, A., and Bismarck, A. (2014). More than meets the eye in bacterial cellulose: Biosynthesis, bioprocessing, and applications in advanced fiber composites. *Macromol. Biosci.* doi:10.1002/mabi.201300298.

(85) Li, H. Z., Chen, S. C., and Wang, Y. Z. (2015). Preparation and characterization of nanocomposites of polyvinyl alcohol/cellulose nanowhiskers/chitosan. *Compos. Sci. Technol.* doi:10.1016/j.compscitech.2015.05.004.

(86) Li, S., Bashline, L., Lei, L., and Gu, Y. (2014a). Cellulose synthesis and its regulation. *Arab. B.* doi:10.1199/tab.0169.

(87) Li, S., Huang, D., Zhang, B., Xu, X., Wang, M., Yang, G., et al. (2014b). Flexible supercapacitors based on bacterial cellulose paper electrodes. *Adv. Energy Mater.* 4, 1301655. doi:10.1002/aenm.201301655.

(88) Li, S., Jasim, A., Zhao, W., Fu, L., Ullah, M. W., Shi, Z., et al. (2018). Fabrication of pH-electroactive bacterial cellulose/polyaniline hydrogel for the development of a controlled drug release system. *ES Mater. Manuf.*, 41–49. doi:10.30919/esmm5f120.

(89) Li, S., Lei, L., Somerville, C. R., and Gu, Y. (2012). Cellulose synthase interactive protein 1 (CSI1) links microtubules and cellulose synthase complexes. *Proc. Natl. Acad. Sci. U. S. A.* doi:10.1073/pnas.1118560109.

(90) Liepman, A. H., and Cavalier, D. M. (2012). The cellulose synthase-like a and cellulose synthase-like C families: Recent advances and future perspectives. *Front. Plant Sci.* doi:10.3389/fpls.2012.00109.

(91) LIN, K. W., and LIN, H. Y. (2006). Quality Characteristics of Chinese-style meatball containing bacterial cellulose (Nata). *J. Food Sci.* doi:10.1111/j.1365-2621.2004.tb13378.x.

(92) Low, L. E., Tan, L. T. H., Goh, B. H., Tey, B. T., Ong, B. H., and Tang, S. Y. (2019). Magnetic cellulose nanocrystal stabilized Pickering emulsions for enhanced bioactive release and human colon cancer therapy. *Int. J. Biol. Macromol.* doi:10.1016/j.ijbiomac.2019.01.037.

(93) Low, L. E., Tey, B. T., Ong, B. H., Chan, E. S., and Tang, S. Y. (2017). Palm olein-in-water Pickering emulsion stabilized by Fe3O4-cellulose nanocrystal nanocomposites and their responses to pH. *Carbohydr. Polym.* doi:10.1016/j.carbpol.2016.08.091.

(94) Luzi, F., Fortunati, E., Giovanale, G., Mazzaglia, A., Torre, L., and Balestra, G. M. (2017). Cellulose nanocrystals from Actinidia deliciosa pruning residues combined with carvacrol in PVA_CH films with antioxidant/antimicrobial properties for packaging applications. *Int. J. Biol. Macromol.* doi:10.1016/j.ijbiomac.2017.05.176.

(95) Ma, X., Cheng, Y., Qin, X., Guo, T., Deng, J., and Liu, X. (2017). Hydrophilic modification of cellulose nanocrystals improves the physicochemical properties of cassava starch-based nanocomposite films. *LWT—Food Sci. Technol.* doi:10.1016/j.lwt.2017.08.012.

(96) Mandal, A., and Chakrabarty, D. (2014). Studies on the mechanical, thermal, morphological and barrier properties of nanocomposites based on poly(vinyl alcohol) and nanocellulose from sugarcane bagasse. *J. Ind. Eng. Chem.* doi:10.1016/j.jiec.2013.05.003.

(97) Mangayil, R., Rajala, S., Pammo, A., Sarlin, E., Luo, J., Santala, V., et al. (2017). Engineering and characterization of bacterial nanocellulose films as low cost and flexible sensor material. *ACS Appl. Mater. Interfaces.* doi:10.1021/acsami.7b04927.

(98) Mao, L., and Wu, T. (2007). Gelling properties and lipid oxidation of kamaboko gels from grass carp (Ctenopharyngodon idellus) influenced by chitosan. *J. Food Eng.* doi:10.1016/j.jfoodeng.2007.01.015.

(99) Marchese, A., Barbieri, R., Coppo, E., Orhan, I. E., Daglia, M., Nabavi, S. F., et al. (2017). Antimicrobial activity of eugenol and essential oils containing eugenol: A mechanistic viewpoint. *Crit. Rev. Microbiol.* doi:10.1080/1040841X.2017.1295225.

(100) Mbituyimana, B., Mao, L., Hu, S., Ullah, M. W., Chen, K., Fu, L., et al. (2021). Bacterial cellulose/glycolic acid/glycerol composite membrane as a system to deliver glycolic acid for anti-aging treatment. *J. Bioresour. Bioprod.* doi:10.1016/j.jobab.2021.02.003.

(101) McCarthy, R. R., Ullah, M. W., Booth, P., Pei, E., and Yang, G. (2019a). The use of bacterial polysaccharides in bioprinting. *Biotechnol. Adv.* 37, 107448. doi:10.1016/j.biotechadv.2019.107448.

(102) McCarthy, R. R., Ullah, M. W., Pei, E., and Yang, G. (2019b). Antimicrobial Inks: The anti-infective applications of bioprinted bacterial polysaccharides. *Trends Biotechnol.* 37, 1153–1155. doi:https://doi.org/10.1016/j.tibtech.2019.05.004.

(103) McClements, D. J. (2012). Nanoemulsions versus microemulsions: Terminology, differences, and similarities. *Soft Matter.* doi:10.1039/c2sm06903b.

(104) McFarlane, H. E., Döring, A., and Persson, S. (2014). The cell biology of cellulose synthesis. *Annu. Rev. Plant Biol.* doi:10.1146/annurev-arplant-050213-040240.

(105) Mesomya, W., Cuptapun, Y., Hengsawadi, D., Tangkanakul, P., Boonvisut, S., and Poosimuang, S. (2001). Protein bioavailability—lowering in rats fed high dietary fiber from cereal and nata de coco. *Kasetsart J. Nat. Sci.* 28, 23–28.

(106) Mikulcová, V., Bordes, R., and Kašpárková, V. (2016). On the preparation and antibacterial activity of emulsions stabilized with nanocellulose particles. *Food Hydrocoll.* doi:10.1016/j.foodhyd.2016.06.031.

(107) Mishra, R. K., Sabu, A., and Tiwari, S. K. (2018). Materials chemistry and the futurist eco-friendly applications of nanocellulose: Status and prospect. *J. Saudi Chem. Soc.* doi:10.1016/j.jscs.2018.02.005.

(108) Moon, R. J., Martini, A., Nairn, J., Simonsen, J., and Youngblood, J. (2011). Cellulose nanomaterials review: Structure, properties and nanocomposites. *Chem. Soc. Rev.* 40, 3941–3994. doi:10.1039/c0cs00108b.

(109) Morgan, J. L. W., Strumillo, J., and Zimmer, J. (2013). Crystallographic snapshot of cellulose synthesis and membrane translocation. *Nature.* doi:10.1038/nature11744.

(110) Mu, R., Hong, X., Ni, Y., Li, Y., Pang, J., Wang, Q., et al. (2019). Recent trends and applications of cellulose nanocrystals in food industry. *Trends Food Sci. Technol.* doi:10.1016/j.tifs.2019.09.013.

(111) Mueller, S. C., and Brown, R. M. (1980). Evidence for an intramembrane component associated with a cellulose microfibril-synthesizing complex in higher plants. *J. Cell Biol.* doi:10.1083/jcb.84.2.315.

(112) Mueller, S. C., Brown, R. M., and Scott, T. K. (1976). Cellulosic microfibrils: Nascent stages of synthesis in a higher plant cell. *Science* 80. doi:10.1126/science.194.4268.949.

(113) Mujtaba, M., Salaberria, A. M., Andres, M. A., Kaya, M., Gunyakti, A., and Labidi, J. (2017). Utilization of flax (Linum usitatissimum) cellulose nanocrystals as reinforcing material for chitosan films. *Int. J. Biol. Macromol.* doi:10.1016/j.ijbiomac.2017.06.127.

(114) Müller, A., Ni, Z., Hessler, N., Wesarg, F., Müller, F. A., Kralisch, D., et al. (2013). The biopolymer bacterial nanocellulose as drug delivery system: Investigation of drug loading and release using the model protein albumin. *J. Pharm. Sci.* doi:10.1002/jps.23385.

(115) Murray, B. S., and Phisarnchananan, N. (2016). Whey protein microgel particles as stabilizers of waxy corn starch + locust bean gum water-in-water emulsions. *Food Hydrocoll.* doi:10.1016/j.foodhyd.2015.11.032.

(116) Nascimento, P., Marim, R., Carvalho, G., and Mali, S. (2016). Nanocellulose produced from rice hulls and its effect on the properties of biodegradable starch films. *Mater. Res.* doi:10.1590/1980-5373-MR-2015-0423.

(117) Neffe-Skocińska, K., Sionek, B., Ścibisz, I., and Kołożyn-Krajewska, D. (2017). Acid contents and the effect of fermentation condition of Kombucha tea beverages on physicochemical, microbiological and sensory properties. *CyTA—J. Food.* doi:10.1080/1947 6337.2017.1321588.

(118) Nguyen, V. T., Gidley, M. J., and Dykes, G. A. (2008). Potential of a nisin-containing bacterial cellulose film to inhibit Listeria monocytogenes on processed meats. *Food Microbiol.* doi:10.1016/j.fm.2008.01.004.

(119) Nicol, F., His, I., Jauneau, A., Vernhettes, S., Canut, H., and Höfte, H. (1998). A plasma membrane-bound putative endo-1,4-β-D-glucanase is required for normal wall assembly and cell elongation in Arabidopsis. *EMBO J.* doi:10.1093/emboj/17.19.5563.

(120) Okiyama, A., Motoki, M., and Yamanaka, S. (1993). Bacterial cellulose IV. Application to processed foods. *Top. Catal.* doi:10.1016/S0268-005X(09)80074-X.

(121) Oun, A. A., and Rhim, J. W. (2016). Isolation of cellulose nanocrystals from grain straws and their use for the preparation of carboxymethyl cellulose-based nanocomposite films. *Carbohydr. Polym.* doi:10.1016/j.carbpol.2016.05.020.

(122) Paximada, P., Kanavou, E., and Mandala, I. G. (2020). Effect of rheological and structural properties of bacterial cellulose fibrils and whey protein biocomposites on electrosprayed food-grade particles. *Carbohydr. Polym.* doi:10.1016/j.carbpol.2020.116319.

(123) Paximada, P., Koutinas, A. A., Scholten, E., and Mandala, I. G. (2016). Effect of bacterial cellulose addition on physical properties of WPI emulsions. Comparison with common thickeners. *Food Hydrocoll.* doi:10.1016/j.foodhyd.2015.10.014.

(124) Pear, J. R., Kawagoe, Y., Schreckengost, W. E., Delmer, D. P., and Stalker, D. M. (1996). Higher plants contain homologs of the bacterial celA genes encoding the catalytic subunit of cellulose synthase. *Proc. Natl. Acad. Sci. U. S. A.* 93, 12637–42. doi:10.1073/pnas.93.22.12637.

(125) Pecoraro, É., Manzani, D., Messaddeq, Y., and Ribeiro, S. J. L. (2007). Bacterial cellulose from glucanacetobacter xylinus: Preparation, properties and applications. In *Monomers, Polymers and Composites from Renewable Resources.* doi:10.1016/B978-0-08-045316-3.00017-X.

(126) Pereira, P. H. F., Waldron, K. W., Wilson, D. R., Cunha, A. P., Brito, E. S. d., Rodrigues, T. H. S., et al. (2017). Wheat straw hemicelluloses added with cellulose nanocrystals and citric acid. Effect on film physical properties. *Carbohydr. Polym.* doi:10.1016/j. carbpol.2017.02.019.

(127) Persson, S., Paredez, A., Carroll, A., Palsdottir, H., Doblin, M., Poindexter, P., et al. (2007). Genetic evidence for three unique components in primary cell-wall cellulose synthase complexes in Arabidopsis. *Proc. Natl. Acad. Sci. U. S. A.* doi:10.1073/pnas.0706592104.

(128) Phisalaphong, M., and Chiaoprakobkij, N. (2016). Applications and products-nata de coco, in *Bacterial NanoCellulose: A Sophisticated Multifunctional Material.* Boca Raton: CRC Press.

(129) Piadozo, M. E. S. (2016). Nata de coco industry in the philippines. In *Bacterial Nanocellulose.* Elsevier, 215–229. doi:10.1016/B978-0-444-63458-0.00013-5.

(130) Pinďáková, L., Kašpárková, V., and Bordes, R. (2019). Role of protein-cellulose nanocrystal interactions in the stabilization of emulsion. *J. Colloid Interface Sci.* doi:10.1016/j.jcis.2019.09.002.

(131) Pitol-Filho, L., Kokol, V., and Voncina, B. (2013). Synthesis and characterization of ethyl cellulose microcapsules containing model active ingredients. *Macromol. Symp.* doi:10.1002/masy.201350605.

(132) Popper, Z. A. (2008). Evolution and diversity of green plant cell walls. *Curr. Opin. Plant Biol.* doi:10.1016/j.pbi.2008.02.012.

(133) Prasertmanakit, S., Praphairaksit, N., Chiangthong, W., and Muangsin, N. (2009). Ethyl cellulose microcapsules for protecting and controlled release of folic acid. *AAPS PharmSciTech.* doi:10.1208/s12249-009-9305-3.

(134) Purwadaria, T., Gunawan, L., and Wydia Gunawan, A. (2010). The production of nata colored by monascus purpureus J1 pigments as functional food. *Microbiol. Indones.* doi:10.5454/mi.4.1.2.

(135) Qazanfarzadeh, Z., and Kadivar, M. (2016). Properties of whey protein isolate nanocomposite films reinforced with nanocellulose isolated from oat husk. *Int. J. Biol. Macromol.* doi:10.1016/j.ijbiomac.2016.06.077.

(136) Reddy, J. P., and Rhim, J. W. (2014). Characterization of bionanocomposite films prepared with agar and paper-mulberry pulp nanocellulose. *Carbohydr. Polym.* doi:10.1016/j.carbpol.2014.04.056.

(137) Riaz, Q. U. A., and Masud, T. (2013). Recent trends and applications of encapsulating materials for probiotic stability. *Crit. Rev. Food Sci. Nutr.* doi:10.1080/10408398.2010.524953.

(138) Richter, S., Voß, U., and Jürgens, G. (2009). Post-Golgi traffic in plants. *Traffic.* doi:10.1111/j.1600-0854.2009.00916.x.

(139) Robson, A. A. (2011). Food nanotechnology: Water is the key to lowering the energy density of processed foods. *Nutr. Health.* doi:10.1177/026010601102000406.

(140) Sajjad, W., He, F., Ullah, M. W., Ikram, M., Shah, S. M., Khan, R., et al. (2020). Fabrication of bacterial cellulose-curcumin nanocomposite as a novel dressing for partial thickness skin burn. *Front. Bioeng. Biotechnol.* 8. doi:10.3389/fbioe.2020.553037.

(141) Salehudin, M. H., Salleh, E., Mamat, S. N. H., and Muhamad, I. I. (2014). Starch based active packaging film reinforced with empty fruit bunch (EFB) cellulose nanofiber. *Procedia Chem.* 9, 23–33. doi:10.1016/j.proche.2014.05.004.

(142) Saxena, I. M., and Brown, R. M. (2005). Cellulose biosynthesis: Current views and evolving concepts. *Ann. Bot.* 96, 9–21. doi:10.1093/aob/mci155.

(143) Saxena, I. M., Kudlicka, K., Okuda, K., and Brown, R. M. (1994). Characterization of genes in the cellulose-synthesizing operon (acs operon) of Acetobacter xylinum: Implications for cellulose crystallization. *J. Bacteriol.* 176, 5735–5752.

(144) Schneider, R., Hanak, T., Persson, S., and Voigt, C. A. (2016). Cellulose and callose synthesis and organization in focus, what's new? *Curr. Opin. Plant Biol.* doi:10.1016/j.pbi.2016.07.007.

(145) Schrecker, S. T., and Gostomski, P. A. (2005). Determining the water holding capacity of microbial cellulose. *Biotechnol. Lett.* 27, 1435–1438. doi:10.1007/s10529-005-1465-y.

(146) Seo, C., Lee, H. W., Suresh, A., Yang, J. W., Jung, J. K., and Kim, Y. C. (2014). Improvement of fermentative production of exopolysaccharides from Aureobasidium pullulans under various conditions. *Korean J. Chem. Eng.* 31, 1433–1437. doi:10.1007/s11814-014-0064-9.

(147) Sethaphong, L., Haigler, C. H., Kubicki, J. D., Zimmer, J., Bonetta, D., DeBolt, S., et al. (2013). Tertiary model of a plant cellulose synthase. *Proc. Natl. Acad. Sci. U. S. A.* doi:10.1073/pnas.1301027110.

(148) Shah, N., Ul-Islam, M., Khattak, W. A., and Park, J. K. (2013). Overview of bacterial cellulose composites: A multipurpose advanced material. *Carbohydr. Polym.* 98, 1585–1598. doi:10.1016/j.carbpol.2013.08.018.

(149) Sharma, G., Sharma, S., Kumar, A., Al-Muhtaseb, A. H., Naushad, M., Ghfar, A. A., et al. (2018). Guar gum and its composites as potential materials for diverse applications: A review. *Carbohydr. Polym.* doi:10.1016/j.carbpol.2018.07.053.

(150) Shoda, M., and Sugano, Y. (2005). Recent advances in bacterial cellulose production. *Biotechnol. Bioprocess Eng.* 10, 1–8. doi:10.1007/bf02931175.

(151) Shoukat, A., Wahid, F., Khan, T., Siddique, M., Nasreen, S., Yang, G., et al. (2019). Titanium oxide-bacterial cellulose bioadsorbent for the removal of lead ions from aqueous solution. *Int. J. Biol. Macromol.* 129, 965–971. doi:10.1016/j.ijbiomac.2019.02.032.

(152) Silvério, H. A., Flauzino Neto, W. P., Dantas, N. O., and Pasquini, D. (2013). Extraction and characterization of cellulose nanocrystals from corncob for application as reinforcing agent in nanocomposites. *Ind. Crops Prod.* doi:10.1016/j.indcrop.2012.10.014.

(153) Singh, S., Gaikwad, K. K., and Lee, Y. S. (2018). Antimicrobial and antioxidant properties of polyvinyl alcohol bio composite films containing seaweed extracted cellulose nano-crystal and basil leaves extract. *Int. J. Biol. Macromol.* doi:10.1016/j. ijbiomac.2017.10.057.

(154) Slavutsky, A. M., and Bertuzzi, M. A. (2014). Water barrier properties of starch films reinforced with cellulose nanocrystals obtained from sugarcane bagasse. *Carbohydr. Polym.* doi:10.1016/j.carbpol.2014.03.049.

(155) Sung, S. H., Chang, Y., and Han, J. (2017). Development of polylactic acid nanocomposite films reinforced with cellulose nanocrystals derived from coffee silverskin. *Carbohydr. Polym.* doi:10.1016/j.carbpol.2017.04.037.

(156) Tang, C., Chen, Y., Luo, J., Low, M. Y., Shi, Z., Tang, J., et al. (2019). Pickering emulsions stabilized by hydrophobically modified nanocellulose containing various structural characteristics. *Cellulose.* doi:10.1007/s10570-019-02648-x.

(157) Tang, W., Jia, S., Jia, Y., and Yang, H. (2010). The influence of fermentation conditions and post-treatment methods on porosity of bacterial cellulose membrane. *World J. Microbiol. Biotechnol.* doi:10.1007/s11274-009-0151-y.

(158) Tanskul, S., Amornthatree, K., and Jaturonlak, N. (2013). A new cellulose-producing bacterium, Rhodococcus sp. MI 2: Screening and optimization of culture conditions. *Carbohydr. Polym.* 92, 421–428. doi:10.1016/j.carbpol.2012.09.017.

(159) Thambiraj, S., and Ravi Shankaran, D. (2017). Preparation and physicochemical characterization of cellulose nanocrystals from industrial waste cotton. *Appl. Surf. Sci.* doi:10.1016/j.apsusc.2017.03.272.

(160) Tomé, L. C., Brandão, L., Mendes, A. M., Silvestre, A. J. D., Neto, C. P., Gandini, A., et al. (2010). Preparation and characterization of bacterial cellulose membranes with tailored surface and barrier properties. *Cellulose.* doi:10.1007/s10570-010-9457-z.

(161) Ul-Islam, M., Ahmad, F., Fatima, A., Shah, N., Yasir, S., Ahmad, M. W., et al. (2021). Ex situ synthesis and characterization of high strength multipurpose bacterial cellulose-aloe vera hydrogels. *Front. Bioeng. Biotechnol.* 9. doi:10.3389/fbioe.2021.601988.

(162) Ul-Islam, M., Ha, J. H., Khan, T., and Park, J. K. (2013). Effects of glucuronic acid oligomers on the production, structure and properties of bacterial cellulose. *Carbohydr. Polym.* 92, 360–366. doi:10.1016/j.carbpol.2012.09.060.

(163) Ul-Islam, M., Khan, S., Ullah, M. W., and Park, J. K. (2019a). Comparative study of plant and bacterial cellulose pellicles regenerated from dissolved states. *Int. J. Biol. Macromol.* 137, 247–252. doi:10.1016/j.ijbiomac.2019.06.232.

(164) Ul-Islam, M., Khan, T., and Park, J. K. (2012). Water holding and release properties of bacterial cellulose obtained by in situ and ex situ modification. *Carbohydr. Polym.* 88, 596–603. doi:10.1016/j.carbpol.2012.01.006.

(165) Ul-Islam, M., Subhan, F., Islam, S. U., Khan, S., Shah, N., Manan, S., et al. (2019b). Development of three-dimensional bacterial cellulose/chitosan scaffolds: Analysis of cell-scaffold interaction for potential application in the diagnosis of ovarian cancer. *Int. J. Biol. Macromol.* 137, 1050–1059. doi:10.1016/j.ijbiomac.2019.07.050.

(166) Ul-Islam, M., Ul-Islam, S., Yasir, S., Fatima, A., Ahmed, M. W., Lee, Y. S., et al. (2020a). Potential applications of bacterial cellulose in environmental and pharmaceutical sectors. *Curr. Pharm. Des.* 26, 5793–5806. doi:10.2174/138161282666620100816 5241.

(167) Ul-Islam, M., Ullah, M. W., Khan, S., Kamal, T., Ul-Islam, S., Shah, N., et al. (2016). Recent advancement in cellulose based nanocomposite for addressing environmental challenges. *Recent Pat. Nanotechnol.* 10, 169–180. doi:10.2174/1872210510666160429144916.

(168) Ul-Islam, M., Ullah, M. W., Khan, S., and Park, J. K. (2020b). Production of bacterial cellulose from alternative cheap and waste resources: A step for cost reduction with positive environmental aspects. *Korean J. Chem. Eng.* 37, 925–937. doi:10.1007/s11814-020-0524-3.

(169) Ul-Islam, M., Ullah, M. W., Khan, S., Shah, N., and Park, J. K. (2017). Strategies for cost-effective and enhanced production of bacterial cellulose. *Int. J. Biol. Macromol.* 102, 1166–1173. doi:10.1016/j.ijbiomac.2017.04.110.

(170) Ullah, H., Santos, H. A., and Khan, T. (2016). Applications of bacterial cellulose in food, cosmetics and drug delivery. *Cellulose.* doi:10.1007/s10570-016-0986-y.

(171) Ullah, M. W., Manan, S., Kiprono, S. J., Ul-Islam, M., and Yang, G. (2019). Synthesis, structure, and properties of bacterial cellulose. In *Nanocellulose* doi:10.1002/9783527807437.ch4.

(172) Ullah, M. W., Ul-Islam, M., Khan, S., Kim, Y., Jang, J. H., and Park, J. K. (2016). In situ synthesis of a bio-cellulose/titanium dioxide nanocomposite by using a cell-free system. *RSC Adv.* 6, 22424–22435. doi:10.1039/C5RA26704H.

(173) Ullah, M. W., Ul-Islam, M., Khan, S., Kim, Y., and Park, J. K. (2015). Innovative production of bio-cellulose using a cell-free system derived from a single cell line. *Carbohydr. Polym.* 132, 286–294. doi:10.1016/j.carbpol.2015.06.037.

(174) Ullah, M. W., Ul-Islam, M., Khan, S., Kim, Y., and Park, J. K. (2016). Structural and physico-mechanical characterization of bio-cellulose produced by a cell-free system. *Carbohydr. Polym.* 136. doi:10.1016/j.carbpol.2015.10.010.

(175) Ullah, M. W., Ul Islam, M., Khan, S., Shah, N., and Park, J. K. (2017). Recent advancements in bioreactions of cellular and cell-free systems: A study of bacterial cellulose as a model. *Korean J. Chem. Eng.* 34, 1591–1599. doi:10.1007/s11814-017-0121-2.

(176) Vain, T., Crowell, E. F., Timpano, H., Biot, E., Desprez, T., Mansoori, N., et al. (2014). The cellulase KORRIGAN is part of the cellulose synthase complex. *Plant Physiol.* doi:10.1104/pp.114.241216.

(177) Velásquez-Riaño, M., and Bojacá, V. (2017). Production of bacterial cellulose from alternative low-cost substrates. *Cellulose.* doi:10.1007/s10570-017-1309-7.

(178) Vieira, D. (2015). Obtenção e caracterização de nanocelulose a partir de fibras de Chorisia speciosa St. Hil. 2015. 60 f. Trabalho de conclusão de curso (Bacharelado - Engenharia de Materiais) - Universidade Estadual Paulista, Faculdade de Engenharia de Guaratinguetá. Available at: http://hdl.handle.net/11449/139202.

(179) Villarreal-Soto, S. A., Beaufort, S., Bouajila, J., Souchard, J. P., and Taillandier, P. (2018). Understanding kombucha tea fermentation: A review. *J. Food Sci.* doi:10.1111/1750-3841.14068.

(180) Wang, J. S., Wang, A. B., Zang, X. P., Tan, L., Ge, Y., Lin, X. E., et al. (2018). Physical and oxidative stability of functional avocado oil high internal phase emulsions collaborative formulated using citrus nanofibers and tannic acid. *Food Hydrocoll.* doi:10.1016/j.foodhyd.2018.02.013.

(181) Wang, L. F., Hu, S., Ullah, M. W., Li, X., Shi, Z., and Yang, G. (2020). Enhanced cell proliferation by electrical stimulation based on electroactive regenerated bacterial cellulose hydrogels. *Carbohydr. Polym.* 249, 116829. doi:https://doi.org/10.1016/j.carbpol.2020.116829.

(182) Wang, L. F., Shankar, S., and Rhim, J. W. (2017). Properties of alginate-based films reinforced with cellulose fibers and cellulose nanowhiskers isolated from mulberry pulp. *Food Hydrocoll.* doi:10.1016/j.foodhyd.2016.08.041.

(183) Wang, W., Du, G., Li, C., Zhang, H., Long, Y., and Ni, Y. (2016). Preparation of cellulose nanocrystals from asparagus (Asparagus officinalis L.) and their applications to palm oil/water Pickering emulsion. *Carbohydr. Polym.* doi:10.1016/j.carbpol.2016.05.052.

(184) Wang, X., Xie, Y., Ge, H., Chen, L., Wang, J., Zhang, S., et al. (2018a). Physical properties and antioxidant capacity of chitosan/epigallocatechin-3-gallate films reinforced with nano-bacterial cellulose. *Carbohydr. Polym.* 179, 207–220. doi:10.1016/j.carbpol.2017.09.087.

(185) Wang, X., Xie, Y., Ge, H., Chen, L., Wang, J., Zhang, S., et al. (2018b). Physical properties and antioxidant capacity of chitosan/epigallocatechin-3-gallate films reinforced with nano-bacterial cellulose. *Carbohydr. Polym.* doi:10.1016/j.carbpol.2017.09.087.

(186) Wonganu, B., and Kongruang, S. (2010). Red bacterial cellulose production by fermentation of monascus purpureus. ICCCE 2010–2010 International Conference on Chemistry and Chemical Engineering, Proceedings. doi:10.1109/ICCCENG.2010.5560376.

(187) Worden, N., Park, E., and Drakakaki, G. (2012). Trans-golgi network-an intersection of trafficking cell wall components. *J. Integr. Plant Biol.* doi:10.1111/j.1744-7909.2012.01179.x.

(188) Wu, J., and Ma, G. H. (2016). Recent studies of Pickering emulsions: Particles make the difference. *Small.* doi:10.1002/smll.201600877.

(189) Wu, W., Xie, J., and Zhang, H. (2016). Dietary fibers influence the intestinal SCFAs and plasma metabolites profiling in growing pigs. *Food Funct.* doi:10.1039/c6fo01406b.

(190) Xiao, J., Li, Y., and Huang, Q. (2016). Recent advances on food-grade particles stabilized Pickering emulsions: Fabrication, characterization and research trends. *Trends Food Sci. Technol.* doi:10.1016/j.tifs.2016.05.010.

(191) Yang, G., Xie, J., Deng, Y., Bian, Y., and Hong, F. (2012). Hydrothermal synthesis of bacterial cellulose/AgNPs composite: A "green" route for antibacterial application. *Carbohydr. Polym.* doi:10.1016/j.carbpol.2011.11.017.

(192) Zhai, X., Lin, D., Zhao, Y., Li, W., and Yang, X. (2018a). Effects of dietary fiber supplementation on fatty acid metabolism and intestinal microbiota diversity in C57BL/6J mice fed with a high-fat diet. *J. Agric. Food Chem.* doi:10.1021/acs.jafc.8b05036.

(193) Zhai, X., Lin, D., Zhao, Y., Li, W., and Yang, X. (2018b). Enhanced anti-obesity effects of bacterial cellulose combined with konjac glucomannan in high-fat diet-fed C57BL/6J mice. *Food Funct.* doi:10.1039/c8fo01211c.

(194) Zhai, X., Lin, D., Zhao, Y., Li, W., and Yang, X. (2018c). Bacterial cellulose relieves diphenoxylate-induced constipation in Rats. *J. Agric. Food Chem.* doi:10.1021/acs.jafc.8b00385.

(195) Zhang, H., Chen, Y., Wang, S., Ma, L., Yu, Y., Dai, H., et al. (2020). Extraction and comparison of cellulose nanocrystals from lemon (Citrus limon) seeds using sulfuric acid hydrolysis and oxidation methods. *Carbohydr. Polym.* doi:10.1016/j.carbpol.2020.116180.

(196) Zhang, H., Jung, J., and Zhao, Y. (2017). Preparation and characterization of cellulose nanocrystals films incorporated with essential oil loaded β-chitosan beads. *Food Hydrocoll.* doi:10.1016/j.foodhyd.2017.01.029.

(197) Zhao, Q., Zaaboul, F., Liu, Y., and Li, J. (2020). Recent advances on protein-based Pickering high internal phase emulsions (Pickering HIPEs): Fabrication, characterization, and applications. *Compr. Rev. Food Sci. Food Saf.* doi:10.1111/1541-4337.12570.

(198) Zhou, Y., Sun, S., Bei, W., Zahi, M. R., Yuan, Q., and Liang, H. (2018). Preparation and antimicrobial activity of oregano essential oil Pickering emulsion stabilized by cellulose nanocrystals. *Int. J. Biol. Macromol.* doi:10.1016/j.ijbiomac.2018.01.102.

7 The Expanding Role of Bacterial Cellulose in Addressing Environmental Challenges

Ajmal Shahzad, Sehrish Manan, Jawad Ali,
Mazhar Ul-Islam, Muhammad Wajid Ullah,
and Guang Yang

CONTENTS

DOI: 10.1201/9781003118756-7

7.1 INTRODUCTION

7.1.1 ENVIRONMENTAL POLLUTION

The earth's natural environment is composed of several interlinking systems, including land, air, water, and living organisms, that make this planet suitable for life. In the present era of industrialization, where the quality of human life is being greatly improved by advanced and diversified technologies, anthropogenic activities are continuously deteriorating the earth's natural environment (e.g., lithosphere, biosphere, hydrosphere, and atmosphere) at an alarming pace (1, 2). Consequently, life on earth is facing serious threats owing to the increasing levels of different types of pollution (e.g., air, water, noise) and emerging contaminants. Many plant and animal species have already become extinct, while several others are on the edge of extinction because of urbanization, pollution, and vanishing natural faunal/floral habitats (3). The terrestrial environment comprising terrestrial and aquatic ecosystems is highly vulnerable to environmental pollution because of its prime biotic systems.

Over time, the burden of numerous emerging contaminants, of both organic and inorganic nature, is increasing and adversely affecting humans and other living biota. Generally, organic contaminants are highly toxic or carcinogenic and have the potential to cause serious problems to human health, such as cancers, physical birth defects, and mental disorders (4, 5). Additionally, such compounds are non-biodegradable and accumulate in the environment as a result of biomagnification, thus increasing the risk of environmental deterioration. The major sources of different organic contaminants include pharmaceuticals, personal care products, endocrine-disrupting compounds, industrial chemicals, and urban and agricultural runoff (6, 7). Toxic and hazardous aquatic pollutants enter the environment by different means. For instance, aquatic pollutants enter the surface and subsurface water matrices directly either from the pollution sources (e.g., the discharge of toxic effluents from industries) or from the wastewater facilities without any adequate treatment (8). Pollutants may also enter the aquatic bodies indirectly, such as from agricultural activities like the disposal of pesticides and fertilizers (Figure 7.1). These pollutants are further transported and distributed to the biosphere via the hydrological cycle and food chain and may produce secondary pollutants, which can be injurious to aquatic biota. Generally, water-soluble contaminants such as heavy metals, organic dyes, phenol and its derivatives, inorganic anions, and pesticides are transported conveniently and cause serious issues to the natural environment and especially to human and animal health.

It is realized that the irreversible rise in population, urbanization, and industrialization has considerably increased environmental pollution and deteriorated the quality of air, water, and soil. Consequently, humans are facing serious health threats, and fauna are losing their natural habitats. To combat or minimize such threats, various pollution control measures have been implemented worldwide, over the past few decades, under environmental laws and obligations. The emerging level of environmental contamination needs to be minimized by using suitable remedial measures. Generally, the appropriate pollutant treatment technologies are selected with respect to the potential contaminants and their physicochemical properties, such as their

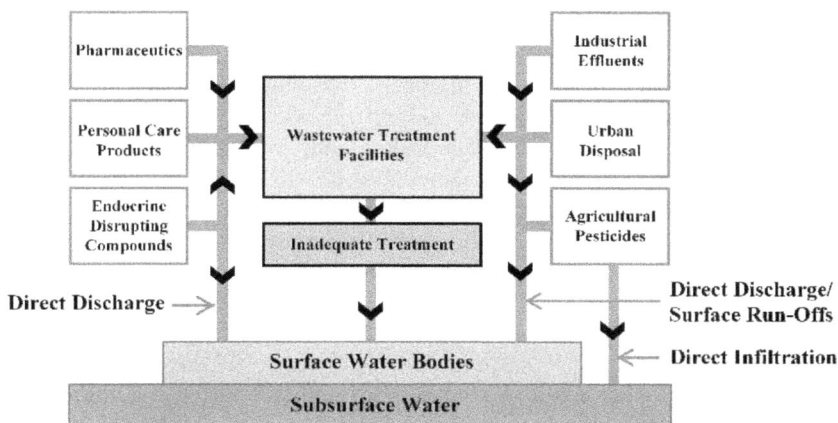

FIGURE 7.1 Sources and routes of aquatic contamination.

composition and concentration. Moreover, technologies used for pollutant removal or treatment should be efficient and economically feasible. Treatment technologies that recover energy and use it efficiently, recover solid reusable matter, and reuse resources (which are all considered green environmental remedial approaches) are highly recommended for effective pollutant treatment. In this regard, different kinds of chemical or biological materials are used for the removal or degradation of a broad spectrum of toxic contaminants. For instance, different kinds of carbon materials such as carbon nanotubes (CNTs), biochar, and graphene, as well as metal-based adsorbents and catalysts, are widely in practice to remediate a variety of toxic pollutants to protect the environment (9–11). Similarly, different microorganisms, specifically bacteria and fungi, are involved in the biological degradation of chemical-oriented pollutants through different metabolic activities carried out by such living organisms (12–14). It is worth mentioning that biotechnology is one of the clean integrated platforms for the conversion of contaminants into less toxic end products (15). Similarly, biological materials such as cellulose nanocrystals (CNCs), cellulose nanofibrils (CNFs), microfibrillated cellulose (MFC), and bacterial nanocellulose (BNCs) (16–19) are regarded as environmentally benign, inexpensive, and efficient materials for environmental applications. Recently, worldwide interest in using cellulosic materials has grown in various sectors. The cellulosic materials are abundantly available, are lightweight, possess high strength and stiffness, and biodegrade easily (20). Mostly, cellulosic materials are obtained from plants (21–23), but they are also produced by a variety of microorganisms such as bacteria, fungi, and algae (22), as well as synthesized enzymatically, such as by cell-free enzyme systems (24, 25).

7.1.2 BACTERIAL CELLULOSE

A variety of microbial species, such as bacteria, fungi, and a few algae, produce cellulosic materials through bioconversion of sugars or other carbon substrates. Such cellulosic materials are termed microbial cellulose but are frequently known as bacterial cellulose (BC) or bacterial nanocellulose (BNC) (22, 26). BC is a natural biomaterial and represents the purest form of cellulose produced by the Gram-negative bacterium *Gluconacetobacter xylinum* (*Komagataeibacter xylinus*) using glucose as the common substrate (27–29). The chemical structure of BC is the same as of plant cellulose (i.e., $C_6H_{10}O_5)_n$; however, its nanoscaled fibrous morphology and physiological behavior differentiate it from plant cellulose (22, 30, 31). Glucose is primarily involved in BC production, whereas some other sugars, such as fructose, sucrose, and galactose, are also be used to produce BC (32).

7.1.2.1 History

BC was first discovered by A.J. Brown in 1886 from *G. xylinum* producing extracellular gelatinous fibers (33), which was later named "bacterial cellulose" (34). Brown employed the acetic acid–producing bacterium on the surface of coconut water media, which produced a gel (33). Until the mid- to late 20th century, BC was mainly used to produce coconut gel through the fermentation of coconut water by bacteria, commonly known as *nata* or *nata de coco* in the Philippines. The demand for this bacterium-produced *nata* as a food supplement with numerous health benefits broke many records in Asian markets, including Vietnam, Indonesia, Japan, and other Southeast Asian countries. The level of its popularity can be imagined through the example that 90% of the Philippine-manufactured *nata* was exported to Japan in 1993. Furthermore, the ease of BC-based *nata* production led to the establishment of many small industries, which ultimately contributed massively to the development of rural areas of many Asian countries (35). Later, different kinds of biological species, media, and equipment/strategies (including static, agitation, and shaking) were brought into practice to produce a variety of BC-based materials (36). To date, a number of microbial species have been discovered that have the potential to produce cost-effective and multidimensional polymeric materials such as BC. Currently, the attraction to BC has been tremendously developed owing to its purity, fibrous network, physicochemical and thermal properties, and nontoxicity. The cost-effective and environmental friendly BC is being researched significantly to substitute or improve the synthetic polymeric materials in different disciplines that pose numerous environmental hazards. The prominent disciplines in which BC offers extraordinary services owing to its diverse properties include biomedical, drug delivery, food processing, textile, environment, electronics, energy, agriculture, and several other daily use products (37–43). Further detailed discussion on the different applications of BC in various fields is provided in Section 7.4.

7.1.2.2 Structure and Properties

Although the chemical structure of BC is similar to that of plant cellulose (Figure 7.2a), it is different in morphological and physiological behavioral aspects, making it a unique and novel functional material (22, 30, 31, 44). It differs from plant cellulose in fibril size, arrangement, geometry, crystallinity, and overall structural morphology (Figure 7.2). Unlike plant cellulose, BC is the purest form of cellulose, as it does not contain lignin, hemicellulose, or pectin. During the typical BC production process, microbial cells such as *G. xylinum* produce cellulose I (ribbon-like polymer) and

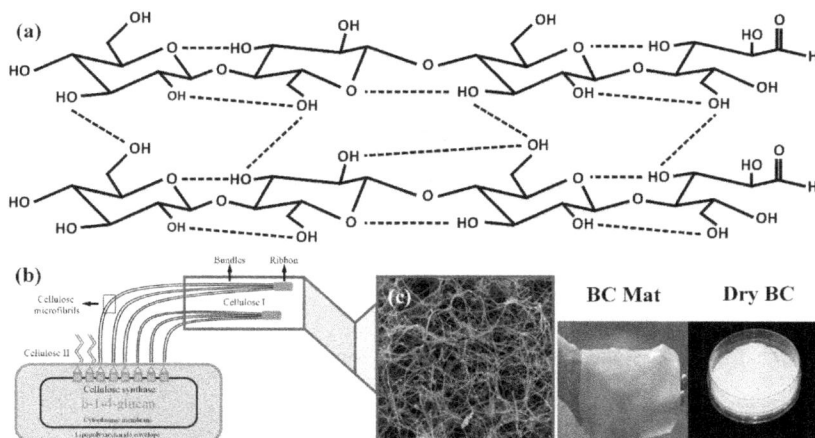

FIGURE 7.2 Chemical structure, inter- and intramolecular hydrogen bonding in BC (a), BC synthesis (production of cellulose microfibrils by *G. xylinum*) (b), and SEM micrographs of BC fibrous network, BC mat, and dried powder (c).

cellulose II (thermodynamically stable polymer) as described in Figure 7.2b, which subsequently results in the formation of a three-dimensional fibrous network structure of BC (Figure 7.2c). During the synthesis process, the protofibrils of glucose chain are secreted across the bacterial cell wall and aggregate to form cellulose ribbons (45, 46). The ribbons further form the web-shaped network structure of BC with a highly porous matrix (47). This well-arranged structure forms a hydrogel sheet at the air-medium interface that contains high surface area and porosity, which can be further dried and powdered if required (Figure 7.2c). The emergence of abundant hydroxyl groups on the cellulose surface governs the hydrophilicity and biodegradability of BC and provides a base for further chemical modifications to be used for environmental and other advanced applications in different fields (15, 48, 49). Pristine BC has stunning properties such as abundant surface functional groups, high water-holding capacity, ultrafine fibrous structure (50 to 80 nm in width and 3 to 8 nm in thickness), excellent tensile strength, biodegradability, and eco-friendly nature. Further, as mentioned earlier, BC can be modified by using a variety of different materials that afterward shows novel physicochemical and mechanical properties. Thus, BC has received enormous attention over the past few decades owing to its stunning properties and applications in diverse fields (29). The common properties of BC are summarized in Figure 7.3.

7.1.2.3 Attractive Features of BC for Environmental Remediation

The rehabilitation of deteriorated environmental entities, such as by removing contaminants from wastewater, requires the use of different materials with unique structural, physicochemical, and mechanical properties. For instance, layered double hydroxide (LDH), a clayey material composed of charged layers and replaceable

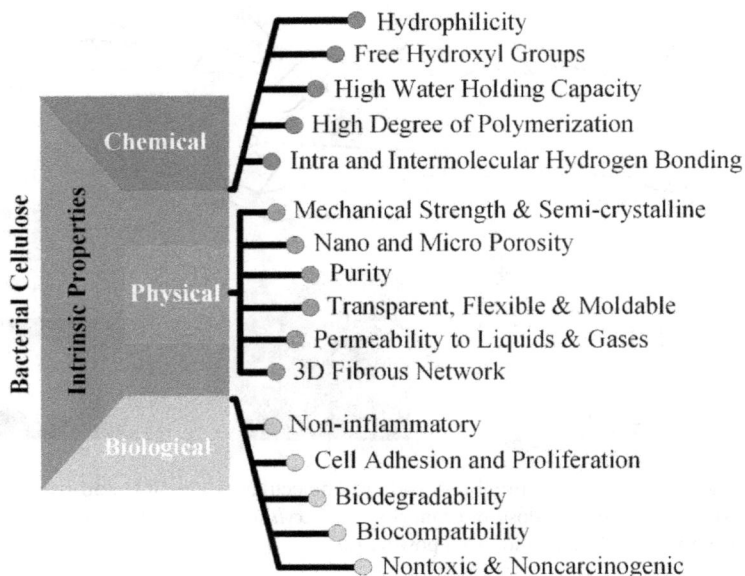

Hydrophilicity
Free Hydroxyl Groups
High Water Holding Capacity
High Degree of Polymerization
Intra and Intermolecular Hydrogen Bonding

Mechanical Strength & Semi-crystalline
Nano and Micro Porosity
Purity
Transparent, Flexible & Moldable
Permeability to Liquids & Gases
3D Fibrous Network

Non-inflammatory
Cell Adhesion and Proliferation
Biodegradability
Biocompatibility
Nontoxic & Noncarcinogenic

FIGURE 7.3   Intrinsic properties of BC.

anions, has been extensively used for different environmental applications. LDH offers characteristics such as tunability in structure, electrostatic interactions, high surface area, and high adsorption capacities for pollutants (9, 50). In a study, Jawad et al. modified the LDHs with Fe-MoS$_4$ for the efficient removal of heavy metals from water. The active MoS$_4$ was compacted inside the hydroxide layers due to which the characteristics of Fe-MoS$_4$ LDH were surprisingly modified, and it exhibited high surface area, enormous capacity, good reusability, and excellent selectivity for a variety of heavy metal ions (51).

In the course of environmental control and remediation, existing materials and techniques are somehow outdated and cause several environmental and health issues for living bodies. For instance, transition metal–based adsorbents and catalysts or nonbiodegradable materials are used to efficiently remove a variety of toxic contaminants. However, these materials themselves cause secondary pollution because of the release of toxic metal ions and nondegradable products that are harmful to the indigenous environment. In this respect, scientists are keen to develop and promote "green" and "biodegradable" materials for environmental remediation, which can reduce the impact of environmental contamination in both primary and secondary aspects of pollution control. From this perspective, BC is an appealing candidate owing to its environmentally friendly nature. During the current decade, BC-based nanomaterials have been studied frequently for their application in environmental remediation. Several studies, as reviewed by Ul-Islam et al., have presented the best pollutant removal outcomes and endorsed such cellulosic materials as the best choice for environmental protection and rehabilitation, as these consume less energy and have inconvertible ability to harmful secondary pollutants or products (19).

As described earlier, BC's potential is due to several stunning characteristics that make it quite feasible for environmental applications, e.g., adsorption, filtration, or degradation of toxic compounds. Other characteristics of BC, such as its nanofibrous and highly porous structure and higher water absorption and holding capacity, could be beneficial for the development of air and water purification or treatment technologies. The properties of BC such as high surface area, purity, high degree of polymerization, superior tensile strength, good biocompatibility, resistance to chemical and heat shock, easy sterilization, selective porosity, and renewable properties, nontoxic nature, environmental benignity, and facile modification ability (15) make BC an attractive candidate for the development of such technologies. Importantly, considering its highly hydrophilic nature, BC can readily interact with water as well as other polar molecules (52). Because of this feature, BC could be an ideal choice in water treatment and the removal of organic pollutants, which can potentially take up several polar molecules (pollutants) due to the presence of hydrogen bonding in its innate structure. Additionally, due to the abundance of surface hydroxyl groups on BC, a variety of organic and inorganic contaminants could interact with BC during the treatment of contaminated waters. Another great attraction of BC is its modification for obtaining synergistically reactive, stable, and environmentally friendly BC-assisted composite materials for the removal of contaminants. For instance, Taha et al. functionalized cellulose acetate with NH_2 and used it for the removal of chromium-VI (Cr-VI) (53). Similarly, Liu et al. modified BC with an isolated soy protein for capturing particulate matter (PMs) from the air (54).

7.1.3 Chapter Summary

The current chapter introduces BC and its characteristics in detail. Primarily, BC is described here as a revolutionary biomaterial that could be applied to remove and degrade a variety of contaminants from air, soil, and water. Further, the role of pristine BC and its composites is critically evaluated, and their prospective applications in environmental rehabilitation are discussed. The fundamentals and salient features of BC, described in this chapter, could help environmental scientists/engineers in modifying BC and provide insights to the modification mechanisms for achieving the desired goals. It is also believed that the contents covered in this chapter could open new avenues to develop state-of-the-art, cost-effective, and environmentally benign BC-assisted materials in the form of adsorbents or catalysts to deal with the different toxic organic contaminants.

7.2 BC PRODUCTION AND WASTE MANAGEMENT

BC is a fermentation product of *Gluconacetobacter xylinus*, an acetic acid bacterial species (55). BC is produced through static or submerged fermentation strategies, where sugar is utilized as the bacterial culture medium. A suitable culture medium and growth conditions are considered crucial for the economical production of BC. Various types of culture media are currently available that contain essential nutrients for microbes to utilize them by different means (degradation or polymerization) and produce BC. A typical culture medium contains essential nutrients for microbes such as carbon, oxygen, hydrogen, and nitrogen, as well as minor elements like iron and zinc, some vitamins, and hormones under special conditions, for the biosynthesis of various metabolites. Generally, production cost is one of the major hurdles in the industrial-scale production of BC. The cost of the culture medium defines the overall pricing, thus necessitating the exploration of alternative cost-effective media for the commercial and sustainable production of BC. The commonly used media for BC production contain chemicals like sugar (glucose, sucrose, fructose, mannitol, and arabitol) as carbon source, complex nitrogen, and vitamin sources such as yeast extract and polypeptone; thus, utilizing wastes from different industries containing such nutrients could be an effective approach for cost-effective production of BC (15, 32, 56).

Nowadays, the generation of industrial wastes and their handling for safe disposal has become a challenging global concern with respect to environmental safety. In this regard, several strategies have been introduced; highly appealing among them is the utilization of industrial waste as a raw material for other industries. By following this strategy, not only the cost of industrial productions could be minimized but it would also protect the environment from massive destruction by reusing or recycling the industrial wastes. Due to the advancement in biotechnologies, BC could be cost-effectively produced from industrial waste, which is a renewable source. Various sources of industrial wastes that could potentially be used for BC production are categorized and listed in Table 7.1.

7.3 MODIFICATIONS IN BC FROM AN ENVIRONMENTAL PERSPECTIVE

Despite the advantages of BC, a few of its deficiencies cannot be ignored. For example, pristine BC is nonconductive (i.e., it does not support electron transfer) and lacks magnetic properties, which are important characteristics for materials used for environmental applications. These limitations impair the direct use of BC in the environmental sector. Fortunately, BC is highly tunable and can be modified using a variety of functional groups and even other active materials, such as polymers and nanomaterials, to develop composites or hybrid materials with different characteristics.

To make BC more attractive and versatile, different kinds of nanomaterials can be incorporated into BC and form composites or hybrids for diverse applications in environmental remediation. In this way, the BC-based composites or hybrids are used to exhibit novel properties and synergistic efficiencies for the removal of contaminants from the air, soil, and water. For instance, different kinds of metallic nanoparticles (NPs) such as Fe^0, Cu^0, TiO_2, and MnO_2 and bimetallic NPs such as Fe/Pd and Fe/Ni are well known for their ability to degrade gaseous/water pollutants. Such nanoparticles are famous for their large surface area and high reactivity (71–73). Although the pollutant removal performance of these inorganic NPs is remarkable, the problems of their separation from aqueous solutions after reaction limit their applications. Therefore, these inorganic NPs are immobilized on different supports like matrix of chitosan, cellulose acetate, BC, and others. For example, Zhu et al. immobilized

TABLE 7.1
Agro-Industrial Wastes Utilized as a Feedstock for the Production of BC

Agro-industrial wastes	Additional nutrients	Microbial strain	Maximum BC production	Ref.
Waste as a complex medium without additional nutrients				
Citrus peels (lemon, mandarin, orange, and grapefruit)	---	*Komagataeibacter hansenii* GA2016	3.92 BC/ 100 g peel	(57)
Sugarcane juice and pineapple residues	---	*Gluconacetobacter medellinensis*	3.24 g/L	(58)
Discarded waste durian shell	---	*G. xylinus* CH001	2.67 g/L	(59)
Waste as a carbon source with additional nutrients				
Enzymatic hydrolysate of wheat straw	Other components are the same as of HS medium	*G. xylinus* ATCC 23770	8.3 g/L	(60)
Coffee cherry husk	Urea and corn steep liquor	*G. hansenii* UAC09	8.2 g/L	(39)
Juice samples watermelon, pineapple, and pawpaw	Other components are the same as that of HS medium	*Gluconacetobacter pasteurianus* PW1	7.7 g/L	(61)
Cashew tree exudates	Other components are the same as of HS medium	*Komagataeibacter rhaeticus*	6.0 g/L	(62)
Extracted data syrup	Other components are the same as of HS medium	*G. xylinus* 0416 MARDI	5.8 g/L	(63)
Pineapple waste medium and pawpaw waste medium	Other components are the same as of HS medium	*G. pasteurianus PW1*	3.9 g/L	(64)
Orange peel fluid and orange peel hydrolysate	Acetate buffer, peptone, and yeast extract	*G. xylinus* BCRC 12334	3.40 g/L	(65)
Cheap agricultural product konjac powder	Yeast extract and tryptone	*G. aceti* ATCC 23770	2.12 g/L	(66)
Sago by-product	Other components are the same as of HS medium	*Beijerinkia fluminensis* WAUPM53 and *G. xylinus* 0416 (reference strain)	0.47 g/L and 1.55 g/L for the reference strain	(61)

(Continued)

Table 7.1 (Continued)

Agro-industrial wastes	Additional nutrients	Microbial strain	Maximum BC production	Ref.
Grape skins aqueous extract, cheese whey, crude glycerol, and sulfite pulping liquor	Organic or inorganic nitrogen	*Gluconacetobacter sacchari*	0.1 g/L	(67)
Waste as a nitrogen source				
Pineapple peel and sugar cane juice	Glucose, fructose, and sucrose	*Gluconacetobacter swingsii*	2.8 g/L	(68)
Others				
Poor-quality apple residues in combination with glycerol	Apple glucose equivalents, glycerol, ammonium sulfate, and citric acid	*G. xylinus* DSMZ-2004	8.6 g/L	(69)
Pineapple and watermelon peels	Sucrose, ammonium sulfate, and cycloheximide	*G. hansenii* MCM B-967	125 g/L (on a wet weight basis)	(70)

Source: Reproduced from Ul-Islam et al. 2020. (32)

bimetallic Cd/S on chitosan for the synergistically superior degradation of Congo red (CR) from water (74).

Similarly, metal organic frameworks (MOFs) are porous hybrid materials composed of metal ions and find applications in numerous disciplines, especially for the removal of gaseous or aquatic pollutants (75). Although MOFs acquire impressive physicochemical characteristics like high surface area, porosity, and thermal stability, they also have shortcomings; for example, their crystalline nature makes them brittle and fragile, which restricts their industrial applications. Recently, several studies have reported the incorporation of MOFs into BC to transform into a new functional material with novel characteristics for use in environmental remediation. For example, Zhu et al. developed a hybrid composite of MOFs-type MIL-100(Fe)@BC for efficient removal of arsenic (As) and Rhodamine B (RhB) from aqueous solution (76).

Besides the association of BC with inorganic NPs, BC is also frequently modified using different natural and organic materials for environmental remediation. For instance, Liu et al. modified BC using the isolated soy protein for efficient air filtration. The soy protein isolate (SPI) possesses several functional groups and strong intermolecular interactions that denaturalized to reduced form for further interaction with a variety of pollutants. A combination of SPI and BC ultimately demonstrated novel properties such as large surface area and abundant active sites for effective capturing of air pollutants (54). In another study, modification of BC with acrylic acid was shown to be useful for the removal of metallic ions such as heavy metals (77).

The nanocellulosic materials (e.g., BC) usually exhibit agglomeration of nanoparticles, thus halting the separation from effluents and recyclability, which restricts

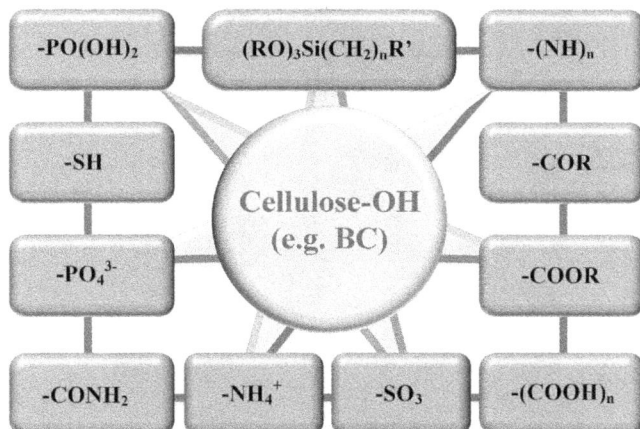

FIGURE 7.4 The major surface chemical functionalization of nanocellulose for application in water purification.

their applications in water treatment. These issues can be rationally addressed by grafting different functional groups on to the cellulose surface using techniques such as carboxylation, oxidation, sulfonation, esterification, phosphorylation, etherification, and amidation (Figure 7.4). Among these, phosphonate, sulfonate, and carboxyl groups can tune BC to make it suitable for selective uptake of contaminants like metallic cations (e.g., heavy metals) (78).

In summary, BC is an adventurous material having great ability to be used for prospective environmental applications. Several reports are available describing how BC and other natural biopolymer composites have been applied for the removal of contaminants from different environmental media, including industrial effluents, wastes, soils, and gases through adsorption and catalytic degradation (26).

7.4 ENVIRONMENTAL PERSPECTIVE OF USING BC IN DIFFERENT INDUSTRIES

The impact of industrialization on natural resources and the environment is increasing at a rapid rate, thus legislation regarding eco-sustainability is mandatory to obey. Eco-friendly and sustainable development could be achieved by considering three factors: (1) use natural resources sensibly to preserve them for coming generation, (2) explore new renewable and cost-effective resources to achieve industrial production goals, and (3) develop and use improved and latest technologies for industrial processes that produce minimum or no waste to protect the environment and its ecosystems. Among these eco-sustainability postulates, exploring new materials needs to be highly considered in this era of industrialization not only to reduce the utilization of natural resources but also to save environmental entities. Abundantly available materials such as BC, which can be cost-effectively synthesized even from wastes

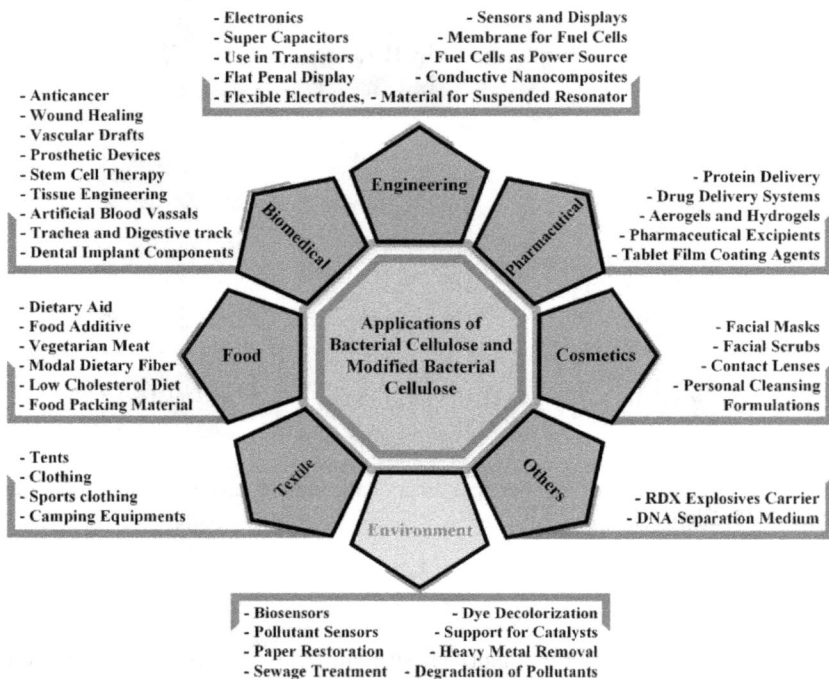

- Anticancer
- Wound Healing
- Vascular Drafts
- Prosthetic Devices
- Stem Cell Therapy
- Tissue Engineering
- Artificial Blood Vassals
- Trachea and Digestive track
- Dental Implant Components

- Electronics
- Super Capacitors
- Use in Transistors
- Flat Penal Display
- Flexible Electrodes,

- Sensors and Displays
- Membrane for Fuel Cells
- Fuel Cells as Power Source
- Conductive Nanocomposites
- Material for Suspended Resonator

- Protein Delivery
- Drug Delivery Systems
- Aerogels and Hydrogels
- Pharmaceutical Excipients
- Tablet Film Coating Agents

- Dietary Aid
- Food Additive
- Vegetarian Meat
- Modal Dietary Fiber
- Low Cholesterol Diet
- Food Packing Material

- Facial Masks
- Facial Scrubs
- Contact Lenses
- Personal Cleansing
Formulations

- Tents
- Clothing
- Sports clothing
- Camping Equipments

- RDX Explosives Carrier
- DNA Separation Medium

Biomedical   Engineering   Pharmaceutical   Food   Applications of Bacterial Cellulose and Modified Bacterial Cellulose   Cosmetics   Textile   Others   Environment

- Biosensors
- Pollutant Sensors
- Paper Restoration
- Sewage Treatment

- Dye Decolorization
- Support for Catalysts
- Heavy Metal Removal
- Degradation of Pollutants

FIGURE 7.5 Schematic illustrations of applications of BC in different fields.

from different industries, are considered renewable resources. BC and its composites have replaced the old and conventional technologies in different disciplines and play a vital role in preserving both natural resources and the environment; they are thus considered "environmentally benign materials." The following section presents the role and versatility of BC in different disciplines owing to its facile synthesis and availability, robust properties, and eco-friendly nature. Figure 7.5 also presents the applications of BC in different fields.

As described previously, BC has remarkable advantages owing to its stunning physicochemical properties, due to which it has found applications in different fields. The following sections describe the potential of BC and its composites to replace the commonly used substances that otherwise cause various environmental hazards.

7.4.1 Pharmaceutical Industry

Different synthetic materials are routinely used in the pharmaceutical industry for different applications such as coatings, dressings, temporary grafts, and several others. Most of these materials are nondegradable in nature and cause environmental hazards when released into the environment after use. A replacement of these materials from natural sources that are biodegradable could largely contribute to minimizing the hazardous effect of wastes released from the pharmaceutical industries.

The practice in modern medicine demands material with advanced characteristics capable of managing temperature and pH, excess exudate, painless wound dressing, appropriate gas diffusion, cost effectiveness, and preventing infections. For instance, BC plays a revolutionary role in the pharmaceutical industry, especially in wound healing (79), antibacterial medicines (80), drug delivery systems (38), diagnostics (81), and tissue regeneration (35, 82, 83), both alone and in the form of composites. The BC membrane and chemically modified BC have been extensively used for drug delivery or controlling drug release. Antimicrobial agents and nonsteroidal anti-inflammatory drugs (NSAIDs) are the most frequent examples of drugs delivered through BC (84). Recently, BC has been used as a scaffold for different types of mammalian cell culturing and tissue engineering (83, 85, 86). Several *in vitro* and *in vivo* studies reveal the role of BC in bone, skin, muscle, and neuronal tissue engineering, chondrocytes growth, stem cell expansion, and differentiation, and more recently in melanoma drug screening (35, 87).

7.4.2 COSMETICS INDUSTRY

Cosmetics are products used for the beautification of the body by altering its appearance; they include products used for skin cleansing and whitening, mostly without affecting the natural functioning of body parts. According to recent studies, the number of cosmetic consumers is growing worldwide (especially in Asia) at an intense rate, which led to a 4.5% rise in the production of cosmetics annually over the last two decades. Overall, the world cosmetic market generated a revenue of US $134 billion in 2019 alone (88; Skin Care Products Market Size, Share & Trends Analysis Report, By Product (Face Cream, Body Lotion), By Region (North America, Central & South America, Europe, APAC, MEA), And Segment Forecasts, 2019–2025 2019). Despite such advantages, the use of cosmetic products has several side effects, mainly damage to human health, and they pollute the environment when washed away from the body due to the presence of several nondegradable compounds used in their manufacturing. Some cosmetic products, for example shampoos, creams, and makeup, contain hazardous or toxic substances (e.g., heavy metals) that cause several adverse health effects like illness, contact dermatitis, skin allergy, and cancer in severe cases (89). For example, a number of cosmetic products (e.g., skin masks and scrubs) are produced from petroleum derivatives that are nonrenewable and cause serious environmental issues. The nondegradable nature of these petroleum residues could risk the lives of many aquatic species through the bioaccumulation of microplastics (90). To combat such issues, continuous research efforts are being implemented to provide cosmetics to end-consumers without any harmful or toxic constituents, preventing diseases to humans and protecting the environment from toxic wastes.

In this regard, the use of BC brought revolutionary change to the cosmeceutical industry. BC has been widely used in *in vivo* skin applicators such as anti-aging agents, moisturizer skin masking, and skin masking with whitening effects because of its hydrophilic nature and excellent water retention ability (91). Chantereau et al. fabricated an ionic liquids-assisted noncytotoxic composite of BC and vitamin B (BC-ILs) and used it for rough dermal care. Herein, BC was used as the nontoxic bioplasticizer for topical delivery of vitamin B in skincare applications (92). Due to its environmental friendly and biodegradability, BC could potentially replace the organic nonbiodegradable plasticizers in the cosmetics industry and thus reduce the load of organic waste (originating from cosmetics) on the environment.

7.4.3 Food Industry

The food industry is top-ranked, being the essential entity of today's life. This industry is also under continuous evolution and introduces healthier, nutritious ingredients and processing of food supplements and additives by using advanced biotechnologies. BC has remarkable significance in food technologies in both the food manufacturing and environmental remedies related to the food industry. Traditionally, BC was used in making *nata de coco*, a fibrous food of Southeast Asia and well-reputed food supplement classified as "generally recognized as safe" (GRAS) by the Food and Drug Administration (FDA) since 1992 (93). A thick sheet of cellulose is washed, boiled, and cooked in sugar syrup to transform it into desserts, and it is used for making fruit cocktails and jellies. Due to the increase of low-calorie, low-fat, and no-cholesterol diets and as a vegetarian substitute for meat, BC could potentially meet today's consumer demand in the food market. In addition, BC has been explored for its use as a food additive (gelling, thickening, and stabilizing agent), as well as a substrate for the immobilization of cells and enzymes for various applications in the food industry, for example, to obtain high-yield food products.

Currently, BC is also used as extraordinarily water-resistant, biodegradable, and environmentally friendly food packaging and preserving material in food industries. BC is becoming a suitable choice for food packaging, which could ultimately minimize the use of nonbiodegradable plastic packaging (94).

7.4.4 Agriculture

Recently, different kinds of cellulosic materials have been introduced in the agricultural area, especially concerning the irrigation sector. The problem of excessive irrigation often leads to the death of seedlings. Although the electro-mechanical aids resolved many of the irrigation-related issues, the use of external resources (machines) may affect the cost and the local environmental conditions. To address such issues, scientists have introduced cellulose-based materials that are not only cost-effective but also reduce the need for electromechanical aids. The switching of conventional technologies to cellulose-based membrane/hydrogel materials in irrigation systems offers high water retention capacity and necessarily controlled release of water, which ultimately improves plant growth and yield (95). Hydrogel is an astonishing cross-linked natural biomaterial exhibiting all the aforementioned qualities; it is thus reported frequently for different agricultural applications. For example, Elbarbary et al. fabricated the polyacrylamide/sodium alginate hydrogel and applied it to a cornfield. It absorbed 200 times more than its dry mass and ultimately improved crop quality and yield up to 50% in sandy soil (96).

The application of fertilizers, herbicides, insecticides, and fungicides to crops is a definite requirement for crop safety and yield. However, the leaching of these essential agro-additives (chemicals) not only affects the crops in different ways but also adversely affects the environment, for example in soil and aquatic bodies through surface runoff and percolations. In this context, the use of cellulose-based hydrogel could be a blessing for both plants and the environment. The safe delivery of essential nutrients to the plants and the release of toxic herbicides or insecticides to the environment could be controlled through the application of such cellulosic biomaterials.

In some cases, materials for the development of hydrogel are nonbiodegradable, which themselves pollute the indigenous environment. This issue could be resolved by using BC with all the other super water-absorbent and nutrient-delivery characteristics and could improve crop quality and yield. Additionally, BC could potentially protect the environment by preventing the unwanted release of toxic metals and other agricultural wastes (18).

7.4.5 Bioelectronics

Recently, BC and its composites with a variety of materials have been extensively used in the fields of sensing, electricity, and energy storage (97, 98). Although pristine BC is electrically inactive, it can be modified with other electrically active materials such as transition metals, carbon allotropes like carbon nanotubes and graphene, and conductive polymers. The functionalized active materials could be transformed into flexible composites after incorporating them into BC without affecting their inherent electrical properties (99). There are a number of studies on developing BC-based conductive composites. For example, the composite of BC with cobalt-ferrite nanowires could be used in electronics as a promising anode for the development of Li-ion batteries (100, 101). Bai et al. fabricated a flexible composite comprised of graphene/carbon nanotube and BC (RGO/CNT/BC) that exhibited high conductivity and thus could be used in the development of supercapacitors (102). Trigona et al. harvested green energy by introducing a conducting composite of BC impregnated with ionic liquids that possessed a property like mechanoelectrical transduction. Due to this property, electric signals were generated during the mechanical deformation, which could thus be used in the development of wireless sensors and wearable and flexible electronics (103). In a recent study, Farooq et al. developed BC/phage-based electrochemical biosensors for rapid detection of *Staphylococcus aureus* in food samples. In their study, BC was surface modified with polyethyleneimine and impregnated with carboxylated MWCNTs for imparting positive charge for phage immobilization and electrical conductivity, respectively. The developed BC/c-MWCNTs-PEI-phage composite supported the immobilization of 11.7±1.2 phage particles·μm^{-2} and detected 5 CFU/mL *S. aureus* within 30 minutes (104).

The modified BC-based sensors are common in the biomedical domain for the detection of several drug molecules. However, from an environmental perspective, BC-based nanopaper optical sensors have been introduced for the detection of different aquatic pollutants such as heavy metals and other recalcitrant pollutants (105). From this discussion, it is clear that BC-based technologies are significantly contributing to the sustainability of nonrenewable natural resources and earth's ecosystems, as well as to environmental remediation and rehabilitation.

7.4.6 Renewable Energy

Energy is an inevitable demand for human life. Usually, a major proportion of worldwide energy production is based on nonrenewable resources like fossil fuels, which are depleting rapidly. Second, during energy production, myriad types of pollutants are produced that deteriorate the earth's natural environment. In this regard, sustainable development is mandatory for providing a healthier environment and natural resources to the forthcoming generations. Therefore, special efforts have been taken to search for renewable and cost-effective energy production sources. BC has astonishing potential in this respect: for example, the cellulose pellicle that originated from BC could be used as an alternative and inexpensive source of sugar sustainable fermentation into bioethanol (106), which could subsequently be used as fuel for energy production. Compared to that of plant cellulose, because BC does not contain lignin and hemicellulose, its purification does not require energy, which reduces the production cost of bioethanol. Additionally, by using industrial wastes for BC production, bioethanol production from BC could be declared the "green production" approach.

7.5 ENVIRONMENTAL APPLICATIONS

Worldwide environmental regulations are strengthening the incentive to exploit novel and state-of-the-art sustainable technologies to protect the environment and prevent the depletion of natural resources. The demand for reformulating and shifting existing technologies and strategies toward eco-friendly or "green" technologies to meeting regulatory changes and pursue more sustainable practices is currently increasing. Concerning this, BC-based renewable technologies and strategies are receiving significant attention; they can potentially replace conventional fossil fuel resources considerably in near future. The innovative BC-based materials are offering vital roles in the environmental domain following different strategies illustrated in Figure 7.6 and detailed in the following sections.

7.5.1 Environmental Management

Prior to implementing a new strategy or material, these need to be evaluated for their environmental impact through their life cycle assessment (LCA). Conducting an LCA ensures that the newly developed strategy or material does not have a greater environmental or socioeconomic impact than the technology to be replaced. Typically, during the production, consumption, and ultimately the disposal of BC, each process requires specific natural inputs and generates different types of contaminants that are potentially harmful to the environment. Thus, LCA covers all these aforementioned aspects and ensures the suitability of the newly developed strategy or material to be employed for the sustainable industrial revolution. Since the last decade, BC has been commercially exploited in a variety of disciplines, as described earlier. Therefore, it is considered eco-friendly and easy to produce, harvest, and modify for broad-spectrum applications. Especially concerning environmental preservation and sustainability, BC has remarkable advantages, as categorized and briefly discussed in the following sections.

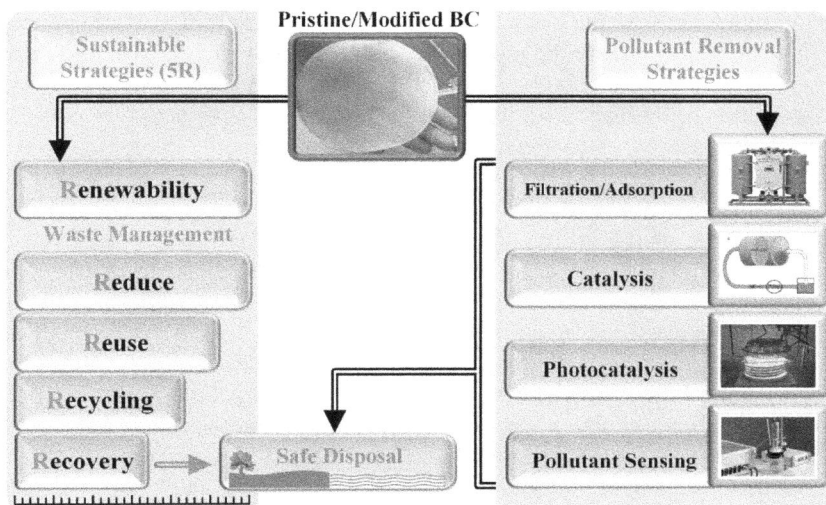

FIGURE 7.6 Application of BC in environmental remediation.

7.5.1.1 Sustainable Strategies

Most of the industrial production line is largely dependent upon the use of non-renewable natural resources, which is leading to the depletion of these resources as well as causing environmental destruction. To address such issues, there is an urgent need to set a balance between the commercial demands and the extraction of natural resources. This could be done by following sustainable strategies through the intervention of novel renewable sources to diminish the full-scale dependency on natural resources. Consequently, the overextraction of natural resources can be controlled, while at the same time protecting the earth's environment from destruction or pollution. There is a huge list of renewable resources that can potentially replace conventional industrial processes with minimum pollution load on the environment. BC is a strong candidate for replacing different nonrenewable and potentially harmful polymers used as industrial inputs. The abundance, easy production, impressive physicochemical properties, and versatility with respect to modification are the characteristics of BC that lead to energy conservation, product improvement, and waste reduction to achieve the sustainability goals.

From this perspective, BC is an ideal choice for the emerging environmental concerns and offers its service in five ways based on its characteristics, which could be enlisted as five **R**s i.e. renewable, reduce, reuse, recycle, and recover. They are detailed as follows:

Renewable: Nowadays, renewable feedstock resources are of prime importance over nonrenewable ones. BC is a renewable raw material that can be easily regenerated from a variety of raw materials, using different methods, and thus serve as the basis for "green processes" for a myriad of applications (107). Owing to the renewability of BC, the extraction load of nonrenewable resources as well as the pollution load on the environment can be considerably reduced, and these entities could be further saved for coming generations.

Reduce: This process refers to the reduction in the input materials, cost, and production of waste in order to reduce the impact on the environment. The application of BC in the industrial sector helps in the reduction of waste (solid, air, and water) generation, which is in fact beneficial for environmental protection. For example, the fabrication and implication of nontoxic BC-based dyes in the textile domain reduces the use of synthetic dyes, which are harmful to humans, animals, and the environment. Moreover, switching from synthetic dyes to BC-based dry dyes could reduce the discharge of toxic synthetic dyes into aquatic bodies and thus could reduce the overuse of water (108). Similarly, owing to the eco-friendly nature of BC, it could reduce the use of synthetic plastics, especially those in daily use such as packaging; thus, it would ultimately reduce the generation of nondegradable solid waste.

Reuse: For materials that cannot be easily reduced, there is a desire to reuse them instead of disposing them directly to the environment or even to the landfill. Many of the BC-based materials or products that are used once could be reused because of their impressive physicochemical and morphological properties. For instance, BC-based catalysts, adsorbents, and membrane filters have been widely explored for environmental remediation, such as for the removal of organic contaminants. Such materials could be easily separated and further used for several cycles (109), which will not only reduce material/processing costs but will also save the environment from pollution. In some other cases, the used material from one process could be reused as a raw material for other processes/applications.

Recycle: Sometimes the spent materials cannot be reused as such but need some extra processing to reform the used materials for another application. The best example of recycling, as discussed in Section 7.2, is BC production from different types of waste materials. With respect to some industrial applications, BC production cost is considered higher, which limits the commercial use of BC. That is why, recently, great measures have been taken to produce BC from the waste of other industries (such as agriculture, food, paper) as a carbon source. In this way, the waste that used to be sent to dumping sites is now being used for BC production, which is further saving the environment from massive wastes. Using waste further minimizes BC production costs and largely replaces the use of chemically defined and expensive cellulose-producing media containing sucrose, yeast extract, and different chemical salts.

Recover: When the used BC material cannot be reused or recycled profitably, then prior to its final disposal, the material could be utilized in the recovery of other useful elements or components like carbon black. The recovered materials could be further functionalized for additional applications (110).

With increasing worldwide environmental concerns and corresponding public awareness and sensitization about environmental safety, these highly appreciated top-to-
bottom practices could sustain the natural environment. BC thus contributes to each
step by lowering costs as well as protecting the environment from the generation of
harmful waste.

7.5.1.2 Pollutant Removal and Environmental Rehabilitation

During industrial processes and manufacturing, countless organic contaminants
are generated, which is largely deteriorating the natural environment. As discussed
earlier, the generation of pollutants and wastes is being managed through physical
sustainable strategies, but only to a small extent. A variety of harmful organic contaminants still originating from and entering the air, water, and soil must be treated
rationally at the source point (i.e., prior to their disposal) by using advanced technologies. Currently, several technologies like adsorption, coagulation, filtration,
catalysis, photocatalysis, and others are in practice for dealing with various toxic
contaminants. In this regard, BC could offer the pollutant removal role through all
the aforementioned pollutant removal strategies. The applications of BC in environmental rehabilitation (air, water, and soil) are described in the following sections.

7.5.1.2.1 Air and Water Pollution

Pollutants originating from different industries and entering the environment are
deteriorating the air and water domains at an intense rate. The application of BC to
curb such pollutants is discussed as follows:

Separation and filtration: The air is being contaminated primarily from the burn-
ing of fossil fuels in industries, automobile emissions, other anthropogenic activities, and natural disasters (e.g., volcanic eruptions). For healthy tropospheric and
stratospheric atmospheres, air pollutants such as sulfur oxides (SOx), nitrogen oxides
(NOx), greenhouse gases (GHGs), volatile organic compounds (VOx), and others
need to be removed at the point of their generation (16). Filtration is an established
technique frequently used to remove such pollutants from discharged gases. For the
last couple of decades, BC-based membrane filters are gaining significant attraction for capturing air pollutants. The BC membrane filters are composed of nanofibers, which have a large surface area that provides excellent pollutant adsorption/
filtration capacity as compared to the conventional fiber mats (54). Moreover, due to
its abundant surface hydroxyl groups, BC can be further functionalized with other
active groups and materials to enhance its reactivity as a catalyst (Figure 7.4) for
degradation of toxic compounds (after absorbance) into nontoxic end products (30,
78). Similarly, BC-based water filtration membranes are in wide use for removing
recalcitrant organic pollutants as well as pathogens. Further details are provided in
the subsequent water treatment section.

7.5.1.2.2 Water Treatment

Different kinds of contaminants from anthropogenic activities are polluting water
bodies and making them unfit for human and animal use. In water remediation, the
decontamination of aquatic pollutants is a global concern and thus included in the
"2030 Agenda for Sustainable Development" (*Transforming Our World: The 2030*

Agenda for Sustainable Development; 118). Currently, the technologies available for removing a variety of contaminants such as dyes, heavy metal ions, polymers, detergents, and pesticides have several advantages and limitations, in terms of budget and effectiveness, thus advancement in this area is continuously in progress. The advanced BC-based water treatment and purification technologies are gaining increasing attention in comparison to other advanced technologies, owing to the renewable, biodegradable, and eco-friendly nature of BC. The 3D fibrous network, abundant OH groups, high water adsorption capacities, and ease of modification of BC with a variety of functional materials (Figure 7.4) make it an ideal choice for water treatment using the techniques of filtration, adsorption, catalysis, and others (18, 111), as discussed in the following sections.

Filtration: Filtration is thought to be a quick and cost-effective technique for treating contaminated water. The membrane filters are designed with well-defined pores and tailored surface chemistry to selectively filter aquatic pollutants. A variety of filters have been implemented, such as microfilters, ultrafilters, and nanofilters, for the purification of water from toxic contaminants. Most of the commercially available water filters are manufactured by using synthetic polymers, which are composed of organic and inorganic solvents such as different acids and chemicals, for example hydrofluoric and hydrochloric acid, alcohols, N,N-dimethylformamide (DMF), and dimethylsulfoxide. Such polymers and chemicals are generally harmful to humans and the environment, owing to their toxic nature. As the promotion of sustainable and ecologically friendly materials has become a global demand, it is mandatory to switch from these conventional synthetic polymer filters to the filters and membranes produced from natural and eco-friendly materials (112).

In this regard, BC and its composites have been largely explored recently as water filtration materials for the selective removal of a variety of aquatic contaminants. Although there are only a few studies of developing and using BC-based filtration materials, it qualifies as the best eco-friendly filtration owing to its impressive features, like its highly porous nature and reticular structure with small pores offering abundant sites for pollutants. The application of BC as a water filter could be further expanded after some modification in the BC domain or its hybridization with other materials. For instance, BC incorporated with graphene oxide (GO) and palladium NPs exhibited advanced properties like water stability, good mechanical strength, and selective ion permeation up to the angstrom scale coming from the GO component. As a result, the BC-GO composite showed outstanding efficiency in wastewater filtration of organic dyes, methylene blue, and 4-nitrophenol, as well as for the removal of pollutants from wastes from the pharmaceutical industries (113, 114). Similarly, BC and chitosan (BC/Ch) composites have also proved to be excellent membrane filters for the removal of heavy metals from water (115).

Adsorption: Adsorption is a known and frequently used technique to adsorb air and water contaminants such as CO_2, SO_x, NO_x, and heavy metals. The BC-based adsorbent materials have gained considerable attraction in wastewater remediation. The fundamental characteristics of BC that make it a suitable adsorbent include its inherent hydrophilicity, high surface area, ease of tunability, and functionalization with a variety of other active materials. Moreover, BC is widely explored as an adsorbent because of its surface functionalities and negative charge as well as highly porous and fibrous morphology, which bestow excellent adsorption capability to BC for metallic cations (116). BC itself has remarkable adsorption efficiency for such pollutants but requires further modification. Fortunately, the negative charge on BC and the abundant −OH groups on its surface allows other active metals to transform it into a variety of reactive BC-based composites for effective pollutant adsorption (117). Inspiring by these facts, BC and its composites have already been developed for the removal of heavy metal ions, such as Hg, Cr, Pb, As, Cu, Co, and others. For instance, recently Derami et al. fabricated polydopamine (PDA) particles incorporated into BC for the removal of several heavy metal ions and organic dyes (118). Further, the combination of zwitterionic poly(2-methacryloyloxyethyl phosphorylcholine) and the hydrophilic BC yielded nanocomposite membranes, which effectively adsorbed anionic and cationic organic dyes (111). Similarly, a BC-based nanocomposite membrane has been developed by modifying it with polydopamine and TiO_2. The developed nanocomposite showed higher adsorption capacity for methyl orange, Rhodamine B, and methylene blue, as well as high photocatalytic activity (119).

7.6 CATALYSIS

Conventional techniques like filtration or absorbance usually fail to effectively remove the organic contaminants such as pharmaceuticals, phenols, and several others. In such cases, the advanced oxidation processes (AOPs) are used to completely degrade the organic molecules through oxidation (50). Different kinds of catalysts are used to activate the oxidants to generate reactive oxygen species, which further attack the organic pollutants and oxidize them into simple nontoxic compounds like CO_2 and water. The conversion of renewable and biodegradable BC into a catalyst is a very recent discovery (120); therefore, as a catalyst, the BC-based materials are only little studied. Intrinsically, BC is not reactive for catalysis; however, it could be used as a support for other catalytically active materials (e.g., metal NPs and carbonaceous materials) to provide well-dispersed active sites and avoid unwanted aggregations (121). BC has several advantages over other types of cellulosic materials owing to its high purity, abundant surface functional group like −OH, and high surface area. For example, recently, Kamal et al. used BC nanofibers as support for carboxymethyl cellulose (CMC) assisted with copper (Cu) NPs and transform a catalytically reactive nanomaterial (CMC-Cu-BC) for degradation of organic pollutants (122).

Photocatalysis: Besides normal chemical-based catalysis, photocatalysis is also receiving immense consideration for the degradation of aquatic organic contaminants. Frequently, the precious metal NPs (e.g., Au, Pt, and Ag) and some metal oxides like TiO_2 are used as photocatalysts. However, the high cost of these materials is one of the factors discouraging their application in photocatalysis. Several studies reveal that the hybridization of such precious metals and their oxides with other active materials like carbon allotropes (CNTs, rGO, MWCNTs, carbon nanodiamonds) is an efficient way not only to reduce the cost but also to enhance the photocatalytic performance of the hybrid composites. Despite the advantages of these hybrid photocatalysts, the low dispersion active sites, small surface area, high metal leaching, self-aggregation, and recyclability and stability are the major issues that still need to be rectified rationally. According to some studies, the versatile BC has been composited with different metallic NPs to originate eco-friendly nanocomposites capable of remarkable optical, electrical, mechanical, and thermal properties and could be used as photocatalysts (19, 123). In such cases, BC usually provides a highly porous support for better dispersion of NPs as active sites and excellent stability, prevents the self-aggregation of NPs, and ultimately results in excellent pollutant removal efficiency and recyclability. For instance, α-Fe_2O_3 nanodisks were dispersed in BC to develop the α-Fe_2O_3/BC nanohybrid as a photocatalyst, which was used for activation of peroxymonosulfate (PMS) for enhanced photodegradation of organic compounds (124).

Pollutant detection: A key strategy of environmental management is the detection of threats to the environment. Major threats to the environment could be detected through the use of sensors. Concerning environmental remediation, pollutant sensing or detection is playing a vital role in curbing such threats. A sensor detects the contaminants up to a certain threshold limit depending upon its sensitivity (104). The availability of highly sensitive and cheap techniques for contaminant detection can definitely improve environmental monitoring systems. Nanomaterials have gained significant attention for sensing applications; however, the low diffusion and aggregation of NPs are the major hurdles in their large-scale commercial applications. Such issues are addressed by immobilizing the NPs on the natural polymers (125). Recently, BC-based sensors are increasingly being considered for their high efficacy and sensitivity toward traces of contaminants (126). It is well known that BC has very low electrochemical activity, because of which its electrochemical applications are very limited. However, after the incorporation of suitable conducting materials like graphene or transition metals, the resulting BC-based composite becomes very reactive and thus could be used for numerous applications, including the detection of contaminants (19, 127). BC-based sensors are commonly used for the detection of pollutants in wastewater and air. For instance, a BC-GO modified electrode was used for the detection of nitrite in tap and pond water. The composite exhibited combined new properties like high surface area, strength, durability, and stability, which came from both BC and GO. The developed BC-GO hybrid efficiently detected nitrite (127). Similarly, there are numerous studies on the sensing applications of BC for the detection of air and water contaminants such as pharma cuticles, organic and inorganic compounds in water, SOx, NOx, and GHGs (CO_2, CH_4) (19). More recently, a BC-based nanopaper optical sensor array was developed for the detection of heavy metals (105).

Waste management and BC production: Direct applications of BC in waste management are rare. The BC production from different industrial wastes is a hot topic nowadays with respect to green environmental approaches. The fabrication of BC from waste definitely reduces the waste sent to dumping sites, thus lowering the pollution load, saving natural resources, and reducing the cost of relevant industrial products. The details of BC production from different industrial wastes are discussed in Section 7.2.

7.6.1 CHALLENGES AND LIMITATIONS

Although BC has tremendous applications from the perspective of environmental protection, numerous challenges and shortcomings still limit its many applications.

Currently, BC-based systems are facing problems, such as their low yield and productivity, which is limited to laboratory scale only. Similarly, although several mechanically disintegrated techniques are useful in scaling up BC production, these techniques are energy dependent, which further increases the production cost (23, 117). Ecotoxicological studies of BC are somewhat obvious because it has been declared as a safe food by competent authorities; nevertheless, its life cycle assessment still needs to be conducted, which is very important for better environmental remediation. To overcome the challenges and shortcomings associated with BC, further research is desired to expand the application of BC for environmental remediation at an elevated pace in the near future as an environmentally benign material.

7.7 CONCLUSIONS AND PROSPECTS

Nanocellulose and nanocellulose-based composites are currently receiving great interest for use in multiple disciplines. BC is the next class of nanocellulosic materials; it does not exhibit many of the deficiencies of other cellulosic materials. BC has tremendous applications in a myriad of areas; it also provides impressive services in the area of environmental safety and rehabilitation. Especially, BC and BC-based composites can be used for applications in sensing and in the removal of a variety of constituents like heavy metals and organic compounds via adsorption, filtration, and catalysis. Moreover, the production of BC from different industrial wastes and the development of packaging materials from BC are considered groundbreaking discoveries directly benefiting solid waste management systems and ultimately our ecosystem.

Currently, the major focus is to upscale BC production and the synthesis of BC-based eco-friendly hybrid materials. Concerning environmental remediation, the fabrication of advanced BC-based hybrid materials is one of the better prospects and will be highly important in dealing with the emerging contaminants from growing industrialization.

REFERENCES

(1) Hill, Marquita K. 2020. *Understanding Environmental Pollution*. Cambridge: Cambridge University Press.
(2) Pöschl, Ulrich. 2020. Air pollution, oxidative stress, and public health in the Anthropocene. In *Health of People, Health of Planet and Our Responsibility*. Cham: Springer.
(3) Khan, Suliman, Ghulam Nabi, Muhammad Wajid Ullah, et al. 2016. Overview on the role of advance genomics in conservation biology of endangered species. *International Journal of Genomics* 2016:3460416.
(4) Alharbi, Omar ML, Rafat A Khattab, and Imran Ali. 2018. Health and environmental effects of persistent organic pollutants. *Journal of Molecular Liquids* 263:442–453.
(5) Mrema, Ezra J, Federico M Rubino, Gabri Brambilla, Angelo Moretto, Aristidis M Tsatsakis, and Claudio Colosio. 2013. Persistent organochlorinated pesticides and mechanisms of their toxicity. *Toxicology* 307:74–88.
(6) Fontenele, EG, MR Martins, AR Quidute, and Renan Magalhães Montenegro Jr. 2010. Environmental contaminants and endocrine disruptors. *Arquivos brasileiros de endocrinologia e metabologia* 54 (1):6.

(7) Melvin, Steven D, and Frederic DL Leusch. 2016. Removal of trace organic contaminants from domestic wastewater: A meta-analysis comparison of sewage treatment technologies. *Environment International* 92:183–188.

(8) Boczkaj, Grzegorz, and André Fernandes. 2017. Wastewater treatment by means of advanced oxidation processes at basic pH conditions: A review. *Chemical Engineering Journal* 320:608–633.

(9) Shahzad, Ajmal, Jawad Ali, Jerosha Ifthikar, Zhulei Chen, and Zhuqi Chen. 2019. The hetero-assembly of reduced graphene oxide and hydroxide nanosheets as superlattice materials in PMS activation. *Carbon* 155:740–755.

(10) Wang, Jianlong, and Shizong Wang. 2019. Preparation, modification and environmental application of biochar: A review. *Journal of Cleaner Production* 227:1002–1022.

(11) Zhu, Shishu, Xiaochen Huang, Fang Ma, Li Wang, Xiaoguang Duan, and Shaobin Wang. 2018. Catalytic removal of aqueous contaminants on N-doped graphitic biochar's: Inherent roles of adsorption and nonradical mechanisms. *Environmental Science & Technology* 52 (15):8649–8658.

(12) Ullah, Muhammad Wajid, Waleed Ahmad Khattak, Mazhar Ul-Islam, Shaukat Khan, and Joong Kon Park. 2015. Encapsulated yeast cell-free system: A strategy for cost-effective and sustainable production of bio-ethanol in consecutive batches. *Biotechnology and Bioprocess Engineering* 20:561–575.

(13) Ullah, Muhammad Wajid, Waleed Ahmad Khattak, Mazhar Ul-Islam, Shaukat Khan, and Joong Kon Park. 2016. Metabolic engineering of synthetic cell-free systems: strategies and applications. *Biochemical Engineering Journal* 105 (B):391–405.

(14) Zhou, H, and Daniel W Smith. 2002. Advanced technologies in water and wastewater treatment. *Journal of Environmental Engineering and Science* 1 (4):247–264.

(15) Hussain, Zohaib, Wasim Sajjad, Taous Khan, and Fazli Wahid. 2019. Production of bacterial cellulose from industrial wastes: A review. *Cellulose* 26 (5):2895–2911.

(16) Ansaloni, Luca, Jesus Salas-Gay, Simone Ligi, and Marco Giacinti Baschetti. 2017. Nanocellulose-based membranes for CO2 capture. *Journal of Membrane Science* 522:216–225.

(17) de Amorim, Julia Didier Pedrosa, Karina Carvalho de Souza, Cybelle Rodrigues Duarte, et al. 2020. Plant and bacterial nanocellulose: Production, properties and applications in medicine, food, cosmetics, electronics and engineering. A review. *Environmental Chemistry Letters*:1–19.

(18) Nascimento, Diego M, Yana L Nunes, Maria CB Figueirêdo, et al. 2018. Nanocellulose nanocomposite hydrogels: Technological and environmental issues. *Green Chemistry* 20 (11):2428–2448.

(19) Ul-Islam, Mazhar, Muhammad Wajid Ullah, Shaukat Khan, et al. 2016. Recent advancement in cellulose based nanocomposite for addressing environmental challenges. *Recent Patents on Nanotechnology* 10 (3):169–180.

(20) Belbekhouche, Sabrina, Julien Bras, Gilberto Siqueira, et al. 2011. Water sorption behavior and gas barrier properties of cellulose whiskers and microfibrils films. *Carbohydrate Polymers* 83 (4):1740–1748.

(21) Khan, Hina, Ashish Kadam, and Dharm Dutt. 2020. Studies on bacterial cellulose produced by a novel strain of Lactobacillus genus. *Carbohydrate Polymers* 229:115513.

(22) Ullah, Muhammad Wajid, Mazhar Ul-Islam, Shaukat Khan, Nasrullah Shah, and Joong Kon Park. 2017. Recent advancements in bioreactions of cellular and cell-free systems: A study of bacterial cellulose as a model. *Korean Journal of Chemical Engineering* 34 (6):1591–1599.

(23) Wang, Jing, Javad Tavakoli, and Youhong Tang. 2019. Bacterial cellulose production, properties and applications with different culture methods—A review. *Carbohydrate Polymers* 219:63–76.

(24) Kim, Yeji, Muhammad Wajid Ullah, Mazhar Ul-Islam, Shaukat Khan, Jae Hyun Jang, and Joong Kon Park. 2019. Self-assembly of bio-cellulose nanofibrils through intermediate phase in a cell-free enzyme system. *Biochemical Engineering Journal* 142:135–144.

(25) Ullah, Muhammad Wajid, Mazhar Ul-Islam, Shaukat Khan, Yeji Kim, and Joong Kon Park. 2015. Innovative production of bio-cellulose using a cell-free system derived from a single cell line. *Carbohydrate Polymers* 132:286–294.

(26) Islam, Mazhar Ul, Muhammad Wajid Ullah, Shaukat Khan, Nasrullah Shah, and Joong Kon Park. 2017. Strategies for cost-effective and enhanced production of bacterial cellulose. *International Journal of Biological Macromolecules* 102:1166–1173.

(27) Czaja, Wojciech, Alina Krystynowicz, Stanislaw Bielecki, and R Malcolm Brown Jr. 2006. Microbial cellulose—the natural power to heal wounds. *Biomaterials* 27 (2):145–151.

(28) Kurosumi, Akihiro, Chizuru Sasaki, Yuya Yamashita, and Yoshitoshi Nakamura. 2009. Utilization of various fruit juices as carbon source for production of bacterial cellulose by Acetobacter xylinum NBRC 13693. *Carbohydrate Polymers* 76 (2):333–335.

(29) Lin, Wen-Chun, Chun-Chieh Lien, Hsiu-Jen Yeh, Chao-Ming Yu, and Shan-hui Hsu. 2013. Bacterial cellulose and bacterial cellulose—chitosan membranes for wound dressing applications. *Carbohydrate Polymers* 94 (1):603–611.

(30) Moniri, Mona, Amin Boroumand Moghaddam, Susan Azizi, et al. 2017. Production and status of bacterial cellulose in biomedical engineering. *Nanomaterials* 7 (9):257.

(31) Yin, Na, Thiago MA Santos, George K Auer, John A Crooks, Piercen M Oliver, and Douglas B Weibel. 2014. Bacterial cellulose as a substrate for microbial cell culture. *Applied and Environmental Microbiology* 80 (6):1926–1932.

(32) Ul-Islam, Mazhar, Muhammad Wajid Ullah, Shaukat Khan, and Joong Kon Park. 2020. Production of bacterial cellulose from alternative cheap and waste resources: A step for cost reduction with positive environmental aspects. *Korean Journal of Chemical Engineering* 37:925–937.

(33) Brown, Adrian J. 1886. XLIII.—On an acetic ferment which forms cellulose. *Journal of the Chemical Society, Transactions* 49:432–439.

(34) Hestrin, Schramm, and MJBJ Schramm. 1954. Synthesis of cellulose by Acetobacter xylinum. 2. Preparation of freeze-dried cells capable of polymerizing glucose to cellulose. *Biochemical Journal* 58 (2):345–352.

(35) Blanco Parte, Francisco German, Shella Permatasari Santoso, Chih-Chan Chou, et al. 2020. Current progress on the production, modification, and applications of bacterial cellulose. *Critical Reviews in Biotechnology* 40 (3):397–414.

(36) Shah, Nasrullah, Mazhar Ul-Islam, Waleed Ahmad Khattak, and Joong Kon Park. 2013. Overview of bacterial cellulose composites: A multipurpose advanced material. *Carbohydrate Polymers* 98 (2):1585–1598.

(37) Jozala, Angela Faustino, Leticia Celia de Lencastre-Novaes, André Moreni Lopes, et al. 2016. Bacterial nanocellulose production and application: A 10-year overview. *Applied Microbiology and Biotechnology* 100 (5):2063–2072.

(38) Li, Sixiang, Ashwak Jasim, Weiwei Zhao, et al. 2018. Fabrication of pH-electroactive bacterial cellulose/polyaniline hydrogel for the development of a controlled drug release system. *ES Materials & Manufacturing* 1 (11):41–49.

(39) Nishi, Y, M Uryu, S Yamanaka, et al. 1990. The structure and mechanical properties of sheets prepared from bacterial cellulose. *Journal of Materials Science* 25 (6):2997–3001.

(40) Sheng, Nan, Shiyan Chen, Jingjing Yao, et al. 2019. Polypyrrole@ TEMPO-oxidized bacterial cellulose/reduced graphene oxide macrofibers for flexible all-solid-state supercapacitors. *Chemical Engineering Journal* 368:1022–1032.

(41) Shi, Zhijun, Yue Zhang, Glyn O Phillips, and Guang Yang. 2014. Utilization of bacterial cellulose in food. *Food Hydrocolloids* 35:539–545.

(42) Shoukat, Annum, Fazli Wahid, Taous Khan, et al. 2019. Titanium oxide-bacterial cellulose bioadsorbent for the removal of lead ions from aqueous solution. *International Journal of Biological Macromolecules* 129:965–971.

(43) Ul-Islam, Mazhar, Shaukat Khan, Muhammad Wajid Ullah, and Joong Kon Park. 2015. Bacterial cellulose composites: Synthetic strategies and multiple applications in biomedical and electro-conductive fields. *Biotechnology Journal* 10 (12):1847–1861.

(44) Ul-Islam, Mazhar, Shaukat Khan, Muhammad Wajid Ullah, and Joong Kon Park. 2019. Comparative study of plant and bacterial cellulose pellicles regenerated from dissolved states. *International Journal of Biological Macromolecules* 137:247–252.

(45) Esa, Faezah, Siti Masrinda Tasirin, and Norliza Abd Rahman. 2014. Overview of bacterial cellulose production and application. *Agriculture and Agricultural Science Procedia* 2:113–119.

(46) Saibuatong, Ong-Ard, and Muenduen Phisalaphong. 2010. Novo aloe vera—bacterial cellulose composite film from biosynthesis. *Carbohydrate Polymers* 79 (2):455–460.

(47) Maria, Luiz, Ana LC Santos, Philippe C Oliveira, et al. 2010. Preparation and antibacterial activity of silver nanoparticles impregnated in bacterial cellulose. *Polimeros* 20 (1):72–77.

(48) Alves, AA, WE Silva, MF Belian, LSG Lins, and A Galembeck. 2020. Bacterial cellulose membranes for environmental water remediation and industrial wastewater treatment. *International Journal of Environmental Science and Technology*:1–12.

(49) Klemm, Dieter, Brigitte Heublein, Hans-Peter Fink, and Andreas Bohn. 2005. Cellulose: Fascinating biopolymer and sustainable raw material. *Angewandte chemie International Edition* 44 (22):3358–3393.

(50) Shahzad, Ajmal, Jawad Ali, Jerosha Ifthikar, et al. 2020. Non-radical PMS activation by the nanohybrid material with periodic confinement of reduced graphene oxide (rGO) and Cu hydroxides. *Journal of Hazardous Materials* 392:122316.

(51) Jawad, Ali, Zhuwei Liao, Zhihua Zhou, et al. 2017. Fe-MoS4: An effective and stable LDH-based adsorbent for selective removal of heavy metals. *ACS Applied Materials & Interfaces* 9 (34):28451–28463.

(52) Napavichayanun, Supamas, Phakdee Amornsudthiwat, Prompong Pienpinijtham, and Pornanong Aramwit. 2015. Interaction and effectiveness of antimicrobials along with healing-promoting agents in a novel biocellulose wound dressing. *Materials Science and Engineering: C* 55:95–104.

(53) Taha, Ahmed A, Yi-na Wu, Hongtao Wang, and Fengting Li. 2012. Preparation and application of functionalized cellulose acetate/silica composite nanofibrous membrane via electrospinning for Cr (VI) ion removal from aqueous solution. *Journal of Environmental Management* 112:10–16.

(54) Liu, Xiaobing, Hamid Souzandeh, Yudong Zheng, Yajie Xie, Wei-Hong Zhong, and Cai Wang. 2017. Soy protein isolate/bacterial cellulose composite membranes for high efficiency particulate air filtration. *Composites Science and Technology* 138:124–133.

(55) Morgan, Jacob LW, Joanna Strumillo, and Jochen Zimmer. 2013. Crystallographic snapshot of cellulose synthesis and membrane translocation. *Nature* 493 (7431):181–186.

(56) Khattak, Waleed Ahmad, Taous Khan, Mazhar Ul-Islam, Muhammad Wajid Ullah, Shaukat Khan, Fazli Wahid, and Joong Kon Park. 2015. Production, characterization and biological features of bacterial cellulose from scum obtained during preparation of sugarcane jaggery (gur). *Journal of Food Science and Technology* 52:8343–8349.

(57) Güzel, Melih, and Özlem Akpınar. 2019. Production and characterization of bacterial cellulose from citrus peels. *Waste and Biomass Valorization* 10 (8):2165–2175.

(58) Algar, Itxaso, Susana CM Fernandes, Gurutz Mondragon, et al. 2015. Pineapple agroindustrial residues for the production of high value bacterial cellulose with different morphologies. *Journal of Applied Polymer Science* 132 (1).

(59) Luo, Mu-Tan, Cheng Zhao, Chao Huang, et al. 2017. Efficient using durian shell hydroly-sate as low-cost substrate for bacterial cellulose production by *Gluconacetobacter xyli-nus*. *Indian Journal of Microbiology* 57 (4):393–399.

(60) Chen, Lin, Feng Hong, Xue-xia Yang, and Shi-fen Han. 2013. Biotransformation of wheat straw to bacterial cellulose and its mechanism. *Bioresource technology* 135:464–468.

(61) Voon, WWY, BJ Muhialdin, NL Yusof, Y Rukayadi, and AS Meor Hussin. 2019. Bio-cellulose Production by Beijerinckia fluminensis WAUPM53 and *Gluconacetobacter xylinus* 0416 in Sago By-product Medium. *Applied Biochemistry and Biotechnology* 187 (1):211–220.

(62) Pacheco, Guilherme, Cláudio R Nogueira, Andreia Bagliotti Meneguin, et al. 2017. Development and characterization of bacterial cellulose produced by cashew tree resi-dues as alternative carbon source. *Industrial Crops and Products* 107:13–19.

(63) Lotfiman, Samaneh, Dayang Radiah Awang Biak, Tey Beng Ti, Suryani Kamarudin, and Saeid Nikbin. 2018. Influence of date syrup as a carbon source on bacterial cel-lulose production by Acetobacter xylinum 0416. *Advances in Polymer Technology* 37 (4):1085–1091.

(64) Adebayo-Tayo, Bukola C, Moyinoluwa O Akintunde, and Jadesola F Sanusi. 2017. Effect of different fruit juice media on bacterial cellulose production by Acinetobacter sp. BAN1 and Acetobacter pasteurianus PW1. *Journal of Advances in Biology & Biotechnology*:1–9.

(65) Kuo, Chia-Hung, Chun-Yung Huang, Chwen-Jen Shieh, Hui-Min David Wang, and Chin-Yin Tseng. 2019. Hydrolysis of orange peel with cellulase and pectinase to produce bacterial cellulose using *Gluconacetobacter xylinus*. *Waste and Biomass Valorization* 10 (1):85–93.

(66) Hong, Feng, and Kaiyan Qiu. 2008. An alternative carbon source from konjac pow-der for enhancing production of bacterial cellulose in static cultures by a model strain Acetobacter aceti subsp. xylinus ATCC 23770. *Carbohydrate Polymers* 72 (3):545–549.

(67) Carreira, Pedro, Joana AS Mendes, Eliane Trovatti, et al. 2011. Utilization of resi-dues from agro-forest industries in the production of high value bacterial cellulose. *Bioresource technology* 102 (15):7354–7360.

(68) Castro, Cristina, Robin Zuluaga, Jean-Luc Putaux, Gloria Caro, Iñaki Mondragon, and Piedad Gañán. 2011. Structural characterization of bacterial cellulose produced by Gluconacetobacter swingsii sp. from Colombian agroindustrial wastes. *Carbohydrate Polymers* 84 (1):96–102.

(69) Casarica, Angela, Gheorghe Campeanu, MIŞU Moscovici, Alexandra Ghiorghita, and Vasilica Manea. 2013. Improvement of bacterial cellulose production by Acetobacter xylinum dsmz-2004 on poor quality horticultural substrates using the taguchi method for media optimization. Part 1. *Cellulose Chemistry Production and Technology* 47 (1–2):61–68.

(70) Kumbhar, Jyoti Vasant, Jyutika Milind Rajwade, and Kishore Madhukar Paknikar. 2015. Fruit peels support higher yield and superior quality bacterial cellulose produc-tion. *Applied Microbiology and Biotechnology* 99 (16):6677–6691.

(71) Jawad, Ali, Kun Zhan, Haibin Wang, et al. 2020. Tuning of persulfate activation from a free radical to a nonradical pathway through the incorporation of non-redox magnesium oxide. *Environmental Science & Technology* 54 (4):2476–2488.

(72) Khan, Aimal, Shuhua Zou, Ting Wang, et al. 2018. Facile synthesis of yolk shell Mn 2 O 3@ Mn 5 O 8 as an effective catalyst for peroxymonosulfate activation. *Physical Chemistry Chemical Physics* 20 (20):13909–13919.

(73) Song, Hocheol, Elizabeth R Carraway, Young Hun Kim, Bill Batchelor, Byong-Hun Jeon, and Jae-gon Kim. 2008. Amendment of hydroxyapatite in reduction of tetrachloroethyl-ene by zero-valent zinc: Its rate enhancing effect and removal of Zn (II). *Chemosphere* 73 (9):1420–1427.

(74) Zhu, Huayue, Ru Jiang, Ling Xiao, et al. 2009. Photocatalytic decolorization and degradation of Congo Red on innovative crosslinked chitosan/nano-CdS composite catalyst under visible light irradiation. *Journal of Hazardous Materials* 169 (1–3):933–940.

(75) Zhu, Li, Xiao-Qin Liu, Hai-Long Jiang, and Lin-Bing Sun. 2017. Metal—organic frameworks for heterogeneous basic catalysis. *Chemical Reviews* 117 (12):8129–8176.

(76) Ashour, Radwa M, Ahmed F Abdel-Magied, Qiong Wu, Richard T Olsson, and Kerstin Forsberg. 2020. Green synthesis of metal-organic framework bacterial cellulose nanocomposites for separation applications. *Polymers* 12 (5):1104.

(77) Choi, Yong-Jin, Yeonghee Ahn, Moon-Sung Kang, Hong-Ki Jun, In Soo Kim, and Seung-Hyeon Moon. 2004. Preparation and characterization of acrylic acid-treated bacterial cellulose cation-exchange membrane. *Journal of Chemical Technology & Biotechnology: International Research in Process, Environmental & Clean Technology* 79 (1):79–84.

(78) Wang, Dong. 2019. A critical review of cellulose-based nanomaterials for water purification in industrial processes. *Cellulose* 26 (2):687–701.

(79) Di, Zeng, Zhijun Shi, Muhammad Wajid Ullah, Sixiang Li, and Guang Yang. 2017. A transparent wound dressing based on bacterial cellulose whisker and poly (2-hydroxyethyl methacrylate). *International Journal of Biological Macromolecules* 105:638–644.

(80) McCarthy, Ronan R, Muhammad Wajid Ullah, Eujin Pei, and Guang Yang. 2019. Antimicrobial Inks: The anti-infective applications of bioprinted bacterial polysaccharides. *Trends in Biotechnology* 37 (11):1155–1159.

(81) Ul-Islam, Mazhar, Fazli Subhan, Salman Ul Islam, et al. 2019. Development of three-dimensional bacterial cellulose/chitosan scaffolds: Analysis of cell-scaffold interaction for potential application in the diagnosis of ovarian cancer. *International Journal of Biological Macromolecules* 137:1050–1059.

(82) Khan, Shaukat, Mazhar Ul-Islam, Muhammad Ikram, et al. 2018. Preparation and structural characterization of surface modified microporous bacterial cellulose scaffolds: A potential material for skin regeneration applications in vitro and in vivo. *International Journal of Biological Macromolecules* 117:1200–1210.

(83) Khan, Shaukat, Mazhar Ul-Islam, Muhammad Wajid Ullah, Muhammad Ikram, Fazli Subhan, Yeji Kim, Jae Hyun Jang, Sik Yoon, and Joong Kon Park. 2015. Engineered regenerated bacterial cellulose scaffolds for application in in vitro tissue regeneration. *RSC Advances* 5 (103):84565–84573.

(84) Abeer, Muhammad Mustafa, Mohd Cairul Iqbal Mohd Amin, and Claire Martin. 2014. A review of bacterial cellulose-based drug delivery systems: Their biochemistry, current approaches and future prospects. *Journal of Pharmacy and Pharmacology* 66 (8):1047–1061.

(85) Khan, Shaukat, Mazhar Ul-Islam, Muhammad Ikram, et al. 2016. Three-dimensionally microporous and highly biocompatible bacterial cellulose—gelatin composite scaffolds for tissue engineering applications. *RSC Advances* 6 (112):110840–110849.

(86) Ul-Islam, Mazhar, Waleed Ahmad Khattak, Muhammad Wajid Ullah, Shaukat Khan, and Joong Kon Park. 2014. Synthesis of regenerated bacterial cellulose-zinc oxide nanocomposite films for biomedical applications. *Cellulose* 21 (1):433–447.

(87) Reis, Emily M dos, Fernanda V Berti, Guilherme Colla, and Luismar M Porto. 2018. Bacterial nanocellulose-IKVAV hydrogel matrix modulates melanoma tumor cell adhesion and proliferation and induces vasculogenic mimicry in vitro. *Journal of Biomedical Materials Research Part B: Applied Biomaterials* 106 (8):2741–2749.

(88) Fischer, Agnieszka, Barbara Brodziak-Dopierała, Krzysztof Loska, and Jerzy Stojko. 2017. The assessment of toxic metals in plants used in cosmetics and cosmetology. *International Journal of Environmental Research and Public Health* 14 (10):1280.

(89) Darbre, Philippa D, and Philip W Harvey. 2008. Paraben esters: Review of recent studies of endocrine toxicity, absorption, esterase and human exposure, and discussion of potential human health risks. *Journal of Applied Toxicology* 28 (5):561–578.

(90) Anderson, AG, Jane Grose, Sabine Pahl, RC Thompson, and Kayleigh J Wyles. 2016. Microplastics in personal care products: Exploring perceptions of environmentalists, beauticians and students. *Marine pollution bulletin* 113 (1–2):454–460.

(91) Amorim, Julia, Andrea Costa, Cláudio Galdino, Gloria Vinhas, Emilia Santos, and Leonie Sarubbo. 2019. Bacterial cellulose production using industrial fruit residues as subtract to industrial application. *Chemical Engineering Transactions* 74:1165–1170.

(92) Chantereau, G, M Sharma, A Abednejad, et al. 2020. Bacterial nanocellulose membranes loaded with vitamin B-based ionic liquids for dermal care applications. *Journal of Molecular Liquids* 302:112547.

(93) Ullah, Hanif, Hélder A Santos, and Taous Khan. 2016. Applications of bacterial cellulose in food, cosmetics and drug delivery. *Cellulose* 23 (4):2291–2314.

(94) Tang, XZ, P Kumar, S Alavi, and KP Sandeep. 2012. Recent advances in biopolymers and biopolymer-based nanocomposites for food packaging materials. *Critical Reviews in Food Science and Nutrition* 52 (5):426–442.

(95) Guilherme, Marcos R, Fauze A Aouada, André R Fajardo, et al. 2015. Superabsorbent hydrogels based on polysaccharides for application in agriculture as soil conditioner and nutrient carrier: A review. *European Polymer Journal* 72:365–385.

(96) Elbarbary, Ahmed M, Hassan A Abd El-Rehim, Naeem M El-Sawy, El-Sayed A Hegazy, and El-Sayed A Soliman. 2017. Radiation induced crosslinking of polyacrylamide incorporated low molecular weights natural polymers for possible use in the agricultural applications. *Carbohydrate Polymers* 176:19–28.

(97) Khan, Shaukat, Mazhar Ul-Islam, Waleed Ahmad Khattak, Muhammad Wajid Ullah, and Joong Kon Park. 2015. Bacterial cellulose—poly (3, 4-ethylenedioxythiophene)—poly (styrenesulfonate) composites for optoelectronic applications. *Carbohydrate Polymers* 127:86–93.

(98) Khan, Shaukat, Mazhar Ul-Islam, Muhammad Wajid Ullah, Yeji Kim, and Joong Kon Park. 2015. Synthesis and characterization of a novel bacterial cellulose—poly (3, 4-ethylenedioxythiophene)—poly (styrene sulfonate) composite for use in biomedical applications. *Cellulose* 22 (4):2141–2148.

(99) Numata, Yukari, Hiroyuki Kono, Minato Tsuji, and Kenji Tajima. 2017. Structural and mechanical characterization of bacterial cellulose—polyethylene glycol diacrylate composite gels. *Carbohydrate Polymers* 173:67–76.

(100) Menchaca-Nal, Sandra, Cesar Leandro Londoño-Calderón, Patricia Cerrutti, et al. 2016. Facile synthesis of cobalt ferrite nanotubes using bacterial nanocellulose as template. *Carbohydrate Polymers* 137:726–731.

(101) Park, Minsung, Dajung Lee, Sungchul Shin, Hyun-Joong Kim, and Jinho Hyun. 2016. Flexible conductive nanocellulose combined with silicon nanoparticles and polyaniline. *Carbohydrate Polymers* 140:43–50.

(102) Bai, Yang, Rong Liu, Enyuan Li, Xiaolong Li, Yang Liu, and Guohui Yuan. 2019. Graphene/carbon nanotube/bacterial cellulose assisted supporting for polypyrene towards flexible supercapacitor applications. *Journal of Alloys and Compounds* 777:524–530.

(103) Trigona, Carlo, Salvatore Graziani, Giovanna Di Pasquale, Antonino Pollicino, Rossella Nisi, and Antonio Licciulli. 2020. Green energy harvester from vibrations based on bacterial cellulose. *Sensors* 20 (1):136.

(104) Farooq, Umer, Muhammad Wajid Ullah, Qiaoli Yang, et al. 2020. High-density phage particles immobilization in surface-modified bacterial cellulose for ultra-sensitive and selective electrochemical detection of Staphylococcus aureus. *Biosensors and Bioelectronics*:112163.

(105) Abbasi-Moayed, Samira, Hamed Golmohammadi, and M Reza Hormozi-Nezhad. 2018. A nanopaper-based artificial tongue: A ratiometric fluorescent sensor array on bacterial nanocellulose for chemical discrimination applications. *Nanoscale* 10 (5):2492–2502.

(106) Jang, Woo Dae, Ji Hyeon Hwang, Hyun Uk Kim, Jae Yong Ryu, and Sang Yup Lee. 2017. Bacterial cellulose as an example product for sustainable production and consumption. *Microbial Biotechnology* 10 (5):1181–1185.

(107) Wang, Bingtao, Kathryn Mireles, Mitch Rock, et al. 2016. Synthesis and Preparation of Bio-Based ROMP Thermosets from Functionalized Renewable Isosorbide Derivative. *Macromolecular Chemistry and Physics* 217 (7):871–879.

(108) Costa, Andréa Fernanda de S, Júlia DP de Amorim, Fabíola Carolina G Almeida, et al. 2019. Dyeing of bacterial cellulose films using plant-based natural dyes. *International Journal of Biological Macromolecules* 121:580–587.

(109) Kamal, Tahseen, Ikram Ahmad, Sher Bahadar Khan, and Abdullah M Asiri. 2019. Bacterial cellulose as support for biopolymer stabilized catalytic cobalt nanoparticles. *International Journal of Biological Macromolecules* 135:1162–1170.

(110) Foresti, María Laura, Analía Vázquez, and Bruno Boury. 2017. Applications of bacterial cellulose as precursor of carbon and composites with metal oxide, metal sulfide and metal nanoparticles: A review of recent advances. *Carbohydrate Polymers* 157:447–467.

(111) Vilela, Carla, Catarina Moreirinha, Adelaide Almeida, Armando JD Silvestre, and Carmen SR Freire. 2019. Zwitterionic nanocellulose-based membranes for organic dye removal. *Materials* 12 (9):1404.

(112) Galdino Jr, Claudio José S, Alexandre D Maia, Hugo M Meira, et al. 2020. Use of a bacterial cellulose filter for the removal of oil from wastewater. *Process Biochemistry* 91:288–296.

(113) Fang, Qile, Xufeng Zhou, Wei Deng, Zhi Zheng, and Zhaoping Liu. 2016. Freestanding bacterial cellulose-graphene oxide composite membranes with high mechanical strength for selective ion permeation. *Scientific Reports* 6 (1):1–11.

(114) Xu, Ting, Qisheng Jiang, Deoukchen Ghim, et al. 2018. Catalytically Active Bacterial Nanocellulose-Based Ultrafiltration Membrane. *Small* 14 (15):1704006.

(115) Urbina, Leire, Olatz Guaresti, Jesús Requies, et al. 2018. Design of reusable novel membranes based on bacterial cellulose and chitosan for the filtration of copper in wastewaters. *Carbohydrate Polymers* 193:362–372.

(116) Cumbo, Cosimo, Giuseppina Tota, Luisa Anelli, Antonella Zagaria, Giorgina Specchia, and Francesco Albano. 2020. TP53 in Myelodysplastic syndromes: Recent biological and clinical findings. *International Journal of Molecular Sciences* 21 (10):3432.

(117) Shak, Katrina Pui Yee, Yean Ling Pang, and Shee Keat Mah. 2018. Nanocellulose: Recent advances and its prospects in environmental remediation. *Beilstein Journal of Nanotechnology* 9 (1):2479–2498.

(118) Derami, Hamed Gholami, Qisheng Jiang, Deoukchen Ghim, et al. 2019. A robust and scalable polydopamine/bacterial nanocellulose hybrid membrane for efficient wastewater treatment. *ACS Applied Nano Materials* 2 (2):1092–1101.

(119) Yang, Luyu, Chuntao Chen, Ying Hu, et al. 2020. Three-dimensional bacterial cellulose/polydopamine/TiO2 nanocomposite membrane with enhanced adsorption and photocatalytic degradation for dyes under ultraviolet-visible irradiation. *Journal of Colloid and Interface Science* 562:21–28.

(120) Ul-Islam, Mazhar, Muhammad Wajid Ullah, Shaukat Khan, et al. 2017. Current advancements of magnetic nanoparticles in adsorption and degradation of organic pollutants. *Environmental Science and Pollution Research* 24:12713–12722.

(121) Ma, Bo, Jai Prakash Chaudhary, Jianguo Zhu, Bianjing Sun, Yang Huang, and Dongping Sun. 2020. Ni nanoparticle-carbonized bacterial cellulose composites for the catalytic reduction of highly toxic aqueous Cr (VI). *Journal of Materials Science-Materials in Electronics* 31 (9):7044–7052.

(122) Kamal, Tahseen, Ikram Ahmad, Sher Bahadar Khan, Mazhar Ul-Islam, and Abdullah M Asiri. 2019. Microwave assisted synthesis and carboxymethyl cellulose stabilized copper nanoparticles on bacterial cellulose nanofibers support for pollutants degradation. *Journal of Polymers and the Environment* 27 (12):2867–2877.

(123) Almasi, Hadi, Laleh Mehryar, and Ali Ghadertaj. 2020. Photocatalytic activity and water purification performance of in situ and ex situ synthesized bacterial cellulose-CuO nanohybrids. *Water Environment Research* 92:1334–1349.

(124) Zhu, Zhong-Shuai, Jin Qu, Shu-Meng Hao, Shuang Han, Kun-Le Jia, and Zhong-Zhen Yu. 2018. α-Fe2O3 Nanodisk/bacterial cellulose hybrid membranes as high-performance sulfate-radical-based visible light photocatalysts under stirring/flowing states. *ACS Applied Materials & Interfaces* 10 (36):30670–30679.

(125) Islam, Mazhar Ul, Shaukat Khan, Muhammad Wajid Ullah, and Joong Kon Park. 2017. Recent advances in biopolymer composites for environmental issues. *Handbook of Composites from Renewable Materials*:673–691.

(126) Thavasi, V, G Singh, and S Ramakrishna. 2008. Electrospun nanofibers in energy and environmental applications. *Energy & Environmental Science* 1 (2):205–221.

(127) Zhang, Ya, Zhifeng Zhou, Fangfang Wen, et al. 2017. Tubular structured bacterial cellulose-based nitrite sensor: Preparation and environmental application. *Journal of Solid State Electrochemistry* 21 (12):3649–3657.

(128) Farooq, Umer, Qiaoli Yang, Muhammad Wajid Ullah, and Shenqi Wang. 2018. Bacterial biosensing: Recent advances in phage-based bioassays and biosensors. *Biosensors and Bioelectronics* 118:204–216.

(129) Khan, Shaukat, Mazhar Ul-Islam, Muhammad Wajid Ullah, Muhammad Israr, Jae Hyun Jang, and Joong Kon Park. 2018. Nano-gold assisted highly conducting and biocompatible bacterial cellulose-PEDOT: PSS films for biology-device interface applications. *International Journal of Biological Macromolecules* 107:865–873.

(130) Rani, M Usha, and KA Anu Appaiah. 2013. Production of bacterial cellulose by Gluconacetobacter hansenii UAC09 using coffee cherry husk. *Journal of Food Science and Technology* 50 (4):755–762.

(131) Skin Care Products Market Size, Share & Trends Analysis Report, By Product (Face Cream, Body Lotion), By Region (North America, Central & South America, Europe, APAC, MEA), and Segment Forecasts, 2019–2025. 2019. In *Market Analysis Report*.

(132) *Transforming our world: The 2030 Agenda for Sustainable Development*. Available from https://sustainabledevelopment.un.org/post2015/transformingourworld

8 Bacterial Cellulose in Drug Delivery

Muhammad Ismail, M.I. Khan, Kalsoom Akhtar, Murad Ali Khan, Tahseen Kamal, and Sher Bahadar Khan

CONTENTS

8.1 INTRODUCTION

Researchers have worked hard to develop efficient, highly selective, biocompatible, and regulated release drug delivery systems. A prolonged therapeutic effect can be achieved through a persistent drug delivery method by releasing drugs over a prolonged period (1, 2). Nevertheless, due to low adsorption of drug, reduced biochemical activity, and complex release conditions, for example in the stomach and intestine, the difficulties of obtaining a lasting drug effect remain (3, 4). To this end, complicated formulations of drugs have been designed using different nanomaterials, like nanoparticles, micelles (5), nanocapsules (6), and aerogels (7). Aerogels are used as a drug carrier due to their open and porous structure, high surface area, and light weight, which promise a good retention and adsorption of water inside a three-dimensional network with no dissolution (8). For such purpose, the most-used classes of aerogel material are polysaccharide, silica, and carbon (9–11). Based on their good stability, low degree of toxicity, and improved drug-loading capability, bio-based polysaccharide polymers such as cellulose (12), chitosan (13), starch (14), and alginate (15) have been commonly used as primary formulation components in drug delivery applications. Cellulose nanofibers (CNFs) have recently provoked a growing curiosity in pharmaceutical and tissue engineering research applications (16). CNF-titania nanocomposites have been used by the Galkina group as a transporter for three different kinds of medicinal drugs: diclofenac sodium, phosphomycin, and penicillamine-D. The research findings showed a high capacity of CNF-based composites for transdermal delivery of drugs and wound curing (17). Similarly, CNF-based foams of varying thicknesses and shapes for riboflavin loading were manufactured by the Svagan group. Materials have demonstrated simulated gastric fluid (SGF) sustained release and its ability as a continuous drug delivery system (18). In addition, polyethylenimine-grafted bamboo nanofibrils that demonstrated temperature and pH tolerance to sodium salicylate of a drug-loading capacity of up to 287.39 mg g^{-1} were synthesized by the Zhao group, exploring the formulation role of cellulose polymer in drug delivery (19). The most common substance on earth is cellulose, which is synthesized by large numbers of the Plantae and Animalia kingdoms in addition to the Eubacteria domain. Urochordates such as *Microcosmus fulcatus* within the animal kingdom have the unusual characteristics of being capable of producing crystals of rod-shaped cellulose and thereby contributing to the production/synthesis of cellulose (20). Many variants in cellulose have contributed to the availability of complex and heterogeneous products, though the varieties share several features. Cellulose is chemically made up of polysaccharide polymer having a β-1,4-glycosidic bond formed by the polymerization of long anhydroglucose unit chains after condensation. In plants, through hydrogen bonding, individual molecules of cellulose chains pack together to eventually form 50 nm diameter "microfibrils, containing both crystalline and amorphous regions" (21).

Tunicate-derived microfibrils of cellulose could be subjected to hydrolysis of sulphuric acid and contribute substantially to nanocrystal formation. Cellulose can exist as "nanocrystalline" cellulose (NCC), frequently getting from acid hydrolysis, with negatively charged and has particular physicochemical characteristics, including tensile purity, surface characteristics, and asymmetry (22). Another type of cellulose that can be collected from a variety of sources is cellulose "nanowhiskers" (CNWs), including tunicates. For certain polymers, the whiskers have quite a unique potential to serve as reinforcement materials and thus confirm a need for selective variation details of cellulose elucidation (23). NCC and CNWs are terms that are occasionally used interchangeably.

Biocellulose, also known as bacterial cellulose (BC), is produced as an extracellular polysaccharide by a number of bacterial genera, including *Sarcina*, *Agrobacter*, *Gluconacetobacter*, among others, in a wet membrane (99% water) form of the interface air/culture medium (25). Cellulose synthesized by bacteria—BC—is extra pure, has a high capacity for water absorption, exceptional mechanical strength, and good permeability, as well as resistance to degradation. It has a nanofibrous structure, as shown in Figure 8.1. Most of these characteristics of BC arise from the unusual three-dimensional nanofibrillar network. Because of its unique characteristics and biocompatibility, BC has generated considerable attention in many fields, but specifically in the field of biomedicine, such as for healing wounds from skin burns (26, 27) and for microsurgery in artificial blood vessels (28). In addition, BC's unique nanofibrillary structure should constitute sufficient supramolecular support for drug incorporation and thus for the design of specific systems for controlled release. Previous experiments have reported the potential of BC membranes to attenuate the

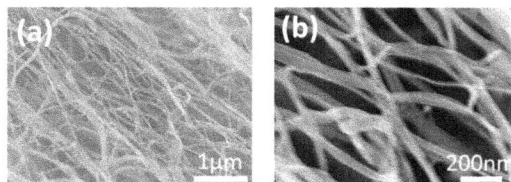

FIGURE 8.1 SEM images at different magnifications of microfibrils cellulose. Image reproduced with permission of (24).

Source: T. Kamal, I. Ahmad, S. B. Khan, and A. M. Asiri, "Bacterial cellulose as support for biopolymer stabilized catalytic cobalt nanoparticles," *International Journal of Biological Macromolecules*, vol. 135, pp. 1162–1170, 2019/08/15/ 2019.

release and bioavailability of percutaneous drug models (29, 30), and have thus been proposed as transdermal or topical drug delivery aids. Therapeutic efficacy relies on the interaction of these supports with the skin, but few details are available on the compatibility with the skin and the irritation ability of BC-based biomaterials. Studies on cytotoxicity have been performed with human cells (31, 32). Reports on *in vivo* biocompatibility, however, are scarce and used primarily in mouse surgery (31, 33). Commercial biocellulose films have been clinically tested specifically for wound healing effects (34), which compensate for their subcutaneous compatibility, but they have not been scientifically characterized by this criterion. Only one study identified the absence of skin discomfort when applying a 24 h biocellulose mask to a human volunteer's arm (35). Due to its distinctive structure and properties, BC synthesized from *Acetobacter xylinum* has attracted growing interest and attention in the field of biomedical devices (36, 37). High purity, good mechanical strength, and high surface area might ensure the good interaction of drug CNFs, improving their ability as adsorbents, drug carriers, and biological scaffolds (38). Nevertheless, the fact that bacterial cellulose is used in medicinal formulations as an active ingredient is not commonly known in the public papers. For the loading of berberine sulphate and berberine hydrochloride, the Huang group studied the controlled discharge behavior of BC membranes and demonstrated the lowest drug release in simulated gastric fluid (SGF) and the maximum in the simulated intestinal fluid (SIF) (39). Similarly, Ahmad's group studied the protein carrier role of BC-g-poly-(acrylic acid) hydrogel, which showed just 10% accumulated release of SGF (40). Studies focusing on the use of BC films to modulate the transdermal system's drug release have also been examined, but the poor loading potential of pristine BC in film form can restrict its additional industrialization. Other researchers have focused on enhancing the properties of bacterial cellulose and biomedical polymer composites, but their extra use in the absorption and release of drugs has not been reported (41).

8.2 BACTERIAL CELLULOSE SOURCES

The history of bacterial cellulose began in 1886, when A.J. Brown first identified it. BC falls under the group of exopolysaccharides, along with polyamides, polyesters, and inorganic polyanhydrides. Alpha-proteobacteria, beta-proteobacteria, and gamma-proteobacteria (Gram-negative) and Gram-positive bacteria are some of the bacteria used for cellulose production (42). In the laboratory, BC can also be produced and processed for food synthesis, such as that of *nata de coco*, an ancient East Asian sweet confectionery. It is noteworthy that using the same laboratory technique, *nata de coco* was synthesized and gained commercial achievement, becoming a well-known product. In the local Philippines language, *nata* means cellulose, and it was called *nata de coco* (43) because it was produced in sugar-based media containing coconut milk. Similar to that generated by bacteria, some algae and fungi manufacture cellulose, and all of three are referred to as microbial cellulose (44). The different species of microbes could be cultivated in different processes of food manufacture (45) and also commercially used in conjunction with various media such as Hestrin and Schramm (peptone, extract of yeast, and buffer) (46), alongside the different factors considered here.

When culture media of bacteria are varied, a number of effects can be detected. This could be practiced in studies conducted to increase the cellulose yield from *Gluconacetobacter xylinus*, where a number of BC concentrations were produced by some media at different incubation periods, composition, and volume. The incubation media used were corn steep liquor and Hestrin and Schramm (HS) media (47). The growth media are categorized for BC synthesis in view of their supply of diverse carbon sources. Based on which polysaccharide is provided, sources of carbon generate differential production of biosynthetic pathways, with galactose, glucose, mannitol, and glycerol as common examples of the different polysaccharides used (48). As a part of the medium, the source of carbon such as fructose also should be closely judged, as not all strains use fructose as a cellulose source. In the same study, Kurosomi et al. identified the related use of HS nitrogen sources in the different fruit juices culture to investigate the effect of carbon sources (49). Currently, microfibrillar BC have been synthesized by using agro-industrial residues. This involves the cultivation of *Gluconacetobacter swingsii* by using sugar cane juice and pineapple peel juice, since all of these compounds serve as concurrent sources of nitrogen and carbon (50). Another research/study has tried to optimize the yield of BC by identifying completely new species of bacteria for its synthesis.

8.3 GENERAL APPLICATIONS OF BACTERIAL CELLULOSE

Bacterial cellulose shows specific characteristics that allow precise applications, of which the cellulose of plants is not appropriate. BC is generated in an extremely pure form, absolutely pectin, hemicellulose, and lignin free (51), which makes it easier to purify than plant cellulose (52). Natural BC is an extremely porous substance with high liquid and gas permeability and also high water absorption potential with water content greater than 90% (46). These characteristics are attributable to the very fragile BC network consisting of micro- and nanofibrils in the form of ribbons (Figure 8.2) (52), and BC is about 100 times thinner than plant.

The density of BC nanofibers is very low, and BC has a high level of polymerization (about ~2,000–6,000) (53). Additionally, their wide aspect ratio and large surface area contribute to significant communication with the adjacent elements, resulting in high water retention, powerful interactions with biomaterials and other polymers, and fixation of nanoparticles of different types (54, 55). In addition, the nanometric

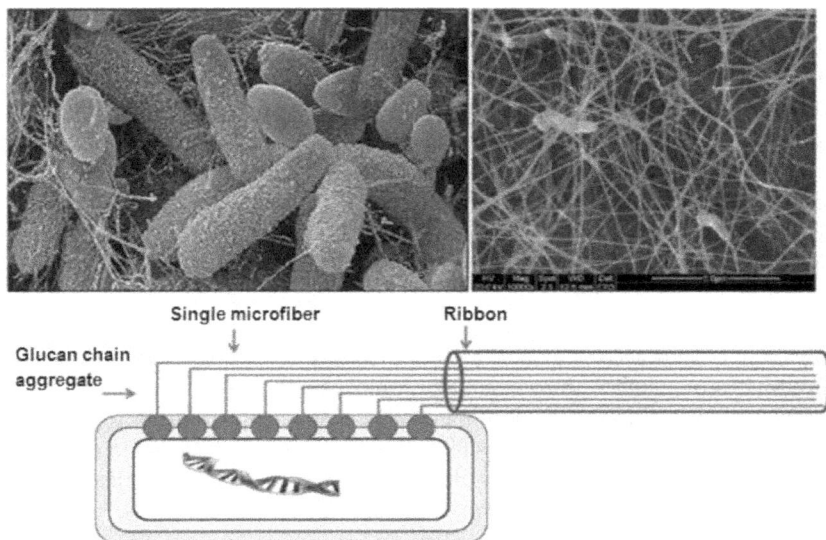

FIGURE 8.2 Scanning electron microscopy image of a BC nanofibrillar network of *Gluconacetobacter xylinus* and schematic representation of bacterial cellulose formation. Image reproduced with permission from (52).

Source: Z. Shi, Y. Zhang, G. O. Phillips, and G. Yang, "Utilization of bacterial cellulose in food," *Food Hydrocolloids*, vol. 35, pp. 539–545, 2014.

diameter of BC fiber gives it a characteristic that is not accessible by plant fibers, that is, transparency, which can be very desirable for many uses. The crystallinity index of BC is also very strong (60%–80%), and BC has high mechanical strength with its 200–300 MPa tensile strength (25, 51). The Young's modulus of BC is 15 GPa (56), and it also has high thermal strength (with a varying decaying temperature from 340 to 370 °C) (57), which is significant for the sterilization processes needed for different products and biomedical instruments.

For certain tissue-engineering applications, BC's tolerance to *in vivo* degradation because of the lack of cellulase in humans and poor solubility may be beneficial (58). Finally, through *in vivo* and *in vitro* studies, the nontoxicity and biocompatibility of BC were also evaluated. Several *in vitro* studies have shown that BC is non-cytotoxic to endothelial cells, Chinese hamster ovaries, and fibroblasts. In addition, BC's *in vivo* cytotoxicity was examined by subcutaneous rat implantation and implant assessment with regard to any signs of inflammation, cell viability, and foreign body responses. The findings obtained showed no major inflammation across the implants and led to the conclusion that BC was useful for the proliferation and attachment of cells (59). The intraperitoneal (within peritoneum) injection of varying BC nanofiber doses in mice gives another solution to these experiments. Blood samples were taken after a few days of exposure, and the findings revealed zero effect on the metabolic profile between the BC-exposed mice and the controls (58). In addition, a recent skin compatibility report showed that BC membranes do not cause skin allergic reactions and induce a positive moisturizing effect when filled with a small proportion of glycerol (60). In addition, BC's highly desirable mechanical properties, *in situ* moldability, biocompatibility, and porosity are also suitable for investigating BC as a tissue-engineering scaffold, including for artificial blood vessels (61), prosthesis of the heart valve (62), artificial cornea (63), artificial cartilage (64), and artificial bone (65). BC also has been identified in a similar vein as an outstanding and nonallergenic biomaterial for the cosmetics sector, where it could be used as a face moisturizer for dry skin treatment (50), as a natural local skin scrub (51), or as a formulating agent for the treatment of personal cleaning compositions (52). Natural BC membranes are also exciting nanostructured systems for drug release, owing to the effectiveness and straightforwardness of the processing of BC membranes loaded with drugs and the fact that it only consists of a monolayer. BC's tendency to absorb exudates and adhere to abnormal skin, along with its consistency, makes it therapeutic for numerous critical problems. In addition to the aforementioned direct applications, research into the production of BC-based nanocomposite materials has been increasing in recent years, with researchers seeking to discover BC's specific properties (especially nanometric parameters and nanostructured networks) for new functional materials.

8.4 BIOCHEMISTRY OF BC

At the heart of BC's existence and properties, like native cellulose, is hydrogen bonding. BC has obtained narrower diameters and enhanced fibril amounts, unlike native cellulose, leading to even stronger hydrogen bonding. Until it was subjected to acid hydrolysis, BC was traditionally synthesized as microfibers. BC's smaller scale of formation also leads to a greater degree of H-bonding. Any alteration such as chemical modification, solubilization, polymerization, or depolymerization can be considered only if the bonding of hydrogen is effectively dealt with. It is used in certain cases as a means of contact with other molecules like glycerol or ethylene glycol, where extensive bonding of hydrogen contributes to high affinity for sorption, making it a hydrophilic object (66). But on the other hand, due to stronger H-bonding, solubility has become a considerable challenge, and only those solvents that decrease the effectiveness of H-bonding have proved effective. N-methylmorpholine N-oxide has been used recently used to dissolve BC where, due to the overwhelming activity of N-methylmorpholine-N-oxide, the crystallinity, a feature of H-bonding, reduced nearly from 79% to 38% (67). However, studies focusing on the pharmaceutical and biomedical applications of BC have not reported at all on the effect of H-bonding. It could not be ruled out that such bonding is accountable for BC's enhanced mechanical properties. Compared to cellulose obtained from other natural sources, BC has greater purity, high biocompatibility, and high mechanical properties. Its strong crystallinity and elastic modulus (EM) signify the mechanical properties of BC (68). Single BC fiber has an EM of 78 GPa reported by atomic force microscopy (AFM) (69) and 114 GPa reported by the Raman spectroscopy (70). Diverse research has given different results for the EM of BC, with AFM showing 78 GPa elastic modulus—slightly higher than the 70 GPa for glass fibers and macroscale native cotton fibers, for example jute and cotton (diameter: 25–200 μm, EM: 13–26.5 GPa and 5.5–12.6 GPa for jute and cotton respectively) (71, 72). Interestingly, the EM is remarkably very poor (15–35 GPa) when BC is rendered into sheets (73). However, the comparative advantages of BC over plant-based cellulose are due to its high degree of polymerization and water-holding capacity and web-like network (55, 74). BC has many chemical characteristics in common with plant-based cellulose, but it remains a different entity. Direct physical alteration of bacterial cellulose was undertaken in a number of experiments, but in terms of certain mechanical properties like burst strain and tensile strength, it needs more improvement. BC's tensile properties generally remain unchanged, but as the ratio of oxygen increases, the burst pressure is increased. In particular, the ratio of oxygen raises BC's burst pressure by increasing the number of layers and the thickness of the tube structure of the internal layer (75). BC's reaction chemistry is becoming significant in this sense. In order to give them modulated properties, Table 8.1 summarizes such processes for BC. Some basic observations are required before explaining the details of BC's chemical interactions. These observations/assumptions may be applicable only to BC's functionality and have not been extended to the distribution of drugs (Table 8.2).

8.5 UNIQUE PROPERTIES OF BNC FOR ITS USE AS DRUG DELIVERY SYSTEM

Exposure to several modern natural materials and quality products is provided by biotechnology. One of these novel biofabricated products developed by bacteria strains of *Komagataeibacter* from glucose in a biotechnological phase is bacterial

---

TABLE 8.1
Bacterial Cellulose Treatment Process and Outcomes.

Treatment type	Result	Reference
Reduction with silver nitrate	Silver nanoparticles adsorption	(76)
Aqueous counter collision	Single nanocellulose fiber formation	(77)
Drying of supercritical fluid	Fine accessed pores	(78)
Gamma radiations	Decreased diffusion coefficient	(79)
Radiation by electron beam	Stimulation of antifungal activity	(79)
Acetylation	Improved BC hygroscopicity	(80)
2,2,6,6-tetramethylpiperidine-1-oxyl-mediated oxidation/sonication	BC microfibril translucent suspension	(81)
Electrospinning	Obtained 1D BC microfibers	(82)
Irradiation by ultrasound	Decreased crystallinity and molecular weight of BC	(83)
In situ action with polyethylene glycol (PEG)	Increased BC fiber poration at lower molecular weight of PEG	(84)
Ionic solvents dissolution	Amorphous character induction	(85)
Agitation through mechanical mixing	Decreased viscosity, increased shear rate of BC suspension	(86)

TABLE 8.2

Bacterial Cellulose Biochemistry Features and Their Possible Applications.

Practical thought	Potential applications
Tensile and mechanical properties' biochemical modification	Mechanically sustainable stuffs for clinical purposes
Hydrophobic modification	Cell adhesion feasibility or transport throughout the phospholipid bilayer of cell increased
Optimized solvent system use	Appropriate testing for distribution and drug delivery
Reaction of bacterial cellulose hydroxyl groups	Novel possible combinations of diverse polymers with BC

nanocellulose (BNC) (62). As regulatory agencies are strengthening the use of xeno-free biomaterials in human treatment and, specifically, in drug delivery, the effectiveness of BNC has increased because of its human-free and animal-free nature. With a wide range of designs, such as spheres, foils, rods, fleeces, fibrous aggregates, and irregularly shaped pulp, BNC may be custom designed (87). In addition, it is possible to obtain strongly crystalline needle-shaped BC nanocrystals by acidic, enzymatic, and oxidative degradation of amorphous fractions (88).

The material shares not only the same chemical structure formula with natural vegetable cellulose, but also characteristics such as sustainability, biodegradation in the environment, easy derivatization, and strong chemical resistance (89). The characteristic 3D nanofibrillary network, however, offers extra material properties than plant cellulose does. These characteristics include extraordinarily high mechanical stability (single-fiber Young's module of 2 GPa) and thermal stability (up to 300 °C), elevated moisture content (>90%), crystallinity (70%–90%), and polymerization degree (up to 10,000 anhydroglucose units), combined with exceptional softness and formability (87, 90). BNC has also been categorized as a biocompatible material ascribed to its biosynthesis process, although extended, biopersistent, and thin nanomaterials as well as nanofibers are subject to intense nano-safety discussions due to their potential shape-dependent health risks (91). *In vitro, in vivo*, and *ex ovo* studies have reported an extraordinarily high biocompatibility (except in long-term applications of over one year) in several species, such as rats, mice, pigs, rabbits, and humans, without chronic inflammation, formation of fibrotic capsules, or genetic or cellular toxicity (58, 60, 92). Even so, although the database of preclinical experiments on *in vitro* cyto- and hemotoxicity in cultured cells and tissues is constantly growing, detailed *in vivo* experiments are still rare, particularly for defined applications. The benefits of BNC, which make it highly attractive for controlled drug delivery, are concluded in Figure 8.3.

8.6 THE ROLE OF BNC PRODUCTION IN DRUG DELIVERY

The mechanization and upscaling of manufacturing processes in the replicable quality of the produced material appear to be a main requirement for their wide use as

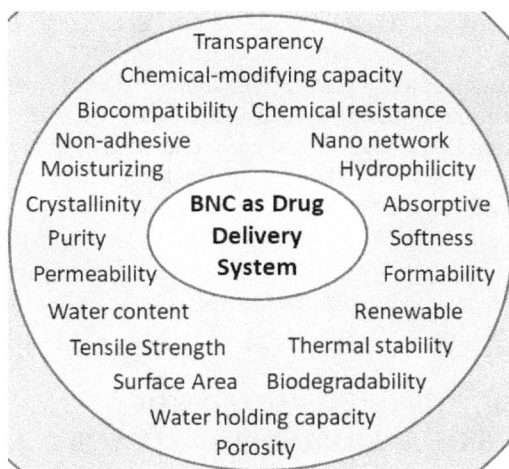

FIGURE 8.3 Overview of advantageous material properties of bacterial nanocellulose for drug delivery.

a system of controlled drug delivery. To date, most BNCs are synthesized by hand in the conventional fermentation industry in Asia of *nata de coco* farms (93, 94); they are cultivated as fleeces in plastic bowls and largely used in the cosmetics and food industries. The use of this form of BNC in the pharmaceutical and medical sectors has so far been hindered by diverse culture situations and feedstock properties, potential impurities, and a relatively low material stability.

A variety of biotechnological methods have been developed over the past few years, varying from batch or fed-batch to continuous cultivation techniques, in addition to this traditional agricultural production technique. As drug delivery systems, flat BNC fleeces in particular have had commercial significance until now. Usually, they are produced in pan-like bioreactors by static cultivation or as a more developed technology in a continuous and integrated biotechnological manufacturing process that could be scaled up for large production (95) with replicable, relatively homogenous fiber networks, greater mechanical strength, and uniform texture. Consequently, in airlift or stirred aerated fermenters, agitated production was often scaled up in submerged cultures. These cultures have a fibrous or pellet-like substance (96). As shear stress encourages bacterial mutation into non-cellulose-producing strains, the BNC yield appears to be lower than it is in static culture (97), but individual strains with high resistance have already been reported. For example, as a drug carrier in oral drug delivery systems, BNC derived from these aggravated cultivation processes may be used and will definitely be researched more widely over the next few years.

For storage, transportation, packaging/labeling, easy handling by doctors and patients, decreased risk of microbial contamination, and manageable exudate absorption, partial or total elimination of moisture from BNC as a drug delivery system may be beneficial. However, since mechanical, thermal, or solvent treatments have been shown to modify the 3D network with a consequent loss of desirable BNC properties, the transformation of wet BNC into xerogels is also a challenge. Freeze-drying (98), critical point-drying after sequentially solvent exchange (99), removal of water under pressure utilizing water-adsorbing materials and/or external heating (29, 100), and air-drying both with and without water-binding additives (101) are standard techniques for dehydrating BNC. The differences found were mainly correlated with the degree of protection of the BNC 3D structure, the loaded drug absorption capability and stability, and the profile of time/cost/environmental hazard. The lower absorption ability of freeze-dried BNC for proteins relative to native BNC has shown the significance of the drying process (102). In contrast with native BNC, air-drying results not only in a systemic breakdown of BNC but also in a very fast release of the model drug azorubine (101).

8.7 HOW TO GET THE DRUG PACKED: THE CHALLENGES OF A HYDROPHILIC NETWORK

The physicochemical properties of the drug, like solubility, molar mass, therapeutic dosage, stability throughout each process step, and the expected drug release profile, should be used to choose the drug-loading system for a given application. In addition, the form of BNC, such as semi-dried (e.g., drained or compressed), native wet, or freeze-dried, BNC alteration, and the bacterial strain used could affect the loading process. Despite major structural differences in the size, lipophilicity, hydrophilicity, and stability of the loaded drugs, the reported drug-loading approaches for BNC are almost comparable and, so far, very limited.

It is possible to differentiate, as summarized in Figure 8.4, between post-synthesis and *in situ* loading approaches. The first is distinguished by the incorporation of active ingredients during the formation of the BNC network and is less commonly used because of the possible impairment of the active ingredient with the growth of BNC-producing bacteria (103). In the latter, which is selected in most experiments, after biosynthesis and purification, drugs are incorporated through the preformed wet fleeces, typically by sorption methods under submerged conditions and gently agitated over 24–48 hours (98, 104). As a significant advantage, this method is quick to conduct and mildly conditioned. As this method is time-consuming, particularly in industrial sectors, two faster alternative approaches have been explored. On the one side, wet BNC was packed, for example, with proteins by vortexing. The very same protein-loading potential obtained by the adsorption process within 24 h was achieved by vortexing within 10 min, relative to the conventional technique. While stability and distribution of protein were comparable, improvements in the fiber network were observed, resulting in a slower release of the drug (105). Conversely, the introduction of drug solutions that favor the benefit of a lower volume of needed drug, specified loading dosage, and better loading efficacy has soaked dry or semi-dried BNC (106). A drug loading by the BNC boiling in the drug solution was also reported, resulting in comparative drug load values of approximately 50% by weight based on the solubilization of the drug (39). The introduction of solutions of the drug into BNC has been proposed to be linked to diffusion forces and capillary action in all of these post-loading approaches. This is focused on the network's wide hydrophilic surface, which consists of up to 99% water, resulting in a homogeneous

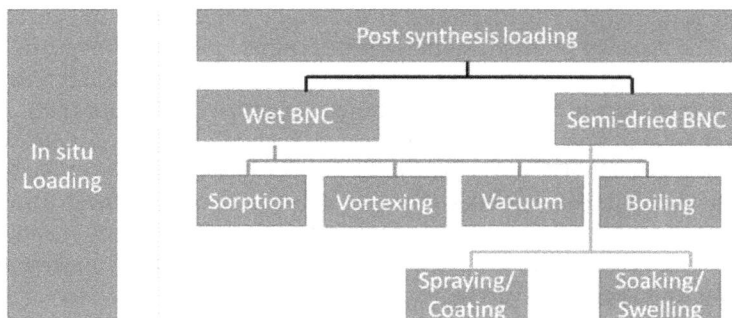

FIGURE 8.4 Summary of common drug-loading preparation applied for bacterial nanocellulose.

distribution of drugs within the BNC (107). Consequently, the resulting release of drugs follows analogous concepts of physical transport that are often described in several articles as the equation of Peppas's semiempirical power law, illustrating the overlay of swelling and diffusion (39, 108). The resultant release profiles are biphasic, with an initial instantaneous release rate within first 0.5–10 h, accompanied by a slow-release phase up to 24 h (98, 108). This well fits, for example, for silk sericin (109), octenidine (108), benzalkonium chloride (106), and diclofenac (110). Loading and release control could be furthermore realized by surface areas and dimensions of BNC, water content of the BNC, drug concentration, incubation time, and shaking velocity (39, 111, 112). Most of the other reports also recorded more or less similar profiles, in particular for small molecules and water-soluble drugs, with changes just in the time periods of both stages. Related but slow profiles were obtained for larger hydrophilic drugs, shown, for example, in povidone-iodine, which demonstrated a slow release mainly due to its high molar mass of drug (104). In addition, protection of their activity and 3D structure due to the normal aqueous phase has been addressed for hydrophilic drugs like proteins (e.g., luciferase).

Conclusively, the design of more advanced loading techniques is urged from a technical perspective, allowing for a fast, effective, and more precise control of the release of the drug from native BNC without additional excipients, particularly with regard to large-scale output.

8.8 BACTERIAL CELLULOSE APPLICATION AS TRANSDERMAL DRUG DELIVERY

Transdermal systems will serve as BC's entrance into the field of drug delivery. It is clear that along with skin substitute products, BC is very effective in wound healing. The characteristics of this wound-healing device can be easily directed toward the delivery of transdermal drugs as it prevents evaporation of moisture, prevents outside contamination, and retains contact with inflamed, exposed, or diseased areas (113). This allows easier localized drug delivery to the target area. Bacterial cellulose dry films were acquired after the exposure to an antimicrobial agent like benzalkonium chloride. The capacity of drug loading was determined to be 0.116 mg/cm^2 per unit surface area, and the drug effects against infected wound flora like *Bacillus subtilis* and *Staphylococcus aureus* (*S. aureus*) lasted for at least 24 hours (114). About a 99.99% inhibition effect against *S. aureus* and *Escherichia coli* (*E. coli*) was successfully developed by BC fibers with silver nanoparticles (Ag-NPs) (115). Bacterial cellulose with Ag-nanocomposites has been approved as an outstanding approach where antibacterial characteristics are essential (116). Propranolol's S-enantiomer, an antihypertensive compound, was released from a methacrylated BC membrane composite layer and applied to transdermal applications where the primary regulation of the release of the drug comes from the parent membrane of BC (76). Molecularly imprinted polymer membrane and gel reservoirs were used to show how a transdermal patch might function for enantiomeric release. Molecularly imprinted polymer membranes alone were examined for *in vitro* drug release and compared to non-imprinting polymer and cellulose membranes. However, *in vitro* skin penetration experiments by dorsal skin excised from Wistar rats of individual males were subjected to the gel reservoir in order that the control imparted on discharge might be defined categorically. The non-imprinting polymer was comprised of a BC membrane with S-propranolol as reactive pore filling. This method was found to impart over the non-imprinting polymer-cellulose membranes the dominant character of enantioselective release to molecularly imprinted polymer. In addition, a combined release of 60% was observed with chitosan and 48% with poloxamer. The improved enantioselectivity of the non-imprinting composite polymer membrane was due to an improvement in the extent of ionization of functional monomer, which enhanced the binding at the imprint site of the S-propranolol enantiomer. Furthermore, the form of gel used has been found to influence the stereoselective propranolol transport, with no selective releases of S-propranolol by the more rheologically formulated poloxamer gel (111). These promising findings showed that BC, which serves as a basis for molecular printing platforms, can be modified for treatments such as reactive pore filling, which enable bacterial cellulose appropriate for chiral uses. A very promising potential for applying BC membranes is in the systems of transdermal delivery that would function bilaterally, to both absorb exudates and deliver drugs. BC membranes' diffusion potential was studied by tetracycline loading in irradiated and non-irradiated electron beam samples. It was found that diffusion was convective from the drug-enriched to the solvent-only cells. In the analysis, the assembly consisted of two compartments of poly-methylmethacrylate, divided by operative membrane. Compared to that with irradiated bacterial cellulose, non-irradiated bacterial cellulose permitted faster movement of the drug, which has been supported by the decline in a coefficient of diffusion of that of the irradiated BC sample. Not only did this research show the potential for transfer via the bacterial cellulose membrane, but in addition suggested a drug adsorption model into the membrane of BC (117). Delivery of drug was tested by putting the film/membrane on uneven surfaces and testing the drug deposition uniformity throughout the epidermis; the rate of permeation was found to be three times higher for the lipophilic drug ibuprofen than for lidocaine hydrochloride (118). Bacterial cellulose could also be used in combination with an electroconductivity-reported conducting polymer, such as polyaniline, which has the ability to serve as a drug delivery system with an electrically induced drug release. The product of aniline polymerization formed in BC matrix on one side and worked as a supercapacitor (119). Therefore, this character may be combined with bacterial cellulose's biodegradability, biocompatibility, and good mechanical properties, even though it is important to adjust the conducting polymers' hydrophobicity for effective protein entrapment. In view of this to expand the application of BC, iontophoresis can be applied. During this observation, multiwalled carbon nanotubes combined with the bacterial cellulose, and an increase of electroconductivity was observed to $1.4 \times 10-1$ S/cm (120, 121). Fang et al. showed that using Na^+ (sodium) and/or ammonium, differently charged moieties of two cellulose lines could be connected to carboxymethyl cellulose. The cellulose-based method for buprenorphine could be complete as compared to a chitosan hydrogel; the ion's bulkiness was the basic factor, with the Na^+ the least bulky with relatively high mobility. Improvement in permeation was therefore attained; although with sodium carboxymethyl cellulose it was lower than with others, the cumulative iontophoretic permeability for CMCNa was the greatest. Compared with the passive transport by the powerful electrophoretic flow of a buprenorphine, ionophoresis permeation enhanced by a factor of 14.27, explaining variations in flux and permeation results. In the same analysis, electroporation was also used, but it proved less successful than iontophoresis. The creation of transient aqueous pores caused this phenomenon to occur. It was also promised that the very same mechanics on the dorsal skin of female mice Balb\c were obtained from the stratum corneum stripped skin (122).

8.9 BC IN HYDROGELS AND AEROGELS

Hydrogels have significant pharmacological and clinical limitations associated with their individual *in vivo* delivery, where limitations are there in cell membranes. In the literature relating the introduction of hydrophobic features to the innately hydrophilic hydrogel, performed through sol-gel chemistry, a different number of approaches are reported.

Hydrophobic inductions could require *in vivo* stimulation or may require additional structures to generate *in situ* hydrogels, such as a linear polymer matrix. Using sol-gel chemistry, the mechanism of hydrogen bonding used in the freeze-thaw method for a starch/cellulose mix could be elicited to synthesize *in vitro* hydrogels that can potentially be used for drug delivery applications (123). Based on polyacrylic acid–grafted BC, an anionic hydrogel was formed that displayed pH-sensitive release at the intestinal pH of a protein bovine serum albumin as a model protein. In this report, hydrogel with pore sizes of 20–110 μm (large size) carried much of the model protein, with higher cumulative release in comparison to SGF in simulated intestinal fluid. At body temperature, a low-swelling pattern was also reported, which enables these hydrogels susceptible to temperature-controlled distribution. For the formulation of 80:20 parts of BC and acrylic acid irradiated with 35 kGy, which contained greater bovine serum albumin due to its bigger pore size, a maximum bovine serum albumin release to 90% was shown. A feature of the radiation dose appears to be porous and hence has the characteristics of trap quality. This hydrogel, according to Peppas's semi-empirical power law, showed non-Fickian diffusion of varying values from 0.46 to 0.76 (124). BC nanocellulose has been shown to properly control the uptake of albumin and release in another study conducted and therefore may be an interesting candidate for the managed delivery of drugs; the studies illustrated retention of protein integrity, which can also be impaired in synthesis methods. The decrease in the uptake of drug's ability of freeze-dried samples, which is a somewhat ordinary feature of BC formulations, is an important finding in this review (102). Using an effective drying technology, aerogels are produced from wet gels, the purpose being to prevent pores from collapsing so as to facilitate the packing/loading of drug ingredients, facilitated by supercritical state fluid drying and the related modifications of such a technology (125). The sterilization and washing of a hydrogel, solvent quantitative exchange, and simultaneous drying of supercritical fluid with carbon dioxide are important processes for making BC aerogel of ultralight weight. This innovative bacterial cellulose aerogel has the outstanding function of maintaining dimensional integrity, which could be a challenge for low-density structures. The low density in this case contributes to the attractiveness and extends the possible uses for this BC aerogel (126). Nanostructures have currently become a promising modality for drug delivery, but a big drawback is their potential for agglomeration. Aerogels not only tackle the aggregation issue, but have enhanced control drug release also. A significant study documenting the discharge from beclomethasone dipropionate nanoparticles coated with proteins hydrophobin use four nanofibrillated celluloses and are correlated with Avicel PH1011 microcrystallinic cellulose. Although TEMPO-oxidized birch cellulose aerogels and quince seed cellulose did not provide sustained release, BC did (127). L-ascorbic acid and dexpanthenol were also loaded into aerogels BC, which were adjusted using solvent rewetting surface tension to possess 100% of their original water content. The appropriateness of BC aerogels as a controlled release matrix is further affirmed by this rehydration characteristics with a related high pore volume (78). If BC is treated as a synthetic hydrogel and made with PEG by freeze thawing, BC can be produced in a composite form. PEG generates pores that support its capacity for their applications in pharmacokinetics within the gel, and it can also enhance BC's properties of drug delivery (128).

8.10 BACTERIAL CELLULOSE ANISOTROPIC PROPERTY FOR DRUG DELIVERY

BC's biocompatibility is a significant advantage and important property in the delivery of drugs and should be studied after some improvements have been made. In the case of composites, this is particularly important. A composite of polyvinyl alcohol-bacterial cellulose (PVA-BC) has been reported on the principle that BC and PVA both are naturally biocompatible. The composite of PVA-BC was engineered for the vascular tissue simulation and was necessary to have a wide variety of mechanical aspects (129). Further, the material has been developed to anisotropy factor, therefore manufacturing the nanocomposite is reliant on the material and processing parameters, a direction-dependent unit, either axially or circumferentially (62). As a stratified structure exhibiting mechanical anisotropy, BC can present a very high tensile modulus in the direction of the fiber layer and a low compressive modulus in the perpendicular direction (130). A current study demonstrates the synthesis of hydrogel by dipping pellicles of BC in a saturated acrylamide. The range of 4.5% to 25% w/w of BC concentration was used. Perpendicular and parallel swelling measurements were performed to the pellicle top and bottom surfaces, with the samples obtained from the BC pellicles. Only the hydrogel composite displayed swelling in thickness, whereas the width and length cut parts were unchanged with the additional condition that no presynthesis pressing was done on the particular subset of hydrogels. Additionally, the subset shrank in thickness-oriented swelling with an increasing amount of ethanol. In order to incorporate carboxylate groups into the hydrogel, partial amide hydrolysis also was undertaken; width swelling was reported for an ionic hydrogel that displayed swelling amounts of 2.6-fold to 2.8-fold. As suggested by Millon et al., it strongly resembles a tunnel-like structure (131). The current emphasis in drug distribution geometry is attributed to anisotropy. A number of *in vivo* metabolic functions like elimination and permeation are thought to be influenced by the shape and size of particles. Anisotropy can also help improve the potential of drug loading, pharmacokinetics, and release profiles of the drug. This involves an analysis of the function of geometry by polymeric nanocarriers' delivery of drugs. In addition, BC nanocomposites can be considered to come into the group of non-spherical transporters to which the anisotropy principle is well adapted. It is known that, due to their nanoscale dimension and massive surface area-to-volume ratio, non-spherical transporters are ideal choices for drug targeting, leading to increased target interactions and improved permeability, respectively (132).

8.11 CONCLUSIONS

After studying published evidence and medical information authorized by the US Food and Drug Administration, it was apparent that bacterial cellulose should be used for equipment and implants, in drug delivery, for transdermal applications, as tissue-engineering scaffolds, and as a therapeutic active ingredient. Oral and transdermal distribution are included in the potential administration routes. In addition, cell therapy for proteins as well as other drugs may also be combined with drug delivery. This suggests that bacterial cellulose can be applicable in even more ways than thought previously, and to completely exploit bacterial cellulose drug delivery ability, a multidisciplinary approach is needed.

REFERENCES

(1) T. M. Allen and P. R. Cullis, "Drug delivery systems: Entering the mainstream," *Science*, vol. 303, pp. 1818–1822, 2004.
(2) C. Shi, D. Guo, K. Xiao, X. Wang, L. Wang, and J. Luo, "A drug-specific nanocarrier design for efficient anticancer therapy," *Nature Communications*, vol. 6, pp. 1–14, 2015.
(3) E. Gultepe, D. Nagesha, S. Sridhar, and M. Amiji, "Nanoporous inorganic membranes or coatings for sustained drug delivery in implantable devices," *Advanced Drug Delivery Reviews*, vol. 62, pp. 305–315, 2010.
(4) M. P. Stewart, A. Sharei, X. Ding, G. Sahay, R. Langer, and K. F. Jensen, "In vitro and ex vivo strategies for intracellular delivery," *Nature*, vol. 538, pp. 183–192, 2016.

(5) Y.-L. Luo, X.-Y. Zhang, Y. Wang, F.-J. Han, F. Xu, and Y.-S. Chen, "Mediating physico-chemical properties and paclitaxel release of pH-responsive H-type multiblock copolymer self-assembly nanomicelles through epoxidation," *Journal of Materials Chemistry B*, vol. 5, pp. 3111–3121, 2017.

(6) J. Wu and M. J. Sailor, "Chitosan hydrogel-capped porous SiO2 as a pH responsive nano-valve for triggered release of insulin," *Advanced Functional Materials*, vol. 19, pp. 733–741, 2009.

(7) F. Tang, L. Li, and D. Chen, "Mesoporous silica nanoparticles: Synthesis, biocompatibility and drug delivery," *Advanced Materials*, vol. 24, pp. 1504–1534, 2012.

(8) A. Bang, A. G. Sadekar, C. Buback, B. Curtin, S. Acar, D. Kolasinac, et al., "Evaluation of dysprosia aerogels as drug delivery systems: A comparative study with random and ordered mesoporous silicas," *ACS Applied Materials & Interfaces*, vol. 6, pp. 4891–4902, 2014.

(9) H. D. Follmann, O. N. Oliveira, D. Lazarin-Bidóia, C. V. Nakamura, X. Huang, T. Asefa, et al., "Multifunctional hybrid aerogels: Hyperbranched polymer-trapped mesoporous silica nanoparticles for sustained and prolonged drug release," *Nanoscale*, vol. 10, pp. 1704–1715, 2018.

(10) H. Cai, X. Wang, H. Zhang, L. Sun, D. Pan, Q. Gong, et al., "Enzyme-sensitive biodegradable and multifunctional polymeric conjugate as theranostic nanomedicine," *Applied Materials Today*, vol. 11, pp. 207–218, 2018.

(11) Y. Dai, H. Cai, Z. Duan, X. Ma, Q. Gong, K. Luo, et al., "Effect of polymer side chains on drug delivery properties for cancer therapy," *Journal of Biomedical Nanotechnology*, vol. 13, pp. 1369–1385, 2017.

(12) E. Pinho and G. Soares, "Functionalization of cotton cellulose for improved wound healing," *Journal of Materials Chemistry B*, vol. 6, pp. 1887–1898, 2018.

(13) N. Bhattarai, J. Gunn and M. Zhang, "Chitosan-based hydrogels for controlled, localized drug delivery," *Advanced Drug Delivery Reviews*, vol. 62, pp. 83–99, 2010.

(14) C. Elvira, J. Mano, J. San Roman, and R. Reis, "Starch-based biodegradable hydrogels with potential biomedical applications as drug delivery systems," *Biomaterials*, vol. 23, pp. 1955–1966, 2002.

(15) Y.-H. Lin, H.-F. Liang, C.-K. Chung, M.-C. Chen, and H.-W. Sung, "Physically crosslinked alginate/N, O-carboxymethyl chitosan hydrogels with calcium for oral delivery of protein drugs," *Biomaterials*, vol. 26, pp. 2105–2113, 2005.

(16) S. Peng, G. Jin, L. Li, K. Li, M. Srinivasan, S. Ramakrishna, et al., "Multi-functional electrospun nanofibres for advances in tissue regeneration, energy conversion & storage, and water treatment," *Chemical Society Reviews*, vol. 45, pp. 1225–1241, 2016.

(17) O. Galkina, V. Ivanov, A. Agafonov, G. Seisenbaeva, and V. Kessler, "Cellulose nanofiber–titania nanocomposites as potential drug delivery systems for dermal applications," *Journal of Materials Chemistry B*, vol. 3, pp. 1688–1698, 2015.

(18) A. J. Svagan, J.-W. Benjamins, Z. Al-Ansari, D. B. Shalom, A. Müllertz, L. Wågberg, et al., "Solid cellulose nanofiber based foams—Towards facile design of sustained drug delivery systems," *Journal of Controlled Release*, vol. 244, pp. 74–82, 2016.

(19) J. Zhao, C. Lu, X. He, X. Zhang, W. Zhang, and X. Zhang, "Polyethylenimine-grafted cellulose nanofibril aerogels as versatile vehicles for drug delivery," *ACS Applied Materials & Interfaces*, vol. 7, pp. 2607–2615, 2015.

(20) S. J. Eichhorn, A. Dufresne, M. Aranguren, N. Marcovich, J. Capadona, S. J. Rowan, et al., "Current international research into cellulose nanofibres and nanocomposites," *Journal of Materials Science*, vol. 45, pp. 1–33, 2010.

(21) C. Zhou and Q. Wu, "Recent development in applications of cellulose nanocrystals for advanced polymer-based nanocomposites by novel fabrication strategies," *Nanocrystals-Synthesis, Characterization and Applications*, pp. 103–120, 2012.

(22) B. L. Peng, N. Dhar, H. Liu, and K. Tam, "Chemistry and applications of nanocrystalline cellulose and its derivatives: A nanotechnology perspective," *The Canadian Journal of Chemical Engineering*, vol. 89, pp. 1191–1206, 2011.

(23) R. Rusli, K. Shanmuganathan, S. J. Rowan, C. Weder, and S. J. Eichhorn, "Stress-transfer in anisotropic and environmentally adaptive cellulose whisker nanocomposites," *Biomacromolecules*, vol. 11, pp. 762–768, 2010.

(24) T. Kamal, I. Ahmad, S. B. Khan, and A. M. Asiri, "Bacterial cellulose as support for bio-polymer stabilized catalytic cobalt nanoparticles," *International Journal of Biological Macromolecules*, vol. 135, pp. 1162–1170, 2019/08/15/ 2019.

(25) D. Klemm, B. Heublein, H. P. Fink, and A. Bohn, "Cellulose: Fascinating biopolymer and sustainable raw material," *Angewandte Chemie International Edition*, vol. 44, pp. 3358–3393, 2005.

(26) W. Czaja, A. Krystynowicz, S. Bielecki, and R. M. Brown, "Microbial cellulose—the natural power to heal wounds," *Biomaterials*, vol. 27, pp. 145–151, 2006/01/01/ 2006.

(27) W. K. Czaja, D. J. Young, M. Kawecki, and R. M. Brown, "The future prospects of micro-bial cellulose in biomedical applications," *Biomacromolecules*, vol. 8, pp. 1–12, 2007.

(28) D. Klemm, D. Schumann, U. Udhardt, and S. Marsch, "Bacterial synthesized cellulose—artificial blood vessels for microsurgery," *Progress in Polymer Science*, vol. 26, pp. 1561–1603, 2001/11/01/ 2001.

(29) E. Trovatti, C. S. R. Freire, P. C. Pinto, I. F. Almeida, P. Costa, A. J. D. Silvestre, et al., "Bacterial cellulose membranes applied in topical and transdermal delivery of lido-caine hydrochloride and ibuprofen: In vitro diffusion studies," *International Journal of Pharmaceutics*, vol. 435, pp. 83–87, 2012/10/01/ 2012.

(30) E. Trovatti, N. H. Silva, I. F. Duarte, C. F. Rosado, I. F. Almeida, P. Costa, et al., "Biocellulose membranes as supports for dermal release of lidocaine," *Biomacromolecules*, vol. 12, pp. 4162–4168, 2011.

(31) L. Fu, Y. Zhang, C. Li, Z. Wu, Q. Zhuo, X. Huang, et al., "Skin tissue repair materi-als from bacterial cellulose by a multilayer fermentation method," *Journal of Materials Chemistry*, vol. 22, pp. 12349–12357, 2012.

(32) P. N. Mendes, S. C. Rahal, O. C. M. Pereira-Junior, V. E. Fabris, S. L. R. Lenharo, J. F. de Lima-Neto, et al., "In vivo and in vitro evaluation of an Acetobacter xylinum synthe-sized microbial cellulose membrane intended for guided tissue repair," *Acta Veterinaria Scandinavica*, vol. 51, p. 12, 2009.

(33) L. S. Scuro, P. Simioni, D. Grabriel, E. E. Saviani, L. V. Modolo, W. M. Tamashiro, et al., "Suppression of nitric oxide production in mouse macrophages by soybean flavonoids accumulated in response to nitroprusside and fungal elicitation," *BMC Biochemistry*, vol. 5, p. 5, 2004.

(34) E. Lenselink and A. Andriessen, "A cohort study on the efficacy of a polyhexanide-con-taining biocellulose dressing in the treatment of biofilms in wounds," *Journal of Wound Care*, vol. 20, pp. 534–539, 2011.

(35) T. Amnuaikit, T. Chusuit, P. Raknam, and P. Boonme, "Effects of a cellulose mask syn-thesized by a bacterium on facial skin characteristics and user satisfaction," *Medical Devices (Auckland, NZ)*, vol. 4, p. 77, 2011.

(36) C. Chen, T. Zhang, Q. Zhang, Z. Feng, C. Zhu, Y. Yu, et al., "Three-dimensional BC/PEDOT composite nanofibers with high performance for electrode—cell interface," *ACS Applied Materials & Interfaces*, vol. 7, pp. 28244–28253, 2015.

(37) C. Chen, T. Zhang, B. Dai, H. Zhang, X. Chen, J. Yang, et al., "Rapid fabrication of composite hydrogel microfibers for weavable and sustainable antibacterial applications," *ACS Sustainable Chemistry & Engineering*, vol. 4, pp. 6534–6542, 2016.

(38) C. Chen, X. Chen, H. Zhang, Q. Zhang, L. Wang, C. Li, et al., "Electrically-responsive core-shell hybrid microfibers for controlled drug release and cell culture," *Acta Biomaterialia*, vol. 55, pp. 434–442, 2017.

(39) L. Huang, X. Chen, T. X. Nguyen, H. Tang, L. Zhang, and G. Yang, "Nano-cellulose 3D-networks as controlled-release drug carriers," *Journal of Materials Chemistry B*, vol. 1, pp. 2976–2984, 2013.

(40) N. Ahmad, M. C. I. M. Amin, S. M. Mahali, I. Ismail, and V. T. G. Chuang, "Biocompatible and mucoadhesive bacterial cellulose-g-poly (acrylic acid) hydrogels for oral protein delivery," *Molecular Pharmaceutics*, vol. 11, pp. 4130–4142, 2014.

(41) M. M. Abeer, M. C. I. Mohd Amin, and C. Martin, "A review of bacterial cellulose-based drug delivery systems: Their biochemistry, current approaches and future prospects," *Journal of Pharmacy and Pharmacology*, vol. 66, pp. 1047–1061, 2014.

(42) B. H. Rehm, "Bacterial polymers: Biosynthesis, modifications and applications," *Nature Reviews Microbiology*, vol. 8, pp. 578–592, 2010.

(43) E. Bernardo, B. Neilan, and I. Couperwhite, "Characterization, differentiation and identification of wild-type cellulose-synthesizing Acetobacter strains involved in Nata de Coco production," *Systematic and Applied Microbiology*, vol. 21, pp. 599–608, 1998.

(44) P. Chen, S. Y. Cho, and H.-J. Jin, "Modification and applications of bacterial celluloses in polymer science," *Macromolecular Research*, vol. 18, pp. 309–320, 2010.

(45) N. Halib, M. Amin, and I. Ahmad, "Physicochemical properties and characterization of nata de coco from local food industries as a source of cellulose," *Sains Malaysiana*, vol. 41, pp. 205–211, 2012.

(46) D. Klemm, D. Schumann, U. Udhardt, and S. Marsch, "Bacterial synthesized cellulose—artificial blood vessels for microsurgery," *Progress in Polymer Science*, vol. 26, pp. 1561–1603, 2001.

(47) D. R. Ruka, G. P. Simon, and K. M. Dean, "Altering the growth conditions of *Gluconacetobacter xylinus* to maximize the yield of bacterial cellulose," *Carbohydrate Polymers*, vol. 89, pp. 613–622, 2012.

(48) D. Mikkelsen, B. M. Flanagan, G. Dykes, and M. Gidley, "Influence of different carbon sources on bacterial cellulose production by *Gluconacetobacter xylinus* strain ATCC 53524," *Journal of Applied Microbiology*, vol. 107, pp. 576–583, 2009.

(49) A. Kurosumi, C. Sasaki, Y. Yamashita, and Y. Nakamura, "Utilization of various fruit juices as carbon source for production of bacterial cellulose by Acetobacter xylinum NBRC 13693," *Carbohydrate Polymers*, vol. 76, pp. 333–335, 2009.

(50) C. Castro, R. Zuluaga, J.-L. Putaux, G. Caro, I. Mondragon, and P. Gañán, "Structural characterization of bacterial cellulose produced by Gluconacetobacter swingsii sp. from Colombian agroindustrial wastes," *Carbohydrate Polymers*, vol. 84, pp. 96–102, 2011.

(51) P. R. Chawla, I. B. Bajaj, S. A. Survase, and R. S. Singhal, "Microbial cellulose: Fermentative production and applications," *Food Technology and Biotechnology*, vol. 47, pp. 107–124, 2009.

(52) Z. Shi, Y. Zhang, G. O. Phillips, and G. Yang, "Utilization of bacterial cellulose in food," *Food Hydrocolloids*, vol. 35, pp. 539–545, 2014.

(53) S. Kalia, A. Dufresne, B. M. Cherian, B. Kaith, L. Avérous, J. Njuguna, et al., "Cellulose-based bio-and nanocomposites: A review," *International Journal of Polymer Science*, vol. 2011, 2011.

(54) L. Fu, J. Zhang, and G. Yang, "Present status and applications of bacterial cellulose-based materials for skin tissue repair," *Carbohydrate Polymers*, vol. 92, pp. 1432–1442, 2013.

(55) D. Klemm, D. Schumann, F. Kramer, N. Heßler, M. Hornung, H.-P. Schmauder, et al., "Nanocelluloses as innovative polymers in research and application," in *Polysaccharides Ii*. Springer, 2006, pp. 49–96.

(56) S. Yamanaka, K. Watanabe, N. Kitamura, M. Iguchi, S. Mitsuhashi, Y. Nishi, et al., "The structure and mechanical properties of sheets prepared from bacterial cellulose," *Journal of Materials Science*, vol. 24, pp. 3141–3145, 1989.

(57) Z. Shi, S. Zang, F. Jiang, L. Huang, D. Lu, Y. Ma, et al., "In situ nano-assembly of bacterial cellulose—polyaniline composites," *Rsc Advances*, vol. 2, pp. 1040–1046, 2012.

(58) S. I. Jeong, S. E. Lee, H. Yang, Y.-H. Jin, C.-S. Park, and Y. S. Park, "Toxicologic evaluation of bacterial synthesized cellulose in endothelial cells and animals," *Molecular & Cellular Toxicology*, vol. 6, pp. 370–377, 2010.

(59) F. Lina, Z. Yue, Z. Jin, and Y. Guang, "Bacterial cellulose for skin repair materials," *Biomedical Engineering—Frontiers and Challenges*, pp. 249–274, 2011.

(60) I. Almeida, T. Pereira, N. Silva, F. Gomes, A. Silvestre, C. Freire, et al., "Bacterial cellulose membranes as drug delivery systems: An in vivo skin compatibility study," *European Journal of Pharmaceutics and Biopharmaceutics*, vol. 86, pp. 332–336, 2014.

(61) Y. Wan, C. Gao, M. Han, H. Liang, K. Ren, Y. Wang, et al., "Preparation and characterization of bacterial cellulose/heparin hybrid nanofiber for potential vascular tissue engineering scaffolds," *Polymers for Advanced Technologies*, vol. 22, pp. 2643–2648, 2011.

(62) L. Millon and W. Wan, "The polyvinyl alcohol—bacterial cellulose system as a new nanocomposite for biomedical applications," *Journal of Biomedical Materials Research Part B: Applied Biomaterials: An Official Journal of The Society for Biomaterials, The Japanese Society for Biomaterials, and The Australian Society for Biomaterials and the Korean Society for Biomaterials*, vol. 79, pp. 245–253, 2006.

(63) J. Wang, C. Gao, Y. Zhang, and Y. Wan, "Preparation and in vitro characterization of BC/ PVA hydrogel composite for its potential use as artificial cornea biomaterial," *Materials Science and Engineering: C*, vol. 30, pp. 214–218, 2010.

(64) L. Nimeskern, H. M. Ávila, J. Sundberg, P. Gatenholm, R. Müller, and K. S. Stok, "Mechanical evaluation of bacterial nanocellulose as an implant material for ear cartilage replacement," *Journal of the Mechanical Behavior of Biomedical Materials*, vol. 22, pp. 12–21, 2013.

(65) K. A. Zimmermann, J. M. LeBlanc, K. T. Sheets, R. W. Fox, and P. Gatenholm, "Biomimetic design of a bacterial cellulose/hydroxyapatite nanocomposite for bone healing applications," *Materials Science and Engineering: C*, vol. 31, pp. 43–49, 2011.

(66) L. K. Pandey, C. Saxena, and V. Dubey, "Studies on pervaporative characteristics of bacterial cellulose membrane," *Separation and Purification Technology*, vol. 42, pp. 213–218, 2005.

(67) G. Shanshan, W. Jianqing, and J. Zhengwei, "Preparation of cellulose films from solution of bacterial cellulose in NMMO," *Carbohydrate Polymers*, vol. 87, pp. 1020–1025, 2012.

(68) W. Czaja, D. Romanovicz, and R. Malcolm Brown, "Structural investigations of microbial cellulose produced in stationary and agitated culture," *Cellulose*, vol. 11, pp. 403–411, 2004.

(69) G. Guhados, W. Wan, and J. L. Hutter, "Measurement of the elastic modulus of single bacterial cellulose fibers using atomic force microscopy," *Langmuir*, vol. 21, pp. 6642–6646, 2005.

(70) Y.-C. Hsieh, H. Yano, M. Nogi, and S. Eichhorn, "An estimation of the Young's modulus of bacterial cellulose filaments," *Cellulose*, vol. 15, pp. 507–513, 2008.

(71) J. Juntaro, M. Pommet, A. Mantalaris, M. Shaffer, and A. Bismarck, "Nanocellulose enhanced interfaces in truly green unidirectional fibre reinforced composites," *Composite Interfaces*, vol. 14, pp. 753–762, 2007.

(72) A. Mohanty, M. a. Misra, and G. Hinrichsen, "Biofibres, biodegradable polymers and biocomposites: An overview," *Macromolecular Materials and Engineering*, vol. 276, pp. 1–24, 2000.

(73) N. Soykeabkaew, C. Sian, S. Gea, T. Nishino, and T. Peijs, "All-cellulose nanocomposites by surface selective dissolution of bacterial cellulose," *Cellulose*, vol. 16, pp. 435–444, 2009.

(74) H. S. Barud, C. Barrios, T. Regiani, R. F. Marques, M. Verelst, J. Dexpert-Ghys, et al., "Self-supported silver nanoparticles containing bacterial cellulose membranes," *Materials Science and Engineering: C*, vol. 28, pp. 515–518, 2008.

(75) A. Bodin, H. Bäckdahl, H. Fink, L. Gustafsson, B. Risberg, and P. Gatenholm, "Influence of cultivation conditions on mechanical and morphological properties of bacterial cellulose tubes," *Biotechnology and Bioengineering*, vol. 97, pp. 425–434, 2007.

(76) C. Bodhibukkana, T. Srichana, S. Kaewnopparat, N. Tangthong, P. Bouking, G. P. Martin, et al., "Composite membrane of bacterially-derived cellulose and molecularly imprinted polymer for use as a transdermal enantioselective controlled-release system of racemic propranolol," *Journal of Controlled Release*, vol. 113, pp. 43–56, 2006.

(77) R. Kose, I. Mitani, W. Kasai, and T. Kondo, ""Nanocellulose" as a single nanofiber prepared from pellicle secreted by *Gluconacetobacter xylinus* using aqueous counter collision," *Biomacromolecules*, vol. 12, pp. 716–720, 2011.

(78) E. Haimer, M. Wendland, K. Schlufter, K. Frankenfeld, P. Miethe, A. Potthast, et al., "Loading of bacterial cellulose aerogels with bioactive compounds by antisolvent precipitation with supercritical carbon dioxide," *Macromolecular Symposia*, vol. 294, pp. 64–74, 2010. https://onlinelibrary.wiley.com/doi/abs/10.1002/masy.201000008.

(79) P. Wanichapichart, W. Taweepreeda, S. Nawae, P. Choomgan, and D. Yasenchak, "Chain scission and anti fungal effect of electron beam on cellulose membrane," *Radiation Physics and Chemistry*, vol. 81, pp. 949–953, 2012.

(80) S. Ifuku, M. Nogi, K. Abe, K. Handa, F. Nakatsubo, and H. Yano, "Surface modification of bacterial cellulose nanofibers for property enhancement of optically transparent composites: Dependence on acetyl-group DS," *Biomacromolecules*, vol. 8, pp. 1973–1978, 2007.

(81) T. Saito, Y. Nishiyama, J.-L. Putaux, M. Vignon, and A. Isogai, "Homogeneous suspensions of individualized microfibrils from TEMPO-catalyzed oxidation of native cellulose," *Biomacromolecules*, vol. 7, pp. 1687–1691, 2006.

(82) R. T. Olsson, R. Kraemer, A. Lopez-Rubio, S. Torres-Giner, M. J. Ocio, and J. M. Lagarón, "Extraction of microfibrils from bacterial cellulose networks for electrospinning of anisotropic biohybrid fiber yarns," *Macromolecules*, vol. 43, pp. 4201–4209, 2010.

(83) S.-S. Wong, S. Kasapis, and Y. M. Tan, "Bacterial and plant cellulose modification using ultrasound irradiation," *Carbohydrate Polymers*, vol. 77, pp. 280–287, 2009.

(84) N. Heßler and D. Klemm, "Alteration of bacterial nanocellulose structure by in situ modification using polyethylene glycol and carbohydrate additives," *Cellulose*, vol. 16, pp. 899–910, 2009.

(85) Y. Wan, L. Hong, S. Jia, Y. Huang, Y. Zhu, Y. Wang, et al., "Synthesis and characterization of hydroxyapatite—bacterial cellulose nanocomposites," *Composites Science and Technology*, vol. 66, pp. 1825–1832, 2006.

(86) T. Kouda, H. Yano, F. Yoshinaga, M. Kaminoyama, and M. Kamiwano, "Characterization of non-Newtonian behavior during mixing of bacterial cellulose in a bioreactor," *Journal of Fermentation and Bioengineering*, vol. 82, pp. 382–386, 1996.

(87) D. Klemm, F. Kramer, S. Moritz, T. Lindström, M. Ankerfors, D. Gray, et al., "Nanocelluloses: A new family of nature-based materials," *Angewandte Chemie International Edition*, vol. 50, pp. 5438–5466, 2011.

(88) N. Duran, A. Paula Lemes, and A. B Seabra, "Review of cellulose nanocrystals patents: Preparation, composites and general applications," *Recent Patents on Nanotechnology*, vol. 6, pp. 16–28, 2012.

(89) K. Schlufter, H. P. Schmauder, S. Dorn, and T. Heinze, "Efficient homogeneous chemical modification of bacterial cellulose in the ionic liquid 1-N-butyl-3-methylimidazolium chloride," *Macromolecular Rapid Communications*, vol. 27, pp. 1670–1676, 2006.

(90) H. Barud, C. Ribeiro, M. Crespi, M. Martines, J. Dexpert-Ghys, R. Marques, et al., "Thermal characterization of bacterial cellulose—phosphate composite membranes," *Journal of Thermal Analysis and Calorimetry*, vol. 87, pp. 815–818, 2007.

(91) N. Lin and A. Dufresne, "Nanocellulose in biomedicine: Current status and future prospect," *European Polymer Journal*, vol. 59, pp. 302–325, 2014.

(92) D. A. Schumann, J. Wippermann, D. O. Klemm, F. Kramer, D. Koth, H. Kosmehl, et al., "Artificial vascular implants from bacterial cellulose: Preliminary results of small arterial substitutes," *Cellulose*, vol. 16, pp. 877–885, 2009.

(93) M. Phisalaphong, T.-K. Tran, S. Taokaew, R. Budiraharjo, G. G. Febriana, D.-N. Nguyen, et al., "Nata de coco industry in Vietnam, Thailand, and Indonesia," in *Bacterial Nanocellulose*. Elsevier, 2016, pp. 231–236.

(94) Y. Pötzinger, D. Kralisch, and D. Fischer, "Bacterial nanocellulose: The future of controlled drug delivery?" *Therapeutic Delivery*, vol. 8, pp. 753–761, 2017.

(95) D. Kralisch, N. Hessler, D. Klemm, R. Erdmann, and W. Schmidt, "White biotechnology for cellulose manufacturing—the HoLiR concept," *Biotechnology and Bioengineering*, vol. 105, pp. 740–747, 2010.

(96) S.-C. Wu and M.-H. Li, "Production of bacterial cellulose membranes in a modified airlift bioreactor by *Gluconacetobacter xylinus*," *Journal of Bioscience and Bioengineering*, vol. 120, pp. 444–449, 2015.

(97) K. Y. Lee, G. Buldum, A. Mantalaris, and A. Bismarck, "More than meets the eye in bacterial cellulose: Biosynthesis, bioprocessing, and applications in advanced fiber composites," *Macromolecular Bioscience*, vol. 14, pp. 10–32, 2014.

(98) W. Shao, H. Liu, S. Wang, J. Wu, M. Huang, H. Min, et al., "Controlled release and antibacterial activity of tetracycline hydrochloride-loaded bacterial cellulose composite membranes," *Carbohydrate Polymers*, vol. 145, pp. 114–120, 2016.

(99) M. Zeng, A. Laromaine, and A. Roig, "Bacterial cellulose films: Influence of bacterial strain and drying route on film properties," *Cellulose*, vol. 21, pp. 4455–4469, 2014.

(100) Y. Qiu, L. Qiu, J. Cui, and Q. Wei, "Bacterial cellulose and bacterial cellulose-vaccarin membranes for wound healing," *Materials Science and Engineering: C*, vol. 59, pp. 303–309, 2016.

(101) A. Müller, M. Zink, N. Hessler, F. Wesarg, F. A. Müller, D. Kralisch, et al., "Bacterial nanocellulose with a shape-memory effect as potential drug delivery system," *Rsc Advances*, vol. 4, pp. 57173–57184, 2014.

(102) A. Müller, Z. Ni, N. Hessler, F. Wesarg, F. A. Müller, D. Kralisch, et al., "The biopolymer bacterial nanocellulose as drug delivery system: Investigation of drug loading and release using the model protein albumin," *Journal of Pharmaceutical Sciences*, vol. 102, pp. 579–592, 2013.

(103) M. L. Cacicedo, K. Cesca, V. E. Bosio, L. M. Porto, and G. R. Castro, "Self-assembly of carrageenin—$CaCO_3$ hybrid microparticles on bacterial cellulose films for doxorubicin sustained delivery," *Journal of Applied Biomedicine*, vol. 13, pp. 239–248, 2015.

(104) C. Wiegand, S. Moritz, N. Hessler, D. Kralisch, F. Wesarg, F. A. Müller, et al., "Antimicrobial functionalization of bacterial nanocellulose by loading with polihexanide and povidone-iodine," *Journal of Materials Science: Materials in Medicine*, vol. 26, p. 245, 2015.

(105) A. Müller, F. Wesarg, N. Hessler, F. A. Müller, D. Kralisch, and D. Fischer, "Loading of bacterial nanocellulose hydrogels with proteins using a high-speed technique," *Carbohydrate Polymers*, vol. 106, pp. 410–413, 2014.

(106) B. V. Mohite, R. K. Suryawanshi, and S. V. Patil, "Study on the drug loading and release potential of bacterial cellulose," *Cellulose Chemistry and Technology*, vol. 50, pp. 219–223, 2016.

(107) Y. Alkhatib, M. Dewaldt, S. Moritz, R. Nitzsche, D. Kralisch, and D. Fischer, "Controlled extended octenidine release from a bacterial nanocellulose/Poloxamer hybrid system," *European Journal of Pharmaceutics and Biopharmaceutics*, vol. 112, pp. 164–176, 2017.

(108) S. Moritz, C. Wiegand, F. Wesarg, N. Hessler, F. A. Müller, D. Kralisch, et al., "Active wound dressings based on bacterial nanocellulose as drug delivery system for octenidine," *International Journal of Pharmaceutics*, vol. 471, pp. 45–55, 2014.

(109) P. Aramwit and N. Bang, "The characteristics of bacterial nanocellulose gel releasing silk sericin for facial treatment," *BMC Biotechnology*, vol. 14, p. 104, 2014.

(110) N. H. Silva, A. F. Rodrigues, I. F. Almeida, P. C. Costa, C. Rosado, C. P. Neto, et al., "Bacterial cellulose membranes as transdermal delivery systems for diclofenac: In vitro dissolution and permeation studies," *Carbohydrate Polymers*, vol. 106, pp. 264–269, 2014.

(111) R. Suedee, C. Bodhibukkana, N. Tangthong, C. Amnuaikit, S. Kaewnopparat, and T. Srichana, "Development of a reservoir-type transdermal enantioselective-controlled delivery system for racemic propranolol using a molecularly imprinted polymer composite membrane," *Journal of Controlled Release*, vol. 129, pp. 170–178, 2008.

(112) M. L. Cacicedo, I. E. León, J. S. Gonzalez, L. M. Porto, V. A. Alvarez, and G. R. Castro, "Modified bacterial cellulose scaffolds for localized doxorubicin release in human colorectal HT-29 cells," *Colloids and Surfaces B: Biointerfaces*, vol. 140, pp. 421–429, 2016.

(113) W. Czaja, A. Krystynowicz, S. Bielecki, and R. M. Brown Jr, "Microbial cellulose— the natural power to heal wounds," *Biomaterials*, vol. 27, pp. 145–151, 2006.

(114) B. Wei, G. Yang, and F. Hong, "Preparation and evaluation of a kind of bacterial cellulose dry films with antibacterial properties," *Carbohydrate Polymers*, vol. 84, pp. 533–538, 2011.

(115) R. Jung, Y. Kim, H.-S. Kim, and H.-J. Jin, "Antimicrobial properties of hydrated cellulose membranes with silver nanoparticles," *Journal of Biomaterials Science, Polymer Edition*, vol. 20, pp. 311–324, 2009.

(116) R. J. Pinto, P. A. Marques, C. P. Neto, T. Trindade, S. Daina, and P. Sadocco, "Antibacterial activity of nanocomposites of silver and bacterial or vegetable cellulosic fibers," *Acta Biomaterialia*, vol. 5, pp. 2279–2289, 2009.

(117) A. Stoica-Guzun, M. Stroescu, F. Tache, T. Zaharescu, and E. Grosu, "Effect of electron beam irradiation on bacterial cellulose membranes used as transdermal drug delivery systems," *Nuclear Instruments and Methods in Physics Research Section B: Beam Interactions with Materials and Atoms*, vol. 265, pp. 434–438, 2007.

(118) E. Trovatti, C. S. Freire, P. C. Pinto, I. F. Almeida, P. Costa, A. J. Silvestre, et al., "Bacterial cellulose membranes applied in topical and transdermal delivery of lidocaine hydrochloride and ibuprofen: In vitro diffusion studies," *International Journal of Pharmaceutics*, vol. 435, pp. 83–87, 2012.

(119) Z. Lin, Z. Guan, and Z. Huang, "New bacterial cellulose/polyaniline nanocomposite film with one conductive side through constrained interfacial polymerization," *Industrial & Engineering Chemistry Research*, vol. 52, pp. 2869–2874, 2013.

(120) Z. Shi, G. O. Phillips, and G. Yang, "Nanocellulose electroconductive composites," *Nanoscale*, vol. 5, pp. 3194–3201, 2013.

(121) S. H. Yoon, H.-J. Jin, M.-C. Kook, and Y. R. Pyun, "Electrically conductive bacterial cellulose by incorporation of carbon nanotubes," *Biomacromolecules*, vol. 7, pp. 1280–1284, 2006.

(122) J. Y. Fang, K. Sung, J. J. Wang, C. C. Chu, and K. T. Chen, "The effects of iontophoresis and electroporation on transdermal delivery of buprenorphine from solutions and hydrogels," *Journal of Pharmacy and Pharmacology*, vol. 54, pp. 1329–1337, 2002.

(123) M. Amin, A. G. Abadi, N. Ahmad, H. Katas, and J. A. Jamal, "Bacterial cellulose film coating as drug delivery system: Physicochemical, thermal and drug release properties," *Sains Malaysiana*, vol. 41, pp. 561–568, 2012.

(124) M. C. I. M. Amin, N. Ahmad, N. Halib, and I. Ahmad, "Synthesis and characterization of thermo-and pH-responsive bacterial cellulose/acrylic acid hydrogels for drug delivery," *Carbohydrate Polymers*, vol. 88, pp. 465–473, 2012.

(125) C. García-González, M. Alnaief, and I. Smirnova, "Polysaccharide-based aerogels—Promising biodegradable carriers for drug delivery systems," *Carbohydrate Polymers*, vol. 86, pp. 1425–1438, 2011.

(126) F. Liebner, E. Haimer, M. Wendland, M. A. Neouze, K. Schlufter, P. Miethe, et al., "Aerogels from unaltered bacterial cellulose: Application of scCO2 drying for the preparation of shaped, ultra-lightweight cellulosic aerogels," *Macromolecular Bioscience*, vol. 10, pp. 349–352, 2010.

(127) H. Valo, S. Arola, P. Laaksonen, M. Torkkeli, L. Peltonen, M. B. Linder, et al., "Drug release from nanoparticles embedded in four different nanofibrillar cellulose aerogels," *European Journal of Pharmaceutical Sciences*, vol. 50, pp. 69–77, 2013.

(128) A. Buyanov, I. Gofman, L. Revel'skaya, A. Khripunov, and A. Tkachenko, "Anisotropic swelling and mechanical behavior of composite bacterial cellulose—poly (acrylamide or acrylamide—sodium acrylate) hydrogels," *Journal of the Mechanical Behavior of Biomedical Materials*, vol. 3, pp. 102–111, 2010.

(129) Z. Cai and J. Kim, "Bacterial cellulose/poly (ethylene glycol) composite: Characterization and first evaluation of biocompatibility," *Cellulose*, vol. 17, pp. 83–91, 2010.

(130) H.-C. Huang, L.-C. Chen, S.-B. Lin, and H.-H. Chen, "Nano-biomaterials application: In situ modification of bacterial cellulose structure by adding HPMC during fermentation," *Carbohydrate Polymers*, vol. 83, pp. 979–987, 2011.

(131) L. E. Millon, G. Guhados, and W. Wan, "Anisotropic polyvinyl alcohol—Bacterial cellulose nanocomposite for biomedical applications," *Journal of Biomedical Materials Research Part B: Applied Biomaterials: An Official Journal of The Society for Biomaterials, The Japanese Society for Biomaterials, and The Australian Society for Biomaterials and the Korean Society for Biomaterials*, vol. 86, pp. 444–452, 2008.

(132) E. A. Simone, T. D. Dziubla, and V. R. Muzykantov, "Polymeric carriers: Role of geometry in drug delivery," in *Expert Opinion on Drug Delivery*, vol. 5, pp. 1283–1300, 2008.

9 Bacterial Cellulose Composites, Synthetic Strategies, and Applications

Zubair Ahmad, Fazal Qayyum, Sher Ali Shah,
Youssef O. Al-Ghamdi, and Shahid Ali Khan

CONTENTS

9.1 INTRODUCTION

Cellulose is an important structural component of the primary cell wall of green plants, algae, and oomycetes and is the most important organic compound on earth. Some species of bacteria also contain cellulose in the form of biofilm. It is a polysaccharide (carbohydrate) with the chemical formula $(C_6H_{10}O_5)_n$ and consists of linear chains of several hundred to thousands of β-(1→4) linked D-glucose units (1–3).

The properties of cellulose differ from that of other polysaccharides and possesses unique properties because of its unique structure. Different glucose subunits also change the properties of cellulose. Cellulose is produced from plant and bacterial sources. It is insoluble in water and most organic solvents, except for a few. It is a semi-crystalline solid and is mostly available in white powder form. The tensile strength of the cellulose molecule is due to the presence of hydrogen bonds among the cellulose microfibrils (4–6).

The cellulose that is synthesized by bacteria is known both as bacterial cellulose (BC) and microbial cellulose. BC is biodegradable and has 20–100 nm microfibrils. It retains a large quantity of water due to its hydrophilic nature and high surface area to mass ratio. The mechanical strength and crystallinity of bacterial cellulose is significant.

BC production is limited, which increases its cost and is therefore not produced commercially (7, 8).

Bacterial cellulose is unable to resist microbial attack and has low biocompatibility, which result in uncertainty in regeneration of tissues and healing of wounds (9, 10).

BC's electrical and magnetic insulation make it unfit in many technological applications. Therefore, some strategies are required to overcome these limitations to make it suitable for effective applications in different fields (11).

As stated earlier, because BC is produced by bacteria in a slow process, various techniques have been developed to overcome these limitations and improve its applications. Some of the techniques studied include the use of nanoparticles and composite materials, which shows significant improvement for specific uses (12). As mentioned, pure BC has several limitations, including its nonconductive nature, lack of antibacterial activity, and poor biocompatibility (13).

To address these shortcomings, BC has been modified chemically and structurally by impregnating it with different types of metal nanoparticles such as Au, Ag, Cu, Ni, and many more, as well as using biocompatible polymers for making composites (14).

The modified BC shows significant improvement in antimicrobial resistance, electrical and magnetic conductance, and biocompatibility, which makes it suitable for different applications in wound healing, tissue regeneration, and in the biomedical, biotechnological, and electrical fields (15).

BC composites impregnated with varieties of metal nanoparticles have wide application, specifically in the biomedical field for artificial organ development, tissue regeneration, and the development of biosensors for targeted uses (16).

BC, with its impressive physicochemical properties such as crystallinity, porous fibrous structure, high mechanical strength, and liquid absorbing capabilities, has emerged as an advanced biomaterial. BC has been synthesized through many routes that produce different structural features. BC's porous geometry and 3D fibrous structure make it an ideal material for synthesizing composites that successfully overcome certain deficiencies of pure BC. Bacterial cellulose and BC composites have the potential to be used in artificial organs, the development of medical devices, and electronic and conducting material (17–20).

9.1.1 BACTERIAL CELLULOSE AND ITS GENERAL FEATURES

Cellulose is a polysaccharide consisting of long chains of glucose molecules (minimum 500 glucose units). These structural chains are arranged in parallel to each other, resulting in the formation of microfibrils. There are strong hydrogen bonds between these microfibrils, which holds them together to form macrofibrils (21). The strong hydrogen bonding makes the crystalline structure inflexible and rigid (2, 22).

Structurally, cellulose is polymer in which monomers are linked to each other by β-(1→4) linkage. The basic units of cellulose comprises two molecules of glucose arranged in such a way that one molecule can be rotated by 180° with respect to the other. BC is similar to plant cellulose in almost all aspects except for the degree of polymerization, which are 13,000–14,000 and 2,000–6,000 respectively (1, 7).

9.1.2 Synthesis of Bacterial Cellulose

Certain bacteria naturally synthesize BC from foodstuffs with a large amount of carbon followed by nitrogen through a complex biochemical procedure using certain enzymes as a catalyst. Besides natural production, BC can be synthesized through a variety of techniques, such as by using synthetic culture media from mainly waste materials. BC synthesized from various media differs from each other in physiology and other properties (23–25).

Agitating culture and static culture techniques are used for the synthesis of BC. In the agitating culture method, the synthesized media is placed in such a way that it remains in contact with the provided suitable environment, and as a result BC takes the form of pellets. Using the static culture method, a broth medium is introduced to empty trays and is left in an incubator for 10 days, resulting in a thick layer of cellulose (26, 27).

9.1.3 Shortcomings in Bacterial Cellulose and Strategies to Overcome Them Using Nanocomposites

As discussed previously, BC has several limitations that restrict its application in many fields. To solve these problems, BC composites are introduced. A variety of composite materials are used with the addition of nanoparticles, which greatly modify their functions.

Nowadays, industries are demanding composite materials, either natural or synthetic, because of their effectiveness as compared to native forms. Polymers are widely used to obtain composites with multifunctional features (28–30).

Pure BC's lack of conductivity and antimicrobial activity restricts its use for numerous applications, so composite materials are used instead. The antimicrobial activity and conducting properties of BC composites make them suitable for biomedical uses (31, 32).

BC composites are synthesized using various materials (Figure 9.1) and differing strategies (Figure 9.2). It has been found that BC composites are more efficient and have better conducting and antimicrobial properties than pure BC, and they play an important role for artificial organ technology, biosensing, and fuel cells (33, 34).

9.1.4 *Ex Situ* and *In Situ* Strategies for BC Composites

In the *In situ* method, the reinforcement materials are introduced to the BC culture media. The reinforcement materials may be polymers and nanomaterials that can

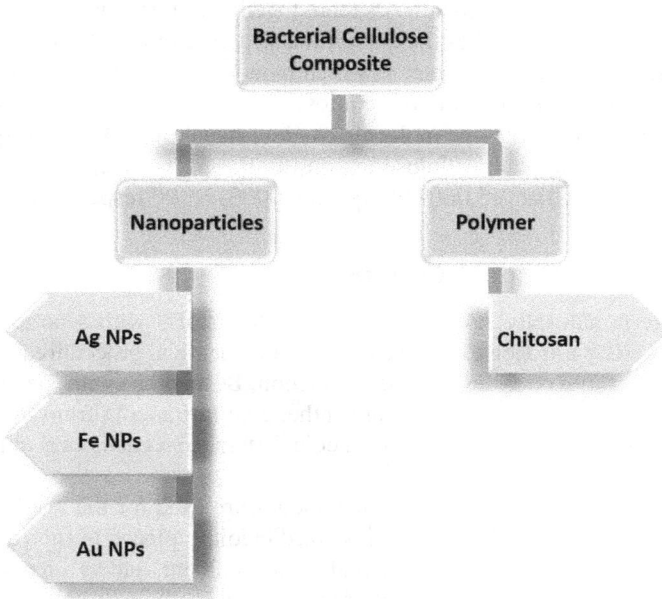

FIGURE 9.1   Various materials impregnated with bacterial cellulose.

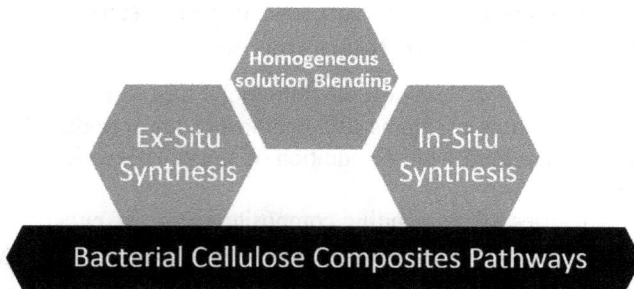

FIGURE 9.2   Strategies for the synthesis of bacterial cellulose composites.

be effectively embedded in the culture media of BC and can efficiently interact both physically and chemically (16).

In the *ex situ* technique, the BC membrane is impregnated with different rein-forcement materials like nanoparticles and polymer solution. Because BC has porous structure, it can accommodate a variety of materials. After penetration, the subjected materials develop chemical or physical interaction, depending on their nature. The –OH moieties in BC provide the opportunity to established hydrogen bonding, while the porous surface traps different types of nanomaterial.

The main issue in the aforementioned process is the toxoid of the resultant composite formed by adding the reinforcement materials, which allows only a specified size of particles to penetrate (35–37).

9.1.5 BACTERIAL CELLULOSE COMPOSITE SYNTHESIS USING HOMOGENEOUS MIXTURE

Because of the many problems associated with synthesized BC composites, a modern and more effective alternative strategy is required to prepare BC composites that are free from the aforementioned shortcomings. For this purpose, solution-blending techniques have been introduced. Solution blending is the latest modern effective technique used to synthesize a wide range of BC composites (38).

In this technique, the BC is dissolved in a specific solvent like ionic liquids, sodium hydroxide, or zinc chloride. After dissolution, the desired materials such as polymers, nanoparticles, and carbon nanotubes are added to the prepared solution to make a homogeneous mixture. The homogeneous mixture is then poured on to a clean glass surface, resulting in composite materials after the solvent is evaporated (38, 39). This technique can used for impregnating BC with a variety of materials to obtain the desired composites in the form of nanocrystal, nanofiber and BC membranes (39).

This strategy is more effective than *in situ* and *ex situ* techniques. However, poor solubility of BC is a drawback to this method (39, 40).

9.1.6 APPLICATION OF BACTERIAL CELLULOSE COMPOSITE

BC and its composites have distinguishing features that help in their applicability in many scientific fields, especially in the biomedical field and for biosensing (41).

9.1.6.1 Application in Wound Healing and Tissue Regeneration

BC, with its large surface area and fine crystalline structure, can retain high amount of water, and can thus work as a hydrogel. Due to these properties, BC composites are widely used in wound dressing, tissue regeneration (especially in epithelial tissues), and for the regeneration of fractured bone. Besides these uses, BC and its composites can be used in biosensing, in electric display devices, and as a source of food (42) (Figure 9.3).

Wound healing is a natural process that can be accelerated by changing the conditions and modifying the involved factors. To date, many conventional therapies have been introduced and show moderate to significant results. However, many problems are associated with conventional therapies, so BC and its composites are used as alternative therapies to accelerate the healing process. BC composites with zinc oxide and silver nanoparticles were found effective against a wide range of infectious bacteria. Mechanistically, BC and its composites possess microfibrils, a nanoporous nature, and the ability to hold a great amount of water. These factors facilitate and accelerate the phenomena occurring during wound healing, such as by enhancing the rate of tissue reconstruction, protecting against invading microorganisms, and providing moisture to the wound, which leads to improved wound healing (43–47).

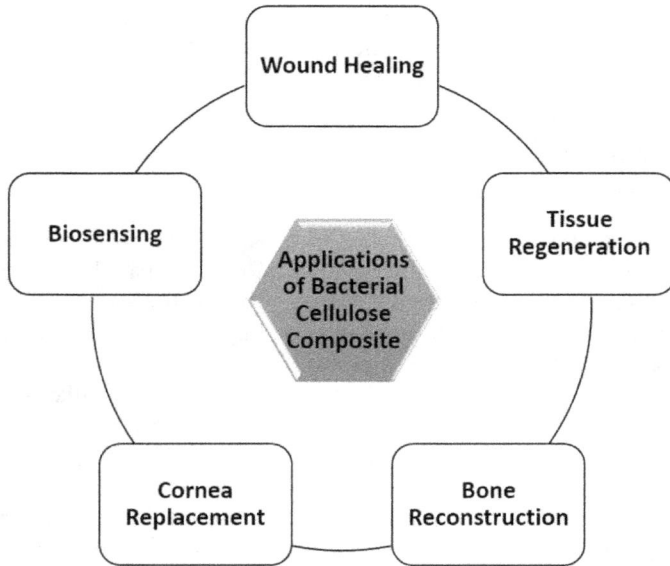

FIGURE 9.3   The generalized applications of bacterial cellulose composites.

9.1.6.2 Cornea Replacement (Artificial Organ Development)

Bacterial cellulose and its composites have been used to develop artificial organs. BC composites with polyvinyl alcohol (BC/PVA) possess high mechanical strength. They allow light to transmit through, providing them with optical properties. Thus, they can used as artificial corneas (48–51).

BC composites can also be used for artificial heart replacement and alternative blood vessels, which decrease the rate of blood clotting and blockage in vessels (52).

Wang et al. reported that BC/PVA contains almost the same amount of water as a natural cornea. The stated composite allowed 95% visible light to transmit and had the ability to block UV light. The BC composite with a fiber ribbon structure of 3–8 nm thickness and 30–100 nm width distorted the scattering of light with resulting excellent transparency (53). Nogi et al. reported that the transmittance of light varies to a very low extent. The wet BC allows transmittance in the range of 400–900 nm up to 60%–80%; this high transmittance is due to high water content. To protect the eye's internal tissue, the ideal artificial cornea should absorb UV light (54, 55).

9.1.6.3 Biosensing

BC composites can attain magnetic and conducting properties by using some nanomaterials as reinforcement materials (56). BC composites with magnetic properties have been used as biosensors to detect humidity and numerous gases. Due to their super magnetic and biosensing properties, these composites have been used in magnetic resonance imaging (MRI) and for the development of nonlinear optics (52, 57, 58).

BC composites are also widely used in biosensing to release antibodies, respond to cardiac infarction, and regenerate tissue by controlling biological signals (59).

It has been reported that BC-Ag nanocomposites are effectively used in bioanalysis using analyte thiosalicylic acid and 2,2-dithiodipyridine. The stated composites were investigated for the bioanalysis of l-phenylalanine, l-glutamine, and l-histidine amino acids.

Wang et al. synthesized BC with Au nanoparticles (BC-Au) for the determination of glucose. The stated enzymatic biosensor had the ability to detect and measure levels as low as 2.3 µM with a linear range of 10–400 µM in human blood (49, 53, 60).

9.1.6.4 Bone Reconstruction

BC has also been used widely for the reconstruction of bone. BC with its high tensile strength is compatible with collagen fiber, and hence is widely used to regenerate infected bone. However, the rate of proliferation of cells in BC is lower than that of collagen but is enough to recover the targeted bone.

Lee et al. prepared a BC composite and compared it with natural collagen and to investigate its use in bone regeneration. Their work shows that both samples had porous structure composed of nanofibers. The SEM images show that the synthesized BC membranes have high tensile strength and modulus. The tensile stress of collagen membrane and BC were 2.34±0.09 MPa and 2.88±0.33 MPa respectively. Similarly, the tensile strain was found to be 1.85±0.10% and 1.85±0.10% respectively. The stated factors show that the BC composite had a high value of tensile strength and tensile strain as compared to collagen membrane.

The BC and collagen membrane substrate were subjected to the CCK-8 assays in order to investigate the adhesion and proliferation of cells. No significant difference in proliferation was found between the two samples. The BC composite was shown to effectively form new bone and did not induced any inflammatory action, which supports the use of BC in the bone reconstruction process (61, 62).

9.2 CONCLUSION

Bacterial cellulose is a biopolymer of glucose units joined by glycosides bond forming a network. BC is a crystalline solid that is soluble in a few organic solvents and insoluble in water. It has many important features, which make it suitable for use in different biomedical industries. However, some limitations are found in pure BC. Therefore, the necessity of creating BC composites has been felt. Many reinforcement materials have been used for BC-based composites by applying different techniques. In this chapter, BC and its composites using a variety of reinforcement materials has been discussed. The uses and applications of BC in different fields has also been discussed. This review provides comprehensive information about pure BC's limitations, its composites, and the techniques to overcome various limitations, and its applications have been discussed in detail.

9.3 ACKNOWLEDGMENTS

The authors highly acknowledge the Department of Chemistry, University of Swabi, Khyber Pakhtunkhwa, Pakistan and Department of Chemistry, College of Science Al-zulfi, Majmaah University, Saudi Arabia for the collaborative work.

REFERENCES

(1) Marchessault, R.; Sundararajan, P. Cellulose. In *The polysaccharides*, Elsevier: 1983; pp. 11–95.

(2) Nevell, T. P.; Zeronian, S. H. *Cellulose chemistry and its applications*. Halsted Press, John Wiley: 1985.

(3) Young, R.; Rowell, R. M. *Cellulose: Structure, modification and hydrolysis*. Wiley-Interscience: 1986.

(4) Klemm, D.; Philpp, B.; Heinze, T.; Heinze, U.; Wagenknecht, W. *Comprehensive cellulose chemistry. Volume 1: Fundamentals and analytical methods*. Wiley-VCH Verlag GmbH: 1998.

(5) Lavanya, D.; Kulkarni, P.; Dixit, M.; Raavi, P. K.; Krishna, L. N. V. Sources of cellulose and their applications—A review, *International Journal of Drug Formulation and Research* **2011**, 2, (6), 19–38.

(6) Trache, D.; Hussin, M. H.; Haafiz, M. M.; Thakur, V. K. Recent progress in cellulose nanocrystals: Sources and production, *Nanoscale* **2017**, 9, (5), 1763–1786.

(7) Iguchi, M.; Yamanaka, S.; Budhiono, A. Bacterial cellulose—A masterpiece of nature's arts, *Journal of Materials Science* **2000**, 35, (2), 261–270.

(8) Shi, Z.; Zhang, Y.; Phillips, G. O.; Yang, G. Utilization of bacterial cellulose in food, *Food Hydrocolloids* **2014**, 35, 539–545.

(9) Nottingham, A. T.; Hicks, L. C.; Ccahuana, A. J.; Salinas, N.; Bååth, E.; Meir, P. Nutrient limitations to bacterial and fungal growth during cellulose decomposition in tropical forest soils, *Biology and Fertility of Soils* **2018**, 54, (2), 219–228.

(10) Roman, M.; Haring, A. P.; Bertucio, T. J. The growing merits and dwindling limitations of bacterial cellulose-based tissue engineering scaffolds, *Current Opinion in Chemical Engineering* **2019**, 24, 98–106.

(11) Ostadhossein, F.; Mahmoudi, N.; Morales-Cid, G.; Tamjid, E.; Navas-Martos, F. J.; Soriano-Cuadrado, B.; Paniza, J. M. L.; Simchi, A. Development of chitosan/bacterial cellulose composite films containing nanodiamonds as a potential flexible platform for wound dressing, *Materials* **2015**, 8, (9), 6401–6418.

(12) Martínez-Sanz, M.; Abdelwahab, M. A.; Lopez-Rubio, A.; Lagaron, J. M.; Chiellini, E.; Williams, T. G.; Wood, D. F.; Orts, W. J.; Imam, S. H. Incorporation of poly (glycidyl-methacrylate) grafted bacterial cellulose nanowhiskers in poly (lactic acid) nanocomposites: Improved barrier and mechanical properties, *European Polymer Journal* **2013**, 49, (8), 2062–2072.

(13) Yamanaka, S.; Sugiyama, J. Structural modification of bacterial cellulose, *Cellulose* **2000**, 7, (3), 213–225.

(14) Dayal, M. S.; Catchmark, J. M. Mechanical and structural property analysis of bacterial cellulose composites, *Carbohydrate Polymers* **2016**, 144, 447–453.

(15) Astley, O. M.; Chanliaud, E.; Donald, A. M.; Gidley, M. J. Tensile deformation of bacterial cellulose composites, *International Journal of Biological Macromolecules* **2003**, 32, (1–2), 28–35.

(16) Shah, N.; Ul-Islam, M.; Khattak, W. A.; Park, J. K. Mechanical and structural property analysis of bacterial cellulose composites, *Carbohydrate Polymers* **2013**, 98, (2), 1585–1598.

(17) Chang, W.-S.; Chen, H.-H. Physical properties of bacterial cellulose composites for wound dressings, *Food Hydrocolloids* **2016**, 53, 75–83.

(18) Hu, Y.; Catchmark, J. M. Integration of cellulases into bacterial cellulose: Toward bioabsorbable cellulose composites, *Journal of Biomedical Materials Research Part B: Applied Biomaterials* **2011**, 97, (1), 114–123.

(19) Kim, J.; Cai, Z.; Chen, Y. Biocompatible bacterial cellulose composites for biomedical application, *Journal of Nanotechnology in Engineering and Medicine* **2010**, 1, (1).

(20) Kiziltas, E. E.; Kiziltas, A.; Rhodes, K.; Emanetoglu, N. W.; Blumentritt, M.; Gardner, D. J. Electrically conductive nano graphite-filled bacterial cellulose composites, *Carbohydrate Polymers* **2016,** 136, 1144–1151.

(21) Turbak, A. F.; Snyder, F. W.; Sandberg, K. R. In *Microfibrillated cellulose, a new cellulose product: Properties, uses, and commercial potential,* Journal of Applied Polymer Science: Applied Polymer Symposium. ITT Rayonier Inc: 1983.

(22) Huang, Y.; Zhu, C.; Yang, J.; Nie, Y.; Chen, C.; Sun, D. Recent advances in bacterial cellulose, *Cellulose* **2014,** 21, (1), 1–30.

(23) Kiziltas, E. E.; Kiziltas, A.; Gardner, D. J. Synthesis of bacterial cellulose using hot water extracted wood sugars, *Carbohydrate Polymers* **2015,** 124, 131–138.

(24) Mohammadkazemi, F.; Azin, M.; Ashori, A. Production of bacterial cellulose using different carbon sources and culture media, *Carbohydrate Polymers* **2015,** 117, 518–523.

(25) Yan, H.; Chen, X.; Song, H.; Li, J.; Feng, Y.; Shi, Z.; Wang, X.; Lin, Q. Synthesis of bacterial cellulose and bacterial cellulose nanocrystals for their applications in the stabilization of olive oil pickering emulsion, *Food Hydrocolloids* **2017,** 72, 127–135.

(26) Cakar, F.; Özer, I.; Aytekin, A. Ö.; Şahin, F. Improvement production of bacterial cellulose by semi-continuous process in molasses medium, *Carbohydrate Polymers* **2014,** 106, 7–13.

(27) Gomes, F. P.; Silva, N. H.; Trovatti, E.; Serafim, L. S.; Duarte, M. F.; Silvestre, A. J.; Neto, C. P.; Freire, C. S. Production of bacterial cellulose by *Gluconacetobacter sacchari* using dry olive mill residue, *Biomass and Bioenergy* **2013,** 55, 205–211.

(28) Clyne, T.; Hull, D. *An introduction to composite materials*. Cambridge University Press: 2019.

(29) Kimberly, L. W. *Composite materials*. Google Patents: 2006.

(30) Vinson, J. R.; Chou, T.-W. *Composite materials and their use in structures*. Elsevier-Applied Science: 1975.

(31) Vasconcelos, N. F.; Feitosa, J. P. A.; da Gama, F. M. P.; Morais, J. P. S.; Andrade, F. K.; de Souza, M. d. S. M.; de Freitas Rosa, M. Bacterial cellulose nanocrystals produced under different hydrolysis conditions: Properties and morphological features, *Carbohydrate Polymers* **2017,** 155, 425–431.

(32) Badshah, M.; Ullah, H.; Khan, A. R.; Khan, S.; Park, J. K.; Khan, T. Surface modification and evaluation of bacterial cellulose for drug delivery, *International Journal of Biological Macromolecules* **2018,** 113, 526–533.

(33) Ul-Islam, M.; Khan, S.; Ullah, M. W.; Park, J. K. Comparative study of plant and bacterial cellulose pellicles regenerated from dissolved states, *International Journal of Biological Macromolecules* **2019,** 137, 247–252.

(34) Choi, S. M.; Shin, E. J. The nanofication and functionalization of bacterial cellulose and its applications, *Nanomaterials* **2020,** 10, (3), 406.

(35) Adepu, S.; Khandelwal, M. Ex-situ modification of bacterial cellulose for immediate and sustained drug release with insights into release mechanism, *Carbohydrate Polymers* **2020,** 249, 116816.

(36) Stumpf, T. R.; Yang, X.; Zhang, J.; Cao, X. In situ and ex situ modifications of bacterial cellulose for applications in tissue engineering, *Materials Science and Engineering: C* **2018,** 82, 372–383.

(37) Ul-Islam, M.; Khan, T.; Park, J. K. Water holding and release properties of bacterial cellulose obtained by in situ and ex situ modification, *Carbohydrate Polymers* **2012,** 88, (2), 596–603.

(38) Zhijiang, C.; Chengwei, H.; Guang, Y. Poly (3-hydroxubutyrate-co-4-hydroxubutyrate)/bacterial cellulose composite porous scaffold: Preparation, characterization and biocompatibility evaluation, *Carbohydrate Polymers* **2012,** 87, (2), 1073–1080.

(39) Qiu, K.; Netravali, A. N. A review of fabrication and applications of bacterial cellulose based nanocomposites, *Polymer Reviews* **2014,** 54, (4), 598–626.

(40) Lai, C.; Zhang, S.; Chen, X.; Sheng, L. Nanocomposite films based on TEMPO-mediated oxidized bacterial cellulose and chitosan, *Cellulose* **2014**, 21, (4), 2757–2772.

(41) Esa, F.; Tasirin, S. M.; Abd Rahman, N. Overview of bacterial cellulose production and application, *Agriculture and Agricultural Science Procedia* **2014**, 2, 113–119.

(42) Jonas, R.; Farah, L. F. Production and application of microbial cellulose, *Polymer Degradation and Stability* **1998**, 59, (1–3), 101–106.

(43) Ahmed, J.; Gultekinoglu, M.; Edirisinghe, M. Bacterial cellulose micro-nano fibres for wound healing applications, *Biotechnology Advances* **2020**, 107549.

(44) Lin, W.-C.; Lien, C.-C.; Yeh, H.-J.; Yu, C.-M.; Hsu, S.-H. Bacterial cellulose and bacterial cellulose–chitosan membranes for wound dressing applications, *Carbohydrate Polymers* **2013**, 94, (1), 603–611.

(45) Pal, S.; Nisi, R.; Stoppa, M.; Licciulli, A. Silver-functionalized bacterial cellulose as antibacterial membrane for wound-healing applications, *ACS Omega* **2017**, 2, (7), 3632–3639.

(46) Qiu, Y.; Qiu, L.; Cui, J.; Wei, Q. Bacterial cellulose and bacterial cellulose-vaccarin membranes for wound healing, *Materials Science and Engineering: C* **2016**, 59, 303–309.

(47) Zmejkoski, D.; Spasojević, D.; Orlovska, I.; Kozyrovska, N.; Soković, M.; Glamočlija, J.; Dmitrović, S.; Matović, B.; Tasić, N.; Maksimović, V. Bacterial cellulose-lignin composite hydrogel as a promising agent in chronic wound healing, *International Journal of Biological Macromolecules* **2018**, 118, 494–503.

(48) Klemm, D.; Schumann, D.; Udhardt, U.; Marsch, S. Bacterial synthesized cellulose— Artificial blood vessels for microsurgery, *Progress in Polymer Science* **2001**, 26, (9), 1561–1603.

(49) Moniri, M.; Boroumand Moghaddam, A.; Azizi, S.; Abdul Rahim, R.; Bin Ariff, A.; Zuhainis Saad, W.; Navaderi, M.; Mohamad, R. Production and status of bacterial cellulose in biomedical engineering, *Nanomaterials* **2017**, 7, (9), 257.

(50) Ul-Islam, M.; Khan, S.; Ullah, M. W.; Park, J. K. Bacterial cellulose composites: Synthetic strategies and multiple applications in bio-medical and electro-conductive fields, *Biotechnology Journal* **2015**, 10, (12), 1847–1861.

(51) Wan, Y.; Huang, Y.; Yuan, C.; Raman, S.; Zhu, Y.; Jiang, H.; He, F.; Gao, C. Biomimetic synthesis of hydroxyapatite/bacterial cellulose nanocomposites for biomedical applications, *Materials Science and Engineering: C* **2007**, 27, (4), 855–864.

(52) Petersen, N.; Gatenholm, P. Bacterial cellulose-based materials and medical devices: Current state and perspectives, *Applied Microbiology and Biotechnology* **2011**, 91, (5), 1277.

(53) Wang, J.; Gao, C.; Zhang, Y.; Wan, Y. Biosensor based on bacterial cellulose-Au nanoparticles electrode modified with laccase for hydroquinone detection, *Materials Science and Engineering: C* **2010**, 30, (1), 214–218.

(54) Nogi, M.; Ifuku, S.; Abe, K.; Handa, K.; Nakagaito, A. N.; Yano, H. Smart films based on bacterial cellulose nanofibers modified by conductive polypyrrole and zinc oxide nanoparticles, *Applied Physics Letters* **2006**, 88, (13), 133124.

(55) Svensson, A.; Nicklasson, E.; Harrah, T.; Panilaitis, B.; Kaplan, D.; Brittberg, M.; Gatenholm, P. Bacterial cellulose as a potential scaffold for tissue engineering of cartilage, *Biomaterials* **2005**, 26, (4), 419–431.

(56) Li, G.; Sun, K.; Li, D.; Lv, P.; Wang, Q.; Huang, F.; Wei, Q. Biosensor based on bacterial cellulose-Au nanoparticles electrode modified with laccase for hydroquinone detection, *Colloids and Surfaces A: Physicochemical and Engineering Aspects* **2016**, 509, 408–414.

(57) Pirsa, S.; Shamusi, T.; Kia, E. M. Smart films based on bacterial cellulose nanofibers modified by conductive polypyrrole and zinc oxide nanoparticles, *Journal of Applied Polymer Science* **2018**, 135, (34), 46617.

(58) Hola, K.; Markova, Z.; Zoppellaro, G.; Tucek, J.; Zboril, R. Tailored functionalization of iron oxide nanoparticles for MRI, drug delivery, magnetic separation and immobilization of biosubstances, *Biotechnology Advances* **2015,** 33, (6), 1162–1176.

(59) Giepmans, B. N.; Adams, S. R.; Ellisman, M. H.; Tsien, R. Y. The fluorescent toolbox for assessing protein location and function, *Science* **2006,** 312, (5771), 217–224.

(60) Wang, W.; Li, H. Y.; Zhang, D. W.; Jiang, J.; Cui, Y. R.; Qiu, S.; Zhou, Y. L.; Zhang, X. X. Fabrication of bienzymatic glucose biosensor based on novel gold nanoparticles-bacteria cellulose nanofibers nanocomposite, *Electroanalysis* **2010,** 22, (21), 2543–2550.

(61) Lee, S.-H.; Lim, Y.-M.; Jeong, S. I.; An, S.-J.; Kang, S.-S.; Jeong, C.-M.; Huh, J.-B. The effect of bacterial cellulose membrane compared with collagen membrane on guided bone regeneration, *The journal of Advanced Prosthodontics* **2015,** 7, (6), 484.

(62) Fang, B.; Wan, Y.-Z.; Tang, T.-T.; Gao, C.; Dai, K.-R. Proliferation and osteoblastic differentiation of human bone marrow stromal cells on hydroxyapatite/bacterial cellulose nanocomposite scaffolds, *Tissue Engineering Part A* **2009,** 15, (5), 1091–1098.

Index

Note: Page numbers in *italic* indicate a figure and page numbers in **bold** indicate a table on the corresponding page.

For Product Safety Concerns and Information please contact our EU
representative GPSR@taylorandfrancis.com
Taylor & Francis Verlag GmbH, Kaufingerstraße 24, 80331 München, Germany

www.ingramcontent.com/pod-product-compliance
Lightning Source LLC
Chambersburg PA
CBHW060551220326
41598CB00024B/3073